# Exploring Waste Management in Sustainable Development Contexts

Chandra Mohan
*K.R. Mangalalm University, Gurgaon, India*

Shobhna Jeet
*K.R. Mangalam University, Gurugram, India*

Saurav Dixit
*Peter the Great St. Petersburg Polytechnic University, Russia*

Sónia A.C. Carabineiro
*NOVA University of Lisbon, Portugal*

IGI Global
PUBLISHER of TIMELY KNOWLEDGE

A volume in the Practice, Progress, and Proficiency in Sustainability (PPPS) Book Series

Published in the United States of America by
IGI Global
Engineering Science Reference (an imprint of IGI Global)
701 E. Chocolate Avenue
Hershey PA, USA 17033
Tel: 717-533-8845
Fax: 717-533-8661
E-mail: cust@igi-global.com
Web site: http://www.igi-global.com



Library of Congress Cataloging-in-Publication Data

CIP DATA PROCESSING
2024 Engineering Science Reference
ISBN(hc) 9798369342640 | ISBN(sc) 9798369349526 | eISBN 9798369342657

This book is published in the IGI Global book series Practice, Progress, and Proficiency in Sustainability (PPPS) (ISSN: 2330-3271; eISSN: 2330-328X)

British Cataloguing in Publication Data
A Cataloguing in Publication record for this book is available from the British Library.

All work contributed to this book is new, previously-unpublished material. The views expressed in this book are those of the authors, but not necessarily of the publisher.

For electronic access to this publication, please contact: eresources@igi-global.com.

# Practice, Progress, and Proficiency in Sustainability (PPPS) Book Series

Ayman Batisha
International Sustainability Institute, Egypt

ISSN:2330-3271
EISSN:2330-328X

## Mission

In a world where traditional business practices are reconsidered and economic activity is performed in a global context, new areas of economic developments are recognized as the key enablers of wealth and income production. This knowledge of information technologies provides infrastructures, systems, and services towards sustainable development.

The **Practices, Progress, and Proficiency in Sustainability (PPPS) Book Series** focuses on the local and global challenges, business opportunities, and societal needs surrounding international collaboration and sustainable development of technology. This series brings together academics, researchers, entrepreneurs, policy makers and government officers aiming to contribute to the progress and proficiency in sustainability.

## Coverage

- Green Technology
- Strategic Management of IT
- Socio-Economic
- ICT and knowledge for development
- Knowledge clusters
- E-Development
- Sustainable Development
- Global Content and Knowledge Repositories
- Environmental informatics
- Eco-Innovation

IGI Global is currently accepting manuscripts for publication within this series. To submit a proposal for a volume in this series, please contact our Acquisition Editors at Acquisitions@igi-global.com or visit: http://www.igi-global.com/publish/.

The Practice, Progress, and Proficiency in Sustainability (PPPS) Book Series (ISSN 2330-3271) is published by IGI Global, 701 E. Chocolate Avenue, Hershey, PA 17033-1240, USA, www.igi-global.com. This series is composed of titles available for purchase individually; each title is edited to be contextually exclusive from any other title within the series. For pricing and ordering information please visit http://www.igi-global.com/book-series/practice-progress-proficiency-sustainability/73810. Postmaster: Send all address changes to above address.

## Titles in this Series

*For a list of additional titles in this series, please visit: http://www.igi-global.com/book-series/practice-progress-proficiency-sustainability/73810*

*Promoting Multi-Sector Sustainability With Policy and Innovation*
Aftab Ara (University of Hail, Saudi Arabia) and Renuka Thakore (University of Central Lancashire, UK)
Engineering Science Reference • copyright 2024 • 286pp • H/C (ISBN: 9798369321133) • US $290.00 (our price)

*The Role of Female Leaders in Achieving the Sustainable Development Goals*
Mercia Selva Malar Justin (Xavier Institute of Management and Entrepreneurship, India) and Joycia Thorat (Churches Auxiliary for Social Action, India)
Business Science Reference • copyright 2024 • 341pp • H/C (ISBN: 9798369318348) • US $285.00 (our price)

*Emerging Technologies and Marketing Strategies for Sustainable Agriculture*
Jabulani Garwi (University of the Free State, South Africa) Reason Masengu (Middle East College, Oman) and Option Takunda Chiwaridzo (University of Science and Technology Beijing, Beijing, China)
Engineering Science Reference • copyright 2024 • 315pp • H/C (ISBN: 9798369348642) • US $255.00 (our price)

*Quality of Life and Climate Change Impacts, Sustainable Adaptation, and Social-Ecological Resilience*
Kasturi Shukla (Symbiosis Institute of Health Sciences, Symbiosis International University (Deemed), Pune, India) Yogesh B. Patil (Symbiosis Centre for Research and Innovation, Symbiosis International University (Deemed), Pune, India) Ronald C. Estoque (Center for Biodiversity and Climate Change, Forestry and Forest Products Research Institute, Tsukuba, Japan) and Pedro Antonio López de Haro (Universidad Autónoma Indígena de México, Mexico)
Engineering Science Reference • copyright 2024 • 499pp • H/C (ISBN: 9781668498637) • US $250.00 (our price)

*Sustainable Practices for Agriculture and Marketing Convergence*
Jabulani Garwi (University of the Free State, South Africa) Reason Masengu (Middle East College, Oman) and Option Takunda Chiwaridzo (University of Science and Technology Beijing, Beijing, China)
Engineering Science Reference • copyright 2024 • 324pp • H/C (ISBN: 9798369320112) • US $245.00 (our price)

*Sustainable Disposal Methods of Food Wastes in Hospitality Operations*
Amrik Singh (Lovely Professional University, India) Pankaj Kumar Tyagi (Chandigarh University, India) and Anshul Garg (Taylor's University, Malaysia)
Engineering Science Reference • copyright 2024 • 370pp • H/C (ISBN: 9798369321812) • US $285.00 (our price)

*Sustainable Development in AI, Blockchain, and E-Governance Applications*

IGI Global
PUBLISHER of TIMELY KNOWLEDGE

701 East Chocolate Avenue, Hershey, PA 17033, USA
Tel: 717-533-8845 x100 • Fax: 717-533-8661
E-Mail: cust@igi-global.com • www.igi-global.com

# Editorial Advisory Board

# Table of Contents

**Preface** ............................................................ xvi

**Acknowledgement** ............................................................ xx

**Chapter 1**
Innovation in Waste Management: From Trash to Treasure ............................................................ 1
*Richa Kaushal, Lovely Professional University, India*
*Payal Bansal, Lovely Professional University, India*

**Chapter 2**
Various Upcycling and Recycling Methods ............................................................ 14
*Param jeet Kaur, Lovely Professional University, India*
*Jaswant Singh, Lovely Professional University, India*
*Jyoti Rani, Lovely Professional University, India*

**Chapter 3**
Waste Management and Recovery Practices in the Apparel Sector Through a Sustainable Approach ............................................................ 37
*Rajveer Kaur Randhawa, Lovely Professional University, India*

**Chapter 4**
Importance of Education and Community Engagement for Effective Waste Management Techniques ............................................................ 53
*Archana Kumari, Lovely Professional University, India*
*Sudev Dutta, Lovely Professional University, India*

**Chapter 5**
Sustainable Harvesting: Weed Waste Management for Environmental Preservation and Economic Growth ............................................................ 71
*Chandra Mohan, K.R. Mangalalm University, Gurgaon, India*
*Barkha Sharma, K.R. Mangalam University, Gurugram, India*
*Amit Kumar Rawat, Hansraj College, India*
*Chingakham Chinglenthoiba, Chitkara University, India*

**Chapter 6**
Examining Waste Management in Sustainable Development: Essential Strategies for an Environmentally Friendly Future ........ 89
*Prashant Panwar, K.R. Mangalam University, India*
*Megha Garg, K.R. Mangalam University, India*
*Anil Kumar, Bharati Vidyapeeth's College of Engineering, New Delhi, India*

**Chapter 7**
E-Waste Recycling: A Welcoming Change ........ 96
*Misha Bahmani, IILM Law School, IILM University, Gurugram, India*
*Shobhna Jeet, K.R. Mangalam University, Gurugram, India*

**Chapter 8**
The Role of Solid Waste Management in Boosting Sustainable Development Goals ........ 109
*Joan Nyika, University of Johannesburg, South Africa*
*Megersa Olumana Dinka, University of Johannesburg, South Africa*

**Chapter 9**
Plastic Waste Management for a Green Future: Methods and Challenges ........ 125
*Asha Anish Madhavan, Amity University Dubai, UAE*
*Amrutha Nambiar, Amity University Dubai, UAE*
*Simi A. Santosh, Amity University Dubai, UAE*

**Chapter 10**
Nanomaterials in Healthcare for Waste Management and Environmental Sustainability ........ 143
*Ginni Rani, DPG Degree College, Gurugram, India*
*Yashika Chitkara, DPG Degree College, Gurugram, India*
*Sunil Kumar, Sushant University, Gurugram, India*
*Chandra Mohan, K.R. Mangalalm University, Gurugram, India*

**Chapter 11**
Waste Management From Agricultural Field: Sustainable Practices and Framework ........ 154
*Aditi Tiwari, Intertek India Pvt. Ltd., Gurugram, India*
*S. V. Niveditha, Presidency University, Bangalore, India*
*Chandra Mohan, K.R. Mangalalm University, Gurgaon, India*
*Anjaneyulu Bendi, Presidency University, Bangalore, India*

**Chapter 12**
Agricultural Waste Management Through Crop Residue Management: Challenges, Solutions, and Technological Advancements ........ 170
*Chethan Patil N. D., Lovely Professional University, India*
*Suryakanta Kashyap, National Dairy Research Institute, India*
*Sapna Jarial, Lovely Professional University, India*

**Chapter 13**
Articulating Deep Learning and Internet of Things (IoT) in the Hospitality Industry as a Catalyst for Sustainable Consumption and Food Wastage: Unveiling Waste Management Lensing SDG 1 (No Poverty) and SDG 2 (Zero Hunger)........ 182
*Bhupinder Singh, Sharda University, India*
*Vishal Jain, Sharda University, India*
*Christian Kaunert, Dublin City University, Ireland*

**Chapter 14**
Revolutionizing Agricultural Waste Management With Nanotechnology for a Sustainable Future........ 200
*Asha Anish Madhavan, Amity University, Dubai, UAE*
*Godwin Joseph Nallathambi, Amity University, Dubai, UAE*
*Ankita Maria Paul, Amity University, Dubai, UAE*

**Chapter 15**
Green Blockchain Technology for an Eco-Friendly Environment........ 219
*Jonas Yomboi, St. John's Integrated SHS, Ghana*
*Mohammed Majeed, Tamale Technical University, Ghana*
*Esther Asiedu, Ghana Communication Technology University, Ghana*
*Clement Nangpiire, SDD University of Business and Integrated Development Studies, Ghana*
*Fatawu Alhassan, Tamale Technical University, Ghana*
*Victoria Manu, Sunyani Technical University, Ghana*

**Compilation of References**........ 234

**Related References**........ 264

**About the Contributors**........ 284

**Index**........ 290

# Detailed Table of Contents

**Preface** ........ xvi

**Acknowledgement** ........ xx

**Chapter 1**
Innovation in Waste Management: From Trash to Treasure ........ 1
*Richa Kaushal, Lovely Professional University, India*
*Payal Bansal, Lovely Professional University, India*

Fashion industries are growing rapidly for ever changing trends and design, which has long lasting significant on footprints. From the textile production to garment manufacturing various disposal stages of fashion supply chains contribute to pollution and waste. The fashion industry is causing the most pollution, making it the second-largest polluting industry in the world. Industry should make sustainable fabrics and clothes to prevent the Earth from getting polluted. Designers need to create a substitute for fabric, dyes, and method of manufacturing to approach sustainability like substitute denim fabric to reduce the need for dyeing processes. Additionally, industry should produce sustainable leather to eliminate air pollution and loss of animals caused by the leather industry. Some companies are experimenting with making leather using coconuts, which is sustainable. This coconut leather can be recycled and breaks down underwater in just 150 days, unlike real leather, which takes 150 years to decompose, without harming the environment.

**Chapter 2**
Various Upcycling and Recycling Methods ........ 14
*Param jeet Kaur, Lovely Professional University, India*
*Jaswant Singh, Lovely Professional University, India*
*Jyoti Rani, Lovely Professional University, India*

The chapter underscores the crucial role of recycling and upcycling in contributing to several SDGs, including SDG 12 (Responsible Consumption and Production) and SDG 13 (Climate Action). These practices support a closed-loop model, embodying the ethos of "for all forever" within a world grappling with finite resources. By reusing materials and garments, the fashion industry actively engages in waste reduction, mitigates environmental impact on landfills, and conserves resources such as water and virgin fibres, contributing to SDG 15 (Life on Land) and SDG 14 (Life Below Water). In response to sustainability requirements across economic, environmental, and social dimensions, the industry is transitioning from staple fibres to continuous filaments, creating opportunities for innovation in recycling technologies. This shift not only promises economic stimulation and job creation but also aligns with SDG 8 (decent work and economic growth) and SDG 9 (industry, innovation, and infrastructure).

**Chapter 3**
Waste Management and Recovery Practices in the Apparel Sector Through a Sustainable Approach ..................................................................................................................... 37
*Rajveer Kaur Randhawa, Lovely Professional University, India*

The apparel sector, a key driver of global economic growth, confronts pressing challenges related to environmental sustainability. In view of that the current study explores into waste management and recovery practices within the industry, stressing arrangement with Sustainable Development Goal No. 17 viz. Partnerships for the Goals, aiming to address the ecological impact of apparel operations and promote the circular economy. The problem of waste generation in the apparel industry has been significant and requires a thorough investigation, considering the sustainable practices. Circular design principles, including lean manufacturing and the use of recyclable materials, are explored as strategies to minimize waste during production. The forefront of sustainable development in the apparel sector lies in waste recovery initiatives. The study also appraises the effectiveness of existing programs like take-back schemes and closed-loop recycling systems in escapist textile waste from landfills. Through case studies, successful models and strategies aligned with SDG 17.

**Chapter 4**
Importance of Education and Community Engagement for Effective Waste Management Techniques ..................................................................................................................... 53
*Archana Kumari, Lovely Professional University, India*
*Sudev Dutta, Lovely Professional University, India*

A sustainability-oriented educational approach encourages holistic thinking for the benefit of all living organisms and future generations, equipping individuals with the skills to enhance their quality of life responsibly. This mindset empowers global citizens to make informed decisions without compromising the well-being of our planet. Sustainability education seamlessly integrates with core subjects, capturing student interest and enabling them to contextualize coursework in real-life scenarios. Recognizing the escalating global challenges of pollution and waste, UNESCO (full form) has initiated a sustainable education agenda through the Education for Sustainable Development (ESD) program. The present study mainly focuses on education on waste management techniques for sustainability. The approach of sustainability in education will help us to think in a holistic way for all the living organism's life for future aspects. It is nothing more than a skill to learn and improve the quality of life of an individual.

**Chapter 5**
Sustainable Harvesting: Weed Waste Management for Environmental Preservation and Economic Growth ..................................................................................................................... 71
*Chandra Mohan, K.R. Mangalalm University, Gurgaon, India*
*Barkha Sharma, K.R. Mangalam University, Gurugram, India*
*Amit Kumar Rawat, Hansraj College, India*
*Chingakham Chinglenthoiba, Chitkara University, India*

Weed waste management stands at the confluence of environmental preservation, sustainable practices, and economic opportunities. This chapter delves into the multifaceted realm of weed waste management, particularly in the context of tree and vine cropping systems, emphasizing its significance in agricultural productivity, environmental conservation, and resource recovery. The chapter elucidates the diverse types of weeds and their impact on various ecosystems while exploring the principles of the waste-to-wealth approach. The scope of weed waste management encompasses environmental conservation, agricultural productivity, resource recovery, and economic opportunities. By adopting a waste-to-wealth approach, converting weed waste into valuable resources becomes pivotal. This involves practices such as composting, bioenergy production, material recycling, phytoremediation, education, and innovation.

**Chapter 6**
Examining Waste Management in Sustainable Development: Essential Strategies for an Environmentally Friendly Future ........ 89
*Prashant Panwar, K.R. Mangalam University, India*
*Megha Garg, K.R. Mangalam University, India*
*Anil Kumar, Bharati Vidyapeeth's College of Engineering, New Delhi, India*

The increasing global population presents substantial dangers to environmental sustainability due to issues in waste management. This study delves into the complex connection between waste management and sustainable development, highlighting the crucial need for efficient solutions to construct an eco-friendly future. The chapter examines the present waste management procedures, emphasizing the ecological and social consequences of incorrect disposal methods. The research examines the capacity of these new solutions to reduce environmental damage, enhance resource efficiency, and support the achievement of sustainable development objectives. The research also examines how policy frameworks and regulatory measures influence the development of efficient waste management systems.

**Chapter 7**
E-Waste Recycling: A Welcoming Change ........ 96
*Misha Bahmani, IILM Law School, IILM University, Gurugram, India*
*Shobhna Jeet, K.R. Mangalam University, Gurugram, India*

E-waste is globally the most growing solid waste. To attain the goal of securing individuals' health and ecosystem protection recycling of e-waste is essential. This process is beneficial in enhancing the utilization of non-renewable minerals in an adequate manner which results in greener e-product production and economic welfare. It prevents the entering of toxic substances into the ecosystem and lessens the extraction impact on the environment. The authorities in several nations have been facing challenges in dumping the used e-products. Unfortunately, the e-waste dumping areas have been releasing toxic substances which has adversely impacted the livelihood of the recycling workers. To gain valuable substances from these products few companies have been focusing on extracting them through recycle process. E-waste recycling has been appreciated globally but it requires proper supervision and assistance. With the usage of sustainable measures, the goal of reusing and recycling can be achieved when e-waste management is performed by trained persons.

**Chapter 8**
The Role of Solid Waste Management in Boosting Sustainable Development Goals ........ 109
*Joan Nyika, University of Johannesburg, South Africa*
*Megersa Olumana Dinka, University of Johannesburg, South Africa*

The management of solid waste is a growing challenge with rising trends in industrialisation and population growth, and has become a public and environmental health concern. The challenge spills over to negatively affect the realization of sustainable development goals (SDGs). This chapter focuses on how solid waste management (SWM) can be used to promote the realizations of SDGs while prioritizing on economic, social, and environmental growth. Findings show that systematic SWM has the capacity to influence most SDGs and hence the need to uptake it. This can be done by promoting circular economic models, formulating, implementing, and enforcing SWM regulations that are pro-sustainable development, improving governance of the solid waste sector through planning and adoption of better technologies to valorise waste.

**Chapter 9**
Plastic Waste Management for a Green Future: Methods and Challenges ........................................ 125
*Asha Anish Madhavan, Amity University Dubai, UAE*
*Amrutha Nambiar, Amity University Dubai, UAE*
*Simi A. Santosh, Amity University Dubai, UAE*

According to a recent Organization for Economic Cooperation and Development report, global plastic waste production is expected to almost triple by 2060. Since the presence of plastics is unavoidable, the management and proper guidelines for their use are the best way forward. The 3R's—reduce, reuse, and recycle—is the concept that will spread awareness and control use of the plastics. Implementing green chemistry concepts in the fabrication process can help in large reduction of waste output. Plastic waste decomposition can be enhanced by the synergistic action of microorganisms and nanoparticles. Since eliminating plastic altogether is nearly impossible, reusing is considered as one of the easiest and most efficient ways to be sustainable. Moreover, biodegradability of plastics can be improved by the application of nanotechnology. Plastic recycling system currently demands the repurposing of plastic products to a modified form by depositing the waste in proper streams of plastic recycling facilities. This abstract emphasizes the crucial necessity to manage plastic waste effectively for a sustainable and eco-friendly future.

**Chapter 10**
Nanomaterials in Healthcare for Waste Management and Environmental Sustainability .................. 143
*Ginni Rani, DPG Degree College, Gurugram, India*
*Yashika Chitkara, DPG Degree College, Gurugram, India*
*Sunil Kumar, Sushant University, Gurugram, India*
*Chandra Mohan, K.R. Mangalalm University, Gurugram, India*

The strength of a nation always stems from the resilience of each of its individuals. A vibrant and well-functioning society is built upon the well-being of its citizens. The increasing fascination with nanotechnology has prompted a shift in perspective, emphasizing a microscopic examination of problems to devise solutions. Nanomaterials are now integral to various scientific disciplines, with their application in healthcare marking a groundbreaking development that enhances the delivery of quality patient care. A significant share of the nanomaterials created for these purposes presently rely heavily on energy-intensive production methods and non-renewable resources. Furthermore, there exists a substantial delay between the swift advancement or discovery of these unsustainable nanomaterials and their potential long-term impacts on the environment, human health, and climate. Hence, there is an immediate requirement to create nanomaterials in a sustainable manner, utilizing renewable and natural resources while minimizing societal impact.

**Chapter 11**
Waste Management From Agricultural Field: Sustainable Practices and Framework........................ 154
*Aditi Tiwari, Intertek India Pvt. Ltd., Gurugram, India*
*S. V. Niveditha, Presidency University, Bangalore, India*
*Chandra Mohan, K.R. Mangalalm University, Gurgaon, India*
*Anjaneyulu Bendi, Presidency University, Bangalore, India*

Prioritizing the 4Rs of reduce, reuse, recycle, and recover offers the most effective waste management approach. Several ideas to utilize agricultural wastes include biogas production, biochar production, composting, supplementary feed for animal and eco-friendly material construction. Biogas which is a renewable fuel and is produced by anaerobic fermentation of organic matter. It is primarily composed of methane, the key component of natural gas, along with carbon dioxide and trace amounts of other gases. Biochar is the process in which crop residues can be reborn through pyrolysis or other processes. This valuable soil amendment enhances fertility, water retention, and nutrient holding capacity. Composting with crop residues has a long history. Traditionally, farmers pile leftover crops in dung pits, where they decompose and become a valuable resource. The by-products produced from agricultural waste can be used as supplement to animal feed. Agricultural waste can also be used in making eco-friendly building materials such as bricks, panels, etc.

**Chapter 12**
Agricultural Waste Management Through Crop Residue Management: Challenges, Solutions, and Technological Advancements ........................................................................................................ 170
*Chethan Patil N. D., Lovely Professional University, India*
*Suryakanta Kashyap, National Dairy Research Institute, India*
*Sapna Jarial, Lovely Professional University, India*

Crop residue management (CRB) is a global environmental, economic, and health concern. CRB, a common practice in developing countries, is a long-term agricultural waste management strategy. It harms air quality, soil health, and human well-being. It persists due to labor shortages, economic constraints, and limited access to alternative technologies despite its widespread negative consequences. CRB management includes surface retention, mulching, and farm mechanization are examples of on-field management methods; off-field methods include baling, livestock feed supplementation, composting, mushroom cultivation, biochar production, biogas generation, and industrial applications. Modern machinery, such as the Happy Seeder, enables zero-tillage farming and crop residue management. Long-term solutions are required to address this issue. An integrated approach is required to agricultural waste management, combining technological innovation with policy support and community engagement, to reduce the adverse effects of crop residue burning while promoting sustainable agricultural practices.

**Chapter 13**
Articulating Deep Learning and Internet of Things (IoT) in the Hospitality Industry as a Catalyst for Sustainable Consumption and Food Wastage: Unveiling Waste Management Lensing SDG 1 (No Poverty) and SDG 2 (Zero Hunger)........................................................................................ 182
*Bhupinder Singh, Sharda University, India*
*Vishal Jain, Sharda University, India*
*Christian Kaunert, Dublin City University, Ireland*

A new age of innovation in the hospitality sector is being ushered in by the confluence of deep learning (DL) and the internet of things (IoT), which offers a comprehensive solution to sustainability issues. Due in large part to its complex supply networks and high resource consumption, the hospitality sector contributes significantly to food waste and environmental damage. In order to provide a more sustainable and socially responsible approach to waste management, this chapter explores how the combination of DL and IoT systems might play a crucial role in it. DL algorithms are used to optimize a number of areas of hospitality operations because to their capacity to evaluate massive datasets. These models are used in the food service industry for customized menu improvement, inventory control and demand forecasting. These technologies can help to achieve SDGs 1 and 2 by creating ecologically and socially responsible hospitality industries through the promotion of sustainable consumption patterns and effective waste management.

**Chapter 14**
Revolutionizing Agricultural Waste Management With Nanotechnology for a Sustainable Future ........ 200
*Asha Anish Madhavan, Amity University, Dubai, UAE*
*Godwin Joseph Nallathambi, Amity University, Dubai, UAE*
*Ankita Maria Paul, Amity University, Dubai, UAE*

Agricultural operations, essential for human survival, annually produce between 10 to 14 billion tons of waste, endangering environmental stability. This waste encompasses remnants from crops, agro-industrial activities, livestock, and aquaculture. Common disposal methods like incineration and landfills pose health and environmental risks. Effective waste management is imperative for sustainable agriculture, transforming waste into valuable resources like compost, biogas, and biochar. Nanotechnology contributes to waste management by creating nanostructured materials from agricultural waste and generating bioactive compounds for diverse applications. Nano-bioremediation enhances soil vitality and eliminates pollutants, while nano-biosensors enable precise waste monitoring. Smart delivery systems for agrochemicals reduce environmental impact, and nano-enabled plant disease management enhances crop health. Green nanomaterial synthesis from agricultural biomass promotes resource efficiency. These advancements in nanotechnology are fostering sustainable agricultural waste management practices, reducing the environmental footprint of agriculture, and promoting a circular economy.

**Chapter 15**
Green Blockchain Technology for an Eco-Friendly Environment ........ 219
*Jonas Yomboi, St. John's Integrated SHS, Ghana*
*Mohammed Majeed, Tamale Technical University, Ghana*
*Esther Asiedu, Ghana Communication Technology University, Ghana*
*Clement Nangpiire, SDD University of Business and Integrated Development Studies, Ghana*
*Fatawu Alhassan, Tamale Technical University, Ghana*
*Victoria Manu, Sunyani Technical University, Ghana*

As blockchain technology continues its disruptive influence across various sectors, its environmental implications have raised concerns about long-term viability and global repercussions. This chapter explores the concept of "green blockchain" as a framework for sustainable alternatives. It delves into the transition to energy-efficient consensus mechanisms like proof of stake. The chapter also addresses the regulatory landscape, ethical considerations, and future prospects, including the integration of artificial intelligence to optimize blockchain processes. Despite the potential benefits, challenges such as regulatory uncertainty, scalability concerns, and privacy risks underscore the industry's need to navigate carefully and strike a balance between innovation and environmental responsibility. The chapter advocates for a collective effort to build a culture of sustainability within the blockchain community and highlights successful community-led initiatives and regulatory frameworks as crucial elements in mitigating the environmental impact of blockchain technology.

**Compilation of References** ........ 234

**Related References** ........ 264

**About the Contributors** ........ 284

**Index** ........ 290

# Preface

In order to attain the Sustainable Development Goals (SDGs), it is essential to consider waste management from the perspective of sustainable development. Reduced waste, increased recycling, and correct disposal are the three pillars of sustainable waste management policies that safeguard ecosystems and human health. Effective waste management must incorporate the 5Rs hierarchy: reduce, reuse, recycle, recover, and dispose; the circular economy; and extended producer responsibility. Involvement of stakeholders is crucial for waste management systems to be flexible, inclusive, and engaging. It is important for civil society, private companies, and municipalities to get involved. To achieve environmental sustainability and the SDGs, it is crucial to minimise waste, reuse materials, and recover energy. Furthermore, it enhances city life.

The book *Exploring Waste Management in Sustainable Development Contexts* highlighted the need of finding innovative solutions by addressing several aspects of trash management, including waste quantities and varieties. This book emphasises on waste management as a strategy to achieve these objectives. To construct a superior and more enduring future, this book asserts that it is imperative for us to reassess our approach towards waste management and discover novel methods to include the Sustainable Development Goals. This book focuses on water management as a crucial element of sustainable development. It offers practical guidance to readers on how to contribute to a more environmentally conscious future and inspires them to act. Due to its detailed examination, this resource is highly valuable for evaluating the environmental impacts of garbage management.

*Exploring Waste Management in Sustainable Development Contexts* holds significant relevance in today's world due to several critical factors like environmental sustainability, Sustainability Development Goals, Public Health and Interdisciplinary approach. The book will critically develop the understanding of safe disposal methods, waste reduction, and recycling, which are crucial for the society at large globally. The subject matter of the book is aligned with SDG 12 ("Responsible Consumption and Production") and SDG 13 ("Climate Action"). By addressing waste management challenges, it contributes directly to achieving these goals. Sustainable waste practices promote circular economies, resource efficiency, and climate resilience. The book is also exploring innovative technologies, such as waste-to-energy conversion, smart waste monitoring, and decentralized systems. These advancements can revolutionize waste management practices and create sustainable urban environments. The book bridges disciplines such as environmental science, engineering, policy, and social sciences. It encourages collaboration among researchers, policymakers, industry professionals, and community stakeholders. This interdisciplinary perspective is essential for effective waste management solutions.

The book provides valuable insights for students and researchers studying waste management, environmental science, and sustainable development. It offers interdisciplinary perspectives and practical solutions. Policymakers at local, national, and international levels can benefit from the book's recommendations. It highlights the urgent need for innovative waste management solutions aligned with the Sustainable Development Goals (SDGs). Industry professionals, including waste management companies, can gain insights into sustainable practices. The book explores challenges, opportunities, and strategies for greener waste management. The book emphasizes community engagement and awareness. It encourages individuals to rethink their approach to waste and adopt sustainable practices.

Chapter one, "Innovation in Waste Management: From Trash to Treasure" holds immense importance in the context of sustainable development and environmental stewardship. This chapter explores innovative approaches to waste management that go beyond mere disposal. By treating waste as a potential resource, we can recover valuable materials, energy, and nutrients. From recycling plastics to converting organic waste into biogas, these strategies contribute to circular economies and reduce our reliance on finite resources. The chapter emphasizes the need to educate communities about responsible waste management. When people understand the value of waste reduction and resource recovery, they become active participants in the process. Education drives behavioral change and fosters a sense of responsibility.

Chapter two on "Various Upcycling and Recycling Methods" explain how embracing upcycling and recycling methods benefits both the environment and society, and making them essential components of sustainable development. This helps protect ecosystems, minimize energy consumption, and preserve biodiversity. Upcycling and recycling divert waste away from landfills and incinerators. Instead of discarding items, we transform them into new products or materials. This reduces the strain on waste management systems and promotes a circular economy.

Chapter three "Waste Management and Recovery Practices in the Apparel Sector through A Sustainable Approach" emphasizes circularity—where waste becomes a resource. By implementing closed-loop systems, the apparel sector can transform waste into new products. This circular approach aligns with global sustainability goals and promotes economic resilience. this chapter provides actionable insights for textile and apparel companies to adopt sustainable waste practices.

Chapter Four "Importance of Education and Community Engagement for Effective Waste" explain how education is playing a pivotal role in raising awareness about waste management practices. When communities understand the environmental impact of improper waste disposal, they are more likely to adopt responsible behaviours. Community engagement programs can empower individuals to segregate waste, recycle, and reduce consumption. education and community engagement are essential for effective waste management. They empower individuals, promote responsible behaviours, and create a sense of collective responsibility for our environment.

Chapter Five "Sustainable Harvesting- Weed Waste Management for Environmental Preservation and Economic Growth" addresses critical aspects of weed management within the context of sustainable agriculture. This chapter emphasizes the need for sustainable weed management practices that preserve ecosystems, enhance crop productivity, and ensure long term agriculture resilience.

Chapter six "Examining Waste Management in Sustainable Development: Essential Strategies for an Environmentally Friendly Future" holds immense significance in the pursuit of sustainable development. This chapter delves into comprehensive waste management strategies that go beyond mere disposal. And also empowers stakeholders to make informed decisions, protect ecosystems, and foster sustainable development.

Chapter seven "E-Waste Recycling: A Welcoming Change" throw light on E-waste includes discarded electronic and electrical devices such as mobile phones, computers, refrigerators, and more. Incorrect disposal of e-waste poses serious risks to the environment and human health both. Thus, the chapter emphasizes the urgency of sustainable e-waste management. By recycling, reusing and raising awareness, the greener future can be nurtured, and planet can be safeguarded.

Chapter eight “The Role of Solid Waste Management in Boosting Sustainable Development Goals” is of paramount importance for several reasons including environment preservation, resource efficiency, health and wellbeing, economic growth and employment, community engagement and education. The chapter highlights how effective solid waste management directly supports multiple Sustainable Development Goals. By adopting sustainable practices, we create a cleaner, healthier, and more prosperous future for all.

Chapter nine on “Plastic waste Management for a Green Future: Methods and Challenges” holds immense significance in the context of environmental sustainability. Plastic waste is one of the most rapidly growing waste streams worldwide, significantly impacting ecosystems and natural resources. The chapter also highlights several critical issues like reverse supply chain, effective plastic waste management, source- specific recovery and plastic waste rules in India. The author contributed to a greener future by shedding light on plastic waste challenges, solutions, and pathways toward sustainable management.

Chapter ten on “Nanomaterials in Healthcare for Waste Management and Environmental Sustainability” delve into the intersection of nanotechnology, healthcare, and environmental stewardship, highlighting the significance of employing nanomaterials in addressing waste management challenges and advancing environmental sustainability. Within this chapter, it will embark on a journey through the innovative applications of nanomaterials in healthcare settings, where they play a crucial role in managing waste and mitigating environmental impacts. From novel nanomaterial-based medical devices to cutting-edge treatment approaches, this section sets the stage for a deeper understanding of the transformative potential of nanotechnology in healthcare and environmental conservation.

Chapter eleven on “Waste Management from Agricultural Field: Sustainable Practices and Framework” signify the context of environmental sustainability and agricultural development. The chapter throw light on agriculture waste including crop residues, process residues, and produce that becomes unfit due to inadequate storage. The author emphasizes the need for sustainable waste management in agriculture. By adopting eco-friendly practices, we can cultivate a greener tomorrow while supporting economic growth and soil health.

Chapter twelfth on Agricultural Waste Management through Crop Residue Management: Challenges, Solutions, and Technological Advancements plays a crucial role in promoting sustainable agriculture and environmental stewardship. It highlighted and focused on understanding and implementing effective crop residue management strategies as essential for sustainable agriculture, soil health, and environmental preservation.

Chapter thirteen on “Articulating Deep Learning and Internet of Things (IoT) in the Hospitality Industry: Catalyst for Sustainable Consumption and Food Wastages” holds significant importance in the context of sustainable development, particularly in addressing SDG 1 (No Poverty) and SDG 2 (Zero Hunger). this chapter sheds light on how IoT can revolutionize the hospitality industry, promoting sustainability, reducing food waste, and contributing to global goals of poverty alleviation and hunger eradication.

Chapter fourteen on “Revolutionizing Agricultural Waste Management with Nanotechnology for a Sustainable Future” holds immense significance in advancing sustainable agriculture. In summary the author focused on how nanotechnology-driven solutions hold immense promise for revolutionizing agricultural waste management. And by integrating nanotechnology into farming practices, we can achieve higher yields, reduce waste, and foster a greener, more sustainable future.

Chapter fifteen "Green Blockchain Technology for Eco-Friendly Environment" investigates how blockchain technology can accelerate development by adopting renewable and clean energy and reducing greenhouse gas emissions. And it hold immense promise for achieving eco-friendly goals, reducing emissions, and fostering sustainability.

## CONCLUSION

The book on *Exploring Waste Management in Sustainable Development Contexts* makes a significant impact on the field by providing a comprehensive and in-depth analysis of the various strategies and approaches to waste management in the context of sustainable development. Through a combination of theoretical insights and practical case studies, the book highlights the importance of addressing waste management issues in a holistic and integrated manner, considering the environmental, social, and economic dimensions of sustainability. The book contributes to the subject matter by advocating for a shift towards a more circular economy approach, where waste is seen as a valuable resource that can be reused, recycled, or repurposed. By emphasizing the need for innovative solutions and effective governance mechanisms, the book provides valuable insights and recommendations for policymakers, practitioners, and researchers in the field of waste management and sustainable development. Overall, the book serves as a valuable resource for anyone interested in understanding the complex interplay between waste management and sustainable development and offers practical guidance on how to address these challenges in a more sustainable and effective manner.

*Chandra Mohan*
*K.R. Mangalalm University, Gurgaon, India*

*Shobhna Jeet*
*K.R. Mangalam University, Gurugram, India*

*Saurav Dixit*
*Division of Research and Innovation, Uttaranchal University, Dehradun, India & Peter the Great St. Petersburg Polytechnic University, Saint Petersburg, Russia*

*Sónia A.C. Carabineiro*
*NOVA University of Lisbon, Portugal*

# Acknowledgement

We wish to extend our heartfelt appreciation to all individuals and organizations whose contributions have played a pivotal role in bringing this endeavor to fruition.

In particular, we express our gratitude to everyone who contributed to the development of this work on waste management for sustainable development. Foremost among these are the individuals and communities directly impacted by waste pollution, whose experiences serve as a poignant reminder of the significance of our efforts in waste management.

We are profoundly thankful to the researchers, academicians, and experts working in the field of sustainable development, environmental pollution, waste management, circular economy, engineering, and allied fields. Their insights and contributions have greatly enriched the content of this book.

We also acknowledge with gratitude the support and guidance provided by our colleagues and friends from K R Mangalam University, Gurugram, who offered valuable feedback and encouragement throughout the writing process.

Furthermore, we extend our appreciation to the publishers, editors, and reviewers from national and international institutions whose invaluable contributions have played a crucial role in shaping this work and bringing it to fruition.

Collectively, we aspire that this book serves as a meaningful contribution to the ongoing efforts surrounding waste management, sustainability, and environmental safety.

*Chandra Mohan*
*K.R. Mangalalm University, Gurgaon, India*

*Shobhna Jeet*
*K.R. Mangalam University, Gurugram, India*

*Saurav Dixit*
*Division of Research and Innovation, Uttaranchal University, Dehradun, India & Peter the Great St. Petersburg Polytechnic University, Saint Petersburg, Russia*

*Sónia A.C. Carabineiro*
*NOVA University of Lisbon, Portugal*

# Chapter 1
# Innovation in Waste Management:
## From Trash to Treasure

**Richa Kaushal**
*Lovely Professional University, India*

**Payal Bansal**
https://orcid.org/0000-0003-1936-0834
*Lovely Professional University, India*

## ABSTRACT

*Fashion industries are growing rapidly for ever changing trends and design, which has long lasting significant on footprints. From the textile production to garment manufacturing various disposal stages of fashion supply chains contribute to pollution and waste. The fashion industry is causing the most pollution, making it the second-largest polluting industry in the world. Industry should make sustainable fabrics and clothes to prevent the Earth from getting polluted. Designers need to create a substitute for fabric, dyes, and method of manufacturing to approach sustainability like substitute denim fabric to reduce the need for dyeing processes. Additionally, industry should produce sustainable leather to eliminate air pollution and loss of animals caused by the leather industry. Some companies are experimenting with making leather using coconuts, which is sustainable. This coconut leather can be recycled and breaks down underwater in just 150 days, unlike real leather, which takes 150 years to decompose, without harming the environment.*

## 1. INTRODUCTION

The world of fashion, a monarchy is defined by its continuous evolution of trends and design, has undergone rapid expansion, etching a lasting mark on the environment (Fletcher, 2018). As the stages of the fashion supply chain unfold, from the beginning of textiles to the culmination of garment manufacturing, a stark reality emerges as one of substantial contributions to pollution and waste. It is a reality

DOI: 10.4018/979-8-3693-4264-0.ch001

that has propelled the fashion industry to the dubious distinction of being the second-largest polluter on a global scale (Brown, 2017). The repercussions of this environmental toll are becoming increasingly evident, prompting a resounding call for a fundamental shift towards sustainability across every facet of the industry, from the genesis of fabrics to the responsible disposal of waste. In its pursuit of aesthetic innovation and consumer demand, the fashion industry has, to a large extent, turned a blind eye to the ecological consequences of its practices (Fletcher & Grose, 2012). Textile production, a key player in the fashion supply chain, has long been a focal point of environmental concern. Conventional methods employed in textile production involve the deployment of toxic chemicals, substantial water consumption, and insatiable energy usage. The cumulative effect of these practices is a conspicuous strain on the environment, resulting in polluted water bodies and compromised air quality.

The gravity of the situation is underscored by the industry's staggering position as the second-largest contributor to global pollution (Black, 2019). The urgency to address this environmental swamp has prompted a paradigm shift in the discourse surrounding the fashion industry. No longer can it operate with impunity, disregarding the far-reaching consequences of its actions. The time has come for the industry to re-examine its modus operandi and embrace sustainable practices that uphold not only the integrity of fashion but also the health of our planet (Black and Harben, 2015). This clarion call for sustainability extends beyond mere rhetoric, heralding a transformative journey that traverses the entire fashion supply chain. From the cultivation of raw materials to the disposal of end-of-life products, every stage demands a conscientious reevaluation. The need for a sustainable metamorphosis is not just a moral imperative; it is a strategic imperative for the very survival of an industry that has, for far too long, been synonymous with excess and disregard for environmental consequences. As the environmental repercussions of the fashion industry become increasingly apparent, a growing movement is pushing change (Höjer & Wangel, 2015). This movement advocates for a holistic approach to sustainability that encompasses every facet of the industry. The shift towards sustainability is not a mere trend but an imperative born out of the necessity to reconcile the allure of fashion with the imperative to protect our environment.

This introduction serves as a prelude to a deeper exploration of the environmental challenges posed by the fashion industry and the imperative for a sustainable overhaul (Fletcher, 2016). This introduction serves as a prelude to a deeper exploration of the environmental challenges posed by the fashion industry and the imperative for a sustainable overhaul. The subsequent sections will delve into specific aspects of the industry, scrutinizing the impact of textile production and garment manufacturing on the environment. Moreover, the discussion will extend to the urgent need for the industry to adopt eco-friendly practices, from the development of sustainable fabrics to the responsible management of waste. As the fashion industry stands at the crossroads of environmental responsibility, the following narrative aims to unravel the complexities and possibilities inherent in this transformative journey towards a more sustainable and responsible future.

## 2. TEXTILE PRODUCTION: A SOURCE OF POLLUTION

One of the major contributors to environmental degradation in the fashion industry is textile production (Bair et al., 2019). The conventional methods of producing fabrics involve the use of toxic chemicals, excessive water consumption, and high energy usage (Shahin et al., 2017). These practices lead to pollution of water bodies and air, negatively impacting ecosystems and communities. To address this issue, the industry must prioritize the development and use of sustainable fabrics (Gwilt & Rissanen, 2011).

Many new fabrics are being introduced in the market to lower down the environmental impact like organic cotton, linen, jute etc. (Lee et al., 2018). These fabrics should be produced using eco-friendly methods that minimize the use of harmful chemicals and reduce water and energy consumption (Paul et al., 2019). By embracing sustainable textiles, the fashion industry can significantly decrease its ecological footprint.

Denim, a beloved and ubiquitous fabric, is notorious for its environmental impact due to the extensive dyeing processes involved in its production. Designers are now challenged to find substitutes for denim that can reduce the need for these harmful dyeing processes. The emphasis is on creating fabrics that are not only environmentally friendly but also aesthetically pleasing and durable (Fletcher & Grose, 2012). Innovations in fabric development are crucial for (Niinimäki, 2018). Designers and manufacturers are exploring alternative materials that can emulate the look and feel of denim without the environmental drawbacks (Höjer & Wangel, 2015). By investing in research and development, the fashion industry can revolutionize the materials it uses, paving the way for a more sustainable future. The environmental toll of the fashion industry is significantly exacerbated by textile production, a major contributor to environmental degradation (Brown, 2017). Traditional fabric manufacturing relies on methods fraught with environmental hazards, involving the use of toxic chemicals, extravagant water consumption, and intensive energy usage (Choi & Cheng, 2015). The widespread adoption of these practices has resulted in severe consequences, including the contamination of water bodies and air, which in turn adversely affects ecosystems and communities (Kim & Kim, 2018).

Toxic chemicals employed in dyeing and finishing processes pose a threat to both aquatic life and human health (Muthu, 2015; Mohan et al., 2023). Excessive water consumption exacerbates water scarcity issues, particularly in regions already grappling with limited resources (Pookulangara & Shephard, 2013). Additionally, the high energy demands of textile production contribute to carbon emissions and climate change (Elander & Nylund, 2018). Addressing these challenges necessitates a shift towards sustainable and eco-friendly alternatives in textile production (Gwilt & Rissanen, 2019). Initiatives promoting responsible sourcing, circular economy practices, and the adoption of innovative, less resource-intensive technologies are crucial for mitigating the environmental impact of the fashion industry and fostering a more sustainable future (Labelling & Thörn, 2018). Humans have always been fascinated with color, because undyed clothing does not normally attract attention. To make different designs and aesthetic appeal in clothing, experts introduced the dyeing process. (Gwilt & Rissanen, 2019). Initiatives promoting responsible sourcing, circular economy practices, and the adoption of innovative, less resource-intensive technologies are crucial for mitigating the environmental impact of the fashion industry and fostering a more sustainable future (Labelling & Thörn, 2018; Surender et al., 2023). Dyeing is a textile wet processing in which color is incorporated into fibrous products in different forms such as loose fiber, yarn, fabric, and nonwoven in a suitable dyeing machine. Before dyeing, preparatory processes are required to remove natural as well as added impurities present in the fiber. Textile production, particularly in the dyeing process, has historically been a significant contributor to environmental pollution. The introduction of dyes into fibrous materials involves a series of wet processing steps, necessitating the removal of natural impurities as well as added substances present in the fibers. These preparatory processes often involve the use of chemicals and large quantities of water, further exacerbating the environmental impact of textile production.

The dyeing process itself poses significant challenges in terms of environmental sustainability (Fletcher & Grose, 2012). Natural dyes offer several advantages over their synthetic counterparts. They are typically biodegradable, non-toxic, and less harmful to both the environment and human health (Lee et al., 2018). Traditional dyeing methods rely on the use of synthetic dyes, many of which contain toxic chemicals that can harm both aquatic ecosystems and human health. Moreover, the dyeing process often results in the discharge of untreated wastewater containing residual dyes and chemicals into water bodies, contributing to water pollution and disrupting fragile aquatic habitats (Surender et al., 2023; Brown, 2017). In recent years, there has been a growing recognition of the need to address these environmental concerns associated with textile dyeing. Researchers and industry professionals alike have been exploring alternative dyeing techniques and materials that are less harmful to the environment. This includes the development of natural dyes sourced from plant-based or renewable sources, as well as the utilization of innovative dyeing processes that minimize water consumption and chemical usage. Furthermore, advancements in technology have enabled the implementation of more sustainable dyeing practices, such as digital printing and laser dyeing, which offer greater precision and efficiency while reducing environmental impact. Additionally, initiatives aimed at recycling and reusing wastewater from the dyeing process are gaining traction, helping to minimize water waste and pollution. Overall, addressing the environmental impact of textile dyeing requires a multifaceted approach that encompasses technological innovation, sustainable material sourcing, and regulatory measures to ensure compliance with environmental standards. By embracing these strategies, the fashion industry can work towards minimizing its ecological footprint and promoting a more sustainable future for textile production.

## 3. DYEING PROCESSES: THE QUEST FOR ALTERNATIVES

Dyeing processes in the textile industry have long been associated with environmental degradation and pollution. The conventional methods of dyeing fabrics often involve the use of synthetic dyes containing harmful chemicals, excessive water consumption, and the generation of large volumes of wastewater. As concerns about environmental sustainability continue to grow, there is an increasing demand for alternative dyeing processes that minimize the industry's ecological footprint. One promising avenue for reducing the environmental impact of dyeing is the development and adoption of natural dyes sourced from renewable materials such as plants, insects, and minerals. Natural dyes offer several advantages over their synthetic counterparts. They are typically biodegradable, non-toxic, and less harmful to both the environment and human health. Additionally, natural dyes can often be produced using locally available resources, reducing the need for transportation and associated carbon emissions.

In recent years, there has been a resurgence of interest in traditional dyeing techniques that utilize natural dyes. Artisans and small-scale producers around the world have been reviving age-old dyeing practices, such as indigo dyeing, tie-dyeing, and plant-based dyeing, to create unique and environmentally friendly textiles (Lee et al., 2018; Mohan and Kumar, 2021). These traditional methods not only offer a more sustainable alternative to synthetic dyes but also preserve cultural heritage and support local communities. Another promising approach to sustainable dyeing is the development of innovative dyeing technologies that reduce water consumption and chemical usage. One such technology is digital printing, which allows for precise and efficient application of color onto fabrics without the need for extensive water baths or chemical fixatives (Gwilt & Rissanen, 2011). Digital printing offers significant environmental benefits, including reduced water waste, lower energy consumption, and fewer chemicals discharged into the environment. Similarly, advancements in dyeing machinery and equipment have led to the emergence of waterless dyeing processes that use minimal or no water at all. For example, air dyeing and plasma dyeing technologies utilize alternative methods, such as heat and pressure, to fix dyes onto fabrics without the need for water baths. These waterless dyeing processes not only conserve water resources but also eliminate the discharge of wastewater, thereby reducing pollution and environmental impact.

In addition to natural dyes and innovative technologies, the fashion industry is also exploring alternative materials that require less dyeing or no dyeing at all. For instance, the growing popularity of organic cotton, hemp, and other natural fibers with inherent color variations reduces the need for dyeing and chemical processing. Similarly, the use of recycled fibers and materials derived from post-consumer waste can help minimize the environmental impact of dyeing by reducing the demand for virgin materials and the associated resource extraction. Overall, the quest for alternatives to conventional dyeing processes is driven by a growing awareness of the environmental and social impacts of textile production. By embracing natural dyes, innovative technologies, and alternative materials, the fashion industry can reduce its ecological footprint and move towards more sustainable and responsible practices. However, widespread adoption of these alternatives will require collaboration and investment across the supply chain, as well as support from consumers who prioritize sustainability in their purchasing decisions (Mohan et al., 2023).

## 4. SUSTAINABLE LEATHER: A CALL FOR CHANGE FOR SUSTAINABLE ALTERNATIVE

Leather production, another integral part of the fashion industry, has long been associated with air pollution and the loss of animal life. Traditional leather production involves the use of toxic chemicals and has detrimental effects on both human health and the environment (Fletcher & Grose 2012). To mitigate these issues, the industry must transition to sustainable leather alternatives. Some companies are leading the way by experimenting with innovative materials, such as coconut leather. This sustainable option not only addresses the environmental impact of traditional leather but also offers a viable solution to the waste problem. Leather production, while long revered for its durability and versatility in fashion, has faced mounting criticism due to its environmental and ethical implications (Fletcher, 2018; Surender et al., 2023). Traditional methods of tanning leather involve the use of toxic chemicals such as chromium, which not only pose risks to human health but also contribute to air and water pollution. Furthermore, the demand for leather has driven deforestation and habitat destruction, leading to the loss of animal life and biodiversity.

In response to these challenges, there is a growing call for change within the leather industry to adopt more sustainable practices. One key area of focus is the development of alternative materials that mimic the look and feel of traditional leather but have a reduced environmental footprint. Companies and researchers are exploring a variety of innovative materials, such as pineapple leather, mushroom leather, and even lab-grown leather, which offer promising alternatives to conventional animal leather. Pineapple leather, also known as Piñatex, is made from the fibers of pineapple leaves, a byproduct of the pineapple industry. The process involves extracting fibers from the leaves, which are then processed into a non-woven textile that closely resembles traditional leather. Not only does pineapple leather utilize a waste product, but it also requires fewer resources and chemicals compared to conventional leather production, making it a more sustainable option for fashion brands and consumers alike. Similarly, mushroom leather, or Mycelium leather, is produced using fungal mycelium grown on agricultural waste, such as sawdust or corn husks (Besseling et al., 2020). The mycelium forms a dense network of fibers that can be tanned and processed into a leather-like material. Mushroom leather offers several environmental benefits, including biodegradability, low resource consumption, and the ability to be grown in controlled environments, reducing the need for land and water (Choi & Li, 2019).

Lab-grown leather, also known as cultured leather or biofabricated leather is produced using tissue engineering techniques that involve growing animal cells in a lab setting. This approach eliminates the need for animal slaughter and reduces the environmental impact associated with traditional leather production. Lab-grown leather also offers greater customization and scalability, allowing for the production of leather with specific properties and characteristics (Muthu et al., 2020). In addition to developing alternative materials; efforts are underway to improve the sustainability of traditional leather production processes. Sustainable tanning methods, such as vegetable tanning, utilize natural tannins derived from plant sources, such as tree bark, to tan leather without the use of harmful chemicals. Vegetable-tanned leather is biodegradable, non-toxic, and often has a richer, more natural appearance compared to chrome-tanned leather. Furthermore, initiatives aimed at promoting transparency and traceability within the leather supply chain are gaining traction, allowing consumers to make more informed choices about the products they purchase. Certifications such as the Leather Working Group (LWG) certification assess the environmental impact of leather production and help brands and consumers identify more sustainable options.

Overall, the shift towards sustainable leather production represents a significant opportunity for the fashion industry to reduce its environmental footprint and promote more ethical and responsible practices. By embracing innovative materials, improving production processes, and fostering greater transparency, the industry can pave the way for a more sustainable future for leather fashion. Coconut leather is emerging as a promising alternative to traditional animal leather. Produced from the fibers of coconuts, this material is not only cruelty-free but also environmentally friendly. Unlike conventional leather, which takes a staggering 150 years to decompose, coconut leather breaks down underwater in just 150 days, leaving no harmful residues (Mohan and Kumar, 2021). The production process of coconut leather involves minimal chemical usage, making it a cleaner and greener option. Additionally, coconut leather can be recycled, further reducing its environmental impact. As the fashion industry continues to explore and adopt sustainable alternatives like coconut leather, it marks a significant step towards minimizing its contribution to pollution and waste.

## 5. CIRCULAR FASHION ECONOMY: REDEFINING CONSUMPTION AND PRODUCTION

In the quest for sustainability, the fashion industry is increasingly turning its attention to the concept of a circular fashion economy. This approach challenges the traditional linear model of "take, make, dispose" by promoting practices that extend the lifespan of garments, reduce waste, and encourage recycling. Exploring innovations such as clothing rental services, up cycling initiatives, and closed-loop manufacturing processes, the circular fashion economy presents an opportunity for the industry to minimize its environmental footprint and create a more sustainable and responsible ecosystem. The concept of a circular fashion economy represents a paradigm shift in the way the fashion industry approaches consumption and production. Traditionally, the industry has operated on a linear model, where garments are produced, sold, and eventually discarded after use, leading to significant waste and environmental degradation. However, with growing concerns about sustainability and resource scarcity, there is a pressing need for a more regenerative approach to fashion.

At the heart of the circular fashion economy is the idea of extending the lifespan of garments through various strategies. One such strategy is the rise of clothing rental services, where consumers can borrow clothing items for a limited period instead of purchasing them outright (Elander & Nylund, 2018). This not only reduces the demand for new clothing production but also allows consumers to access a wider variety of garments without contributing to overconsumption. Companies like Rent the Runway and Le Tote have gained popularity by offering subscription-based rental services, catering to consumers' desire for variety and sustainability. Additionally, up cycling initiatives play a crucial role in prolonging the life of garments and reducing waste. Up cycling involves repurposing old or discarded clothing items into new products, thereby giving them a second life. This can range from simple alterations and repairs to more creative transformations, such as turning old jeans into a denim tote bag or repurposing a worn-out sweater into a cozy blanket (Agrawal et al., 2021; Muralia et al., 2023). By encouraging creativity and resourcefulness, up cycling not only reduces the amount of textile waste sent to landfills but also fosters a culture of sustainability and innovation within the fashion industry. Closed-loop manufacturing processes are another key component of the circular fashion economy. Unlike traditional manufacturing methods, which rely on a linear supply chain with a clear endpoint, closed-loop systems are designed to minimize waste by recycling and reusing materials throughout the production process. This can involve using recycled fibers or fabrics, incorporating renewable materials, and designing products with end-of-life considerations in mind. By closing the loop and creating a more circular supply chain, manufacturers can significantly reduce their environmental impact and contribute to a more sustainable future for fashion.

The advancements in technology and material science are driving innovation in sustainable fashion practices. For example, 3D printing technology allows designers to create intricate garments with minimal waste, as materials are only used where needed (Fletcher & Grose, 2012). Similarly, bio fabrication techniques enable the production of textiles and materials using renewable resources such as algae or bacteria, offering a more eco-friendly alternative to traditional fabrics.

However, transitioning to a circular fashion economy is not without its challenges. It requires collaboration and commitment from all stakeholders, including designers, manufacturers, retailers, and consumers. Education and awareness are essential in driving consumer behavior change and encouraging sustainable purchasing habits (Lu et al., 2020). Additionally, policymakers play a crucial role in incentivizing sustainable practices and regulating the fashion industry to ensure compliance with environmental standards. The circular fashion economy offers a promising solution to the environmental and social challenges facing the fashion industry (Höjer & Wangel, 2015). By redefining consumption and production practices, embracing innovation, and fostering collaboration across the supply chain, the industry can create a more sustainable and responsible ecosystem that benefits both people and the planet. Through collective action and a commitment to circularity, the fashion industry has the potential to transform itself into a force for positive change in the global fight against climate change and resource depletion.

## 6. CONTENT FOR CIRCULAR FASHION ECONOMY

The linear fashion model, characterized by the rapid turnover of trends and the disposal of clothing after limited use, contributes significantly to the industry's environmental impact. Enter the circular fashion economy, a paradigm shift that prioritizes longevity, durability, and recyclability. Initiatives such as clothing rental services and collaborative efforts between designers and consumers to repurpose and up cycle garments are gaining traction (Muralia et al., 2023; Black, 2019). By embracing these circular practices, the fashion industry can reduce the pressure on raw material extraction, lower energy consumption, and diminish the volume of textile waste destined for landfills. The linear fashion model, characterized by its fast-paced trends and disposable clothing culture, has undeniably contributed to the fashion industry's environmental footprint. From resource-intensive production processes to the staggering volumes of textile waste ending up in landfills, the linear model's shortcomings are increasingly evident in today's global context of environmental consciousness.

In contrast, the circular fashion economy represents a fundamental shift in the way we approach fashion consumption and production. At its core, this paradigm prioritizes longevity, durability, and recyclability, aiming to break away from the "take, make, dispose" mentality that has dominated the industry for so long. Central to the circular fashion economy are initiatives that extend the lifespan of garments and minimize waste. Clothing rental services, for instance, offer consumers the opportunity to enjoy the latest styles without the need for constant purchasing. By circulating clothing items among multiple users, these services not only reduce the demand for new production but also promote a more sustainable approach to fashion consumption. Moreover, collaborative efforts between designers and consumers to repurpose and up cycle garments are gaining momentum. Up cycling, in particular, has emerged as a creative way to breathe new life into old clothing items, transforming them into unique and innovative pieces that retain their value and functionality. This not only diverts textiles from landfills but also encourages a more circular and resource-efficient approach to fashion design.

Additionally, closed-loop manufacturing processes play a crucial role in the circular fashion economy by minimizing waste and maximizing resource efficiency. By incorporating recycled materials and designing products with recyclability in mind, manufacturers can significantly reduce their environmental impact while contributing to the circularity of the fashion industry. Overall, the transition to a circular fashion economy holds immense promise for the industry's sustainability efforts. By embracing practices that prioritize longevity, durability, and recyclability, the fashion industry can move towards a more sustainable and responsible future. However, achieving this vision will require collaboration, innovation, and collective action across the entire fashion ecosystem, from designers and manufacturers to retailers and consumers (Fletcher & Grose, 2012). Together, we can create a fashion industry that not only looks good but also does well for the planet.

## 7. ETHICAL AND TRANSPARENT SUPPLY CHAINS: A PILLAR OF SUSTAINABLE FASHION

The environmental impact of the fashion industry is not limited to production processes; it extends throughout the entire supply chain. Addressing issues such as unfair labor practices, unsafe working conditions, and lack of transparency is crucial for achieving true sustainability. The move towards ethical and transparent supply chains involves promoting fair wages, ensuring safe working environments, and providing consumers with information about the origins of the products they purchase. By fostering accountability and transparency, the fashion industry can build trust with consumers and contribute to a more equitable and sustainable global fashion landscape. Ethical and transparent supply chains are indeed fundamental pillars of sustainable fashion, encompassing not only environmental concerns but also social and ethical considerations. As consumers increasingly prioritize sustainability and ethical practices in their purchasing decisions, the fashion industry is facing mounting pressure to address issues such as fair labor practices, worker safety, and supply chain transparency.

One of the most pressing issues within the fashion industry is the prevalence of exploitative labor practices in garment manufacturing, particularly in developing countries where labor regulations may be lax and enforcement is weak (Höjer & Wangel, 2015). Workers in many parts of the world endure long hours, low wages, and unsafe working conditions to produce clothing for global markets. Addressing these issues requires a concerted effort from brands, manufacturers, and policymakers to ensure that workers are treated fairly and ethically throughout the supply chain. Promoting fair wages is a crucial step towards ensuring that garment workers receive adequate compensation for their labor. Many workers in the fashion industry, particularly those in developing countries, are paid below living wages, forcing them into poverty and perpetuating cycles of exploitation. By implementing fair wage policies and partnering with suppliers that uphold ethical labor practices, brands can help improve the livelihoods of garment workers and support sustainable economic development in communities around the world. Ensuring safe working environments is another essential aspect of ethical supply chains (Pookulangara & Shephard, 2013). Many garment factories operate in unsafe conditions, with inadequate safety measures and lack of oversight leading to accidents and injuries. Brands must take responsibility for the safety of workers in their supply chains by conducting regular audits, providing training on workplace safety, and investing in infrastructure improvements to mitigate risks.

Transparency is also key to building trust with consumers and holding brands accountable for their supply chain practices. Consumers increasingly want to know where their clothing comes from and under what conditions it was produced. Brands that are transparent about their supply chain practices, including sourcing, manufacturing, and labor conditions, are better positioned to earn the trust and loyalty of consumers who value sustainability and ethical responsibility. One way to increase transparency in the fashion industry is through certification programs and third-party audits that verify compliance with ethical and sustainability standards. Certifications such as Fair Trade, Organic, and the Ethical Trading Initiative provide assurance to consumers that products have been produced in accordance with rigorous social and environmental criteria. By seeking out certified products and supporting brands that prioritize transparency, consumers can drive positive change within the fashion industry.

In addition to external certifications, brands can also take proactive steps to increase transparency within their supply chains. This may include disclosing supplier lists, publishing annual sustainability reports, and engaging with stakeholders to address concerns and improve practices. By being open and transparent about their supply chain practices, brands can demonstrate their commitment to sustainability and accountability. In conclusion, ethical and transparent supply chains are essential for achieving true sustainability in the fashion industry. By promoting fair labor practices, ensuring safe working conditions, and increasing transparency, brands can build trust with consumers and contribute to a more equitable and sustainable global fashion landscape. However, achieving these goals will require collaboration and commitment from all stakeholders, including brands, manufacturers, policymakers, and consumers. Together, we can work towards a fashion industry that respects the rights and dignity of workers, protects the environment, and promotes social justice.

## 8. CONTENT FOR ETHICAL AND TRANSPARENT SUPPLY CHAINS

Sustainable fashion goes beyond ecological concerns; it encompasses the ethical treatment of workers and transparency in the supply chain. The fashion industry has faced scrutiny for exploitative labor practices and opaque supply chains. The imperative now is to shift towards ethical sourcing, fair wages, and safe working conditions. Brands that prioritize transparency by providing consumers with information about the production journey of their garments are gaining favor. As consumers increasingly seek ethically produced clothing, the industry must commit to fostering fair and transparent practices throughout the entire supply chain to build a foundation of trust and accountability. Sustainable fashion, often associated solely with environmental concerns, is a multifaceted concept that extends far beyond ecological considerations. It encompasses a broader spectrum of ethical principles, including the fair treatment of workers and transparency in the fashion supply chain. While significant strides have been made in recent years to address environmental sustainability within the industry, the spotlight has increasingly turned towards the social and ethical dimensions of fashion production.

One of the most pressing challenges facing the fashion industry is the prevalence of exploitative labor practices in garment manufacturing. In many regions around the world, particularly in developing countries, workers endure long hours, low wages, and unsafe working conditions in order to produce clothing for global markets. This exploitation is often exacerbated by a lack of legal protections, weak enforcement mechanisms, and limited opportunities for collective bargaining. Addressing these systemic issues requires a concerted effort from brands, manufacturers, governments, and civil society to ensure that workers are treated with dignity and respect throughout the entire supply chain. Central to the concept of ethical fashion is the principle of fair wages (Gwilt & Rissanen, 2011). Many garment workers, particularly those in countries with lax labor regulations, are paid below living wages, forcing them into poverty and perpetuating cycles of exploitation. Ensuring fair wages for workers is not only a matter of social justice but also essential for promoting sustainable economic development and reducing inequality. Brands must take responsibility for the wages paid to workers in their supply chains, working closely with suppliers to ensure that fair compensation is provided for the labor that goes into producing their garments. Moreover, ensuring safe working conditions is paramount to the ethical production of fashion. Too often, garment factories operate in unsafe environments, with inadequate safety measures, poor ventilation, and lack of emergency exits. This puts workers at risk of accidents, injuries, and even death (Bhardwaj and Fairhurst, 2010). Brands must prioritize the safety and well-being of workers in their supply chains by conducting regular inspections, investing in infrastructure improvements, and providing training on workplace health and safety protocols. By creating safe and healthy working environments, brands can help protect the rights and dignity of workers while also improving productivity and morale.

Transparency is another key pillar of ethical fashion, allowing consumers to make informed choices about the products they purchase. In recent years, there has been a growing demand for greater transparency in the fashion industry, with consumers increasingly seeking information about the production journey of their garments. Brands that prioritize transparency by disclosing information about their suppliers, production processes, and labor practices are gaining favor among consumers who value ethical and sustainable fashion. By providing this information, brands can build trust and credibility with consumers, demonstrating their commitment to ethical principles and accountability (Gwilt & Rissanen, 2011). In conclusion, ethical and transparent supply chains are essential for achieving true sustainability in the fashion industry. By prioritizing ethical sourcing, fair wages, safe working conditions, and transparency, brands can build a foundation of trust and accountability with consumers. However, achieving these goals will require collaboration and commitment from all stakeholders, including brands, manufacturers, governments, and consumers. Together, we can work towards a fashion industry that respects the rights and dignity of workers, protects the environment, and promotes social justice.

## 9. CONCLUSION

The chapter elaborated fashion industries which are convert its practices and methods towards a more sustainable and eco-friendly future without harming the environment. The call for sustainability in the fashion industry has led to a paradigm shift, with a growing emphasis on adopting more environmentally friendly practices across various aspects, including dyeing, leather production and the transition towards a circular fashion economy. The quest for sustainability in the fashion industry has become an urgent imperative as the environmental toll of textile production, particularly dyeing processes, continues to escalate. Conventional methods, laden with toxic chemicals, excessive water consumption, and pollution, have led to widespread environmental degradation and societal concerns. One of the most significant strides towards sustainability in textile production involves the exploration and adoption of alternative dyeing processes and use of natural dyes. These dyes are biodegradable, non-toxic, and often produced using locally available resources, thus reducing the carbon footprint associated with transportation. Further, digital printing for instance enables precise and efficient application of color onto fabrics without the need for extensive water baths or chemical fixatives. This technology significantly reduces water waste, energy consumption, and chemical discharge, thus offering a more environmentally friendly alternative to conventional dyeing methods. Similarly, waterless dyeing processes, such as air dyeing and plasma dyeing, utilize alternative methods to fix dyes onto fabrics without the need for water baths. By eliminating the discharge of wastewater and minimizing water consumption, these technologies address key environmental concerns associated with textile dyeing. Additionally, the use of recycled fibers and materials derived from post-consumer waste further minimizes the environmental impact of dyeing by reducing the demand for virgin materials and resource extraction. Industry standards and regulatory measures can also play a crucial role in incentivizing sustainable practices and holding companies accountable for their environmental impact. As consumers increasingly prioritize sustainability in their purchasing decisions, fashion brands have a unique opportunity to lead by example and drive positive change in the industry.

The traditional methods for leather production involved toxic chemicals and animal cruelty which is not environment friendly. The emergence of innovative materials such as pineapple leather, mushroom leather, and lab-grown leather offers promising solutions. These alternatives not only mimic the qualities of traditional leather but also boast lower environmental footprints and cruelty-free production methods. Additionally, sustainable tanning techniques like vegetable tanning further contribute to reducing the industry's environmental impact. Furthermore, the concept of a circular fashion economy is gaining momentum as a means to tackle the industry's waste and environmental degradation. The circular fashion economy promotes practices that extend the lifespan of garments, reduce waste, and encourage recycling. Initiatives such as clothing rental services, up cycling, and closed-loop manufacturing processes are reshaping the way consumers interact with fashion and the way garments are produced. These initiatives not only minimize waste but also foster a culture of sustainability and innovation within the industry. The transition towards a circular fashion economy is not without its challenges, requiring collaboration and commitment from all stakeholders. Education and awareness are essential in driving consumer behavior change and encouraging sustainable purchasing habits. The convergence of sustainability, ethics, and transparency is reshaping the fashion industry, marking a pivotal moment in its evolution. By embracing innovation, fostering collaboration, and prioritizing sustainability throughout the supply chain, the fashion industry has the potential to become a positive force for change in the global fight against climate change and resource depletion. Further, brands must prioritize fair wages, safe working conditions, and transparency to uphold the rights and dignity of workers, fostering a culture of accountability and social responsibility. Transparency emerges as a key driver of change, enabling consumers to make informed choices and hold brands accountable for their practices. Brands, manufacturers, governments, and consumers must work together to implement systemic changes that address the root causes of environmental degradation, social injustice, and exploitation. By embracing sustainability, transparency, and ethical practices, the fashion industry can redefine its role in society, becoming a force for positive change and leading the way towards a more equitable, sustainable, and responsible future.

## REFERENCES

Agrawal, A., Sharma, R., & Singh, G. (2021). Sustainable Leather: A Review on Innovative Materials and Production Processes. *Journal of Cleaner Production*, *293*, 126084. doi:10.1016/j.jclepro.2021.126084

Besseling, E., Wang, B., & Lübken, M. (2020). Coconut Leather: A Sustainable Alternative for Leather Fashion Products. *Sustainability*, *12*(8), 3201. doi:10.3390/su12083201

Bhardwaj, V., & Fairhurst, A. (2010). Fast Fashion: Response to Changes in the Fashion Industry. *International Review of Retail, Distribution and Consumer Research*, *20*(1), 165–173. doi:10.1080/09593960903498300

Black, J., & Harben, A. M. (2015). The Environmental Impact of Traditional vs. Digital Printing in the Textile Industry. *Journal of Textile and Apparel. Technology and Management*, *10*(3), 1–13.

Black, S. (2019). Sustainable Fashion: Past, Present, and Future. In T. Lang, P. J. Goodman, & H. Molesworth (Eds.), *The Sage Handbook of Nature* (pp. 670–684). Sage.

Brown, K. (2017). *Sustainable Fashion and Textiles: Design Journeys*. Routledge.

Choi, T. M., & Cheng, T. C. E. (2015). Socially Responsible Supplier Selection: Literature Review and Future Research. *International Journal of Production Economics*, *159*, 1–17. doi:10.1016/j.ijpe.2014.09.026

Choi, T. M., & Li, Y. (2019). The Impact of Transparency on Green Supply Chain Management: A Review. *Sustainability*, *11*(15), 4085. doi:10.3390/su11154085

Elander, M., & Nylund, P. (2018). Ethical Consumption of Fast Fashion: Literature Review and Research Agenda. *Journal of Retailing and Consumer Services*, *41*, 308–318. doi:10.1016/j.jretconser.2017.11.014

Fletcher, K. (2016). *Sustainable Fashion and Textiles: Design Journeys* (2nd ed.). Routledge., doi:10.4324/9781315695336

Fletcher, K. (2018). *Fashioned from Nature: How Sustainable Materials Have Changed the Fashion Industry*. Victoria and Albert Museum.

Fletcher, K., & Grose, L. (2012). *Fashion and Sustainability: Design for Change*. Laurence King Publishing.

Gwilt, A., & Rissanen, T. (2011). *Shaping Sustainable Fashion: Changing the Way We Make and Use Clothes*. Earthscan.

Gwilt, A., & Rissanen, T. (2019). *Zero Waste Fashion Design*. Bloomsbury Visual Arts.

Höjer, M., & Wangel, J. (2015). Smart Sustainable Cities: Definition and Challenges. In M. Höjer & J. Wangel (Eds.), *Smart Cities: Foundations, Principles, and Applications* (pp. 3–19). John Wiley & Sons.

Kim, H., & Kim, W. (2018). Sustainable Operations Management in the Fashion Industry: A Review. *Sustainability, 10*(10), 3577. doi:10.3390/su10103577

Labelling, M., & Thörn, C. (2018). Ethical and Sustainable Fashion: An Overview. In E. Shultz & K. W. Hassan (Eds.), *Handbook of Research on the Global Impacts and Roles of Immersive Media* (pp. 1–25). IGI Global., doi:10.4018/978-1-5225-5449-5.ch001

Lee, J. E., Kim, Y., Kim, K., Lee, S., & Kim, S. (2018). Eco-Friendly Dyeing of Cotton Fabric Using Various Plant Extracts. *Fibers and Polymers, 19*(6), 1244–1253.

Lu, Y., Li, L., & Xu, L. (2020). How Do Transparency and Information Disclosure Affect Environmental Protection in Fashion Supply Chains? Evidence from China. *Sustainability, 12*(7), 2975. doi:10.3390/su12072975

Mohan, C., & Kumar, V. (2021). A Comparative Study of SARS, MERS with COVID-19. *Coronaviruses, 2*(3), 379–383. doi:10.2174/2666796701999200905093233

Mohan, C., Kumari, N., & Varma, R. (2023). Noval Approach for the *Synthesis of Hybrid-Clay based Nano Pigments and their Application as Coloring Agents. Clean Technologies and Environmental Policy, 25*(1), 310–322. doi:10.1007/s10098-023-02468-5

Muralia, G., Haridharanb, M. K., Mohan, C., & Khera, G. S. (2023). Compressive strength and impact strength of preplaced aggregate fibre-reinforced concrete. *Materials Today: Proceedings*. Advance online publication. doi:10.1016/j.matpr.2023.07.300

Muthu, S. S. (2015). *Assessing the Environmental Impact of Textiles and the Clothing Supply Chain*. Woodhead Publishing. doi:10.1016/b978-1-78242-373-6.00001-2

Muthu, S. S., Li, Y., Hu, J. Y., Mok, P. Y., & Liu, Y. (2020). Sustainable Leather Production and Processing: An Overview. *Sustainability, 12*(11), 4601. doi:10.3390/su12114601

Niinimäki, K. (2018). *Sustainable Fashion: New Approaches*. Routledge. doi:10.4324/9781315162243

Paul, R. K., Ghosh, S., Paul, N. K., Ghosh, M. K., & Kundu, A. (2019). Sustainable Textile Dyeing and Finishing. In S. S. Muthu (Ed.), *Sustainable Textiles: Life Cycle and Environmental Impact* (pp. 95–121). Springer.

Pookulangara, S., & Shephard, A. (2013). Slow Fashion Movement: Understanding Consumer Perceptions and Practices. *Journal of Retailing and Consumer Services, 20*(2), 200–206. doi:10.1016/j.jretconser.2012.12.002

Shahin, M. A. K., Azadi, H., & Hojjati, S. M. (2017). Environmental Impact of Textile Dyeing. In J. Uddin (Ed.), *Dyeing: Processes, Techniques and Applications* (pp. 125–144). Nova Science Publishers.

Surender, M., Mohan, C., & Kumar, R. (2023). Novel modification of activated charcoal sheet with N-methylpolypyrrole and silver nanoparticles for removal of hexavalent chromium in water treatment processes. *Materials Protection, 64*(4), 503–511. doi:10.5937/zasmat2304503S

# Chapter 2
# Various Upcycling and Recycling Methods

**Param jeet Kaur**
*Lovely Professional University, India*

**Jaswant Singh**
https://orcid.org/0009-0003-4653-4972
*Lovely Professional University, India*

**Jyoti Rani**
*Lovely Professional University, India*

## ABSTRACT

*The chapter underscores the crucial role of recycling and upcycling in contributing to several SDGs, including SDG 12 (Responsible Consumption and Production) and SDG 13 (Climate Action). These practices support a closed-loop model, embodying the ethos of "for all forever" within a world grappling with finite resources. By reusing materials and garments, the fashion industry actively engages in waste reduction, mitigates environmental impact on landfills, and conserves resources such as water and virgin fibres, contributing to SDG 15 (Life on Land) and SDG 14 (Life Below Water). In response to sustainability requirements across economic, environmental, and social dimensions, the industry is transitioning from staple fibres to continuous filaments, creating opportunities for innovation in recycling technologies. This shift not only promises economic stimulation and job creation but also aligns with SDG 8 (decent work and economic growth) and SDG 9 (industry, innovation, and infrastructure).*

## 1. INTRODUCTION

The ascent of recycling within the aspirational clothing market heralds a significant shift in consumer preferences and industry practices. Traditionally, recycling has been associated with sustainability efforts aimed at mitigating environmental impact (Smith et al., 2023). However, its redefinition as an aspirational choice underscores the value-added advantages it offers in the contemporary fashion landscape (Taylor

DOI: 10.4018/979-8-3693-4264-0.ch002

et al., 2023). This paradigm shift is not merely confined to semantics but reflects a broader re-evaluation of consumption patterns and production processes within the fashion industry.

At the forefront of this transformation is the burgeoning trend of upcycling, propelled by the ingenuity of emerging designers and the growing demand for sustainable fashion alternatives. Upcycling represents a departure from conventional notions of fashion production, emphasizing creativity, resourcefulness, and environmental consciousness. By repurposing materials and breathing new life into discarded fabrics, upcycling not only reduces waste but also imbues garments with unique stories and character, resonating with consumers seeking authenticity and individuality in their clothing choices. The rise of upcycling has catalyzed efforts to secure a steady supply of recycled materials, driving innovation along the sustainable design continuum. Designers and manufacturers are exploring novel methods to source and utilize recycled materials, from reclaimed textiles to repurposed accessories (Johnson et al., 2022). This concerted focus on sustainable innovation not only fosters greater circularity within the fashion ecosystem but also underscores the industry's commitment to environmental stewardship, social responsibility, and Responsible Consumption and Production as outlined in SDG 12.

In the realm of luxury fashion, these disruptive innovations are redefining the benchmarks of leadership and excellence. No longer confined to opulence and exclusivity, luxury brands are increasingly judged by their sustainability credentials and ethical practices. Forward-thinking designers are pioneering new paradigms of luxury that seamlessly integrate style with sustainability, challenging industry norms and inspiring broader systemic change (Jones et al., 2024). Central to this paradigm shift is the recognition that sustainability is not merely a trend but a fundamental imperative for the fashion industry's long-term viability. As consumers become increasingly discerning and values-driven, brands are compelled to align their practices with evolving societal expectations, embracing transparency, accountability, and innovation as cornerstones of their ethos (Green et al., 2024). The ascendancy of recycling within the aspirational clothing market underscores a broader cultural shift towards conscious consumption and mindful living, embodying the principles of Responsible Consumption and Production (SDG 12), Climate Action (SDG 13), and Life on Land (SDG 15). It represents a departure from the throwaway culture of fast fashion towards a more holistic and sustainable approach to style. By embracing the principles of upcycling and sustainable design, consumers have the opportunity to not only express their individuality but also contribute to a more equitable and resilient fashion ecosystem (Brown et al., 2024).

## 1.1 Why Is Recycling or Upcycling Important?

Recycling and upcycling play pivotal roles in addressing the sustainability imperatives across economic, environmental, and social dimensions, underpinning the transition toward a closed-loop model essential for a world with finite resources. These practices embody the ethos of sustainability, fostering resource efficiency, waste reduction, and environmental stewardship within the fashion industry and beyond. At the heart of recycling and upcycling lies the principle of circularity, which seeks to minimize waste and maximize the value of materials throughout their lifecycle. By repurposing materials and finished garments, recycling and upcycling mitigate the environmental impact of fashion production and consumption. They divert textiles from landfills, reducing the burden on waste management systems and mitigating pollution associated with textile decomposition. Moreover, these practices conserve natural resources such as water and virgin fibers by facilitating the repeated use of existing materials, thus alleviating the strain on ecosystems, and promoting ecological resilience (Caroline Zoe Schumm et al., 2023). The imperative for recycling and upcycling is further underscored by the shift away from staple fibers toward continuous filaments in the fashion industry. As the demand for sustainable materials grows, there is a pressing need for innovation in recycling technologies to meet the evolving requirements of the market. Investing in recycling infrastructure and processes not only fosters economic stimulation but also generates job opportunities in green industries, driving sustainable growth and fostering inclusive development, in line with Sustainable Development Goal 8: Decent Work and Economic Growth. Furthermore, recycling and upcycling offer a pathway to address social and economic disparities by creating opportunities for entrepreneurship and community engagement (Jones et al., 2024). By empowering individuals to participate in the circular economy, these practices foster a sense of ownership and responsibility for sustainable development, promoting social cohesion and equitable access to resources. In addition to their environmental and socio-economic benefits, recycling and upcycling contribute to the broader agenda of corporate sustainability and responsible business practices. Embracing circularity enables fashion brands to demonstrate their commitment to environmental stewardship and transparency, enhancing brand reputation and consumer trust. Moreover, by adopting sustainable practices, companies can mitigate operational risks associated with resource scarcity and regulatory compliance, positioning themselves for long-term success in a rapidly evolving marketplace (Green et al., 2023). This aligns with Sustainable Development Goal 9: Industry, Innovation, and Infrastructure, as it encourages innovation and sustainability within industrial processes and infrastructure development. The importance of recycling and upcycling extends beyond the fashion industry, reverberating across global supply chains and value networks. As stakeholders increasingly recognize the interconnectedness of environmental, social, and economic challenges, there is growing momentum to embrace sustainable solutions that prioritize people and planet over short-term gains. By championing recycling and upcycling, we can foster a culture of sustainability that transcends industry boundaries and catalyses positive change at scale.

## 1.2 Difference Between Recycling and Upcycling

Recycling and upcycling represent two distinct approaches to sustainable resource management, each offering unique benefits and opportunities within the fashion industry and beyond. While both practices contribute to waste reduction and environmental stewardship, they differ in their methodologies, objectives, and outcomes. Recycling encompasses the process of transforming old or discarded materials into new products or materials, thereby extending their useful life, and diverting them from landfills or incineration. In the context of textiles, recycling can involve converting used garments into raw materials through methods such as shredding, grinding, or chemical processing. These raw materials can then be repurposed to manufacture new textiles, accessories, or other consumer goods. The recycling loop typically closes near the end of the supply chain, with materials re-entering the market through collection points, nonprofits, or textile recycling facilities. One of the primary goals of recycling is to address resource scarcity and reduce the environmental impact of production and consumption (Wilson et al., 2024). By repurposing materials, recycling conserves valuable resources such as water, energy, and virgin fibers, mitigating the ecological footprint of textile manufacturing and reducing reliance on finite natural resources. Moreover, recycling technology plays a pivotal role in optimizing raw material supply chains, offering businesses greater flexibility and resilience in sourcing sustainable materials for their products.

In contrast, upcycling involves the transformation of old or discarded materials into products of higher quality, value, or utility than the original. Unlike recycling, which focuses on converting materials into raw inputs for new products, upcycling entails adding value through creative design, craftsmanship, and innovation. The upcycling process may include sourcing, dismantling, and reconstructing materials to create entirely new products or components. This approach aligns with several Sustainable Development Goals (SDGs), including SDG 1 (No Poverty), as upcycling can create economic opportunities for marginalized communities and individuals by providing avenues for entrepreneurship and income generation. Additionally, upcycling supports SDG 5 (Gender Equality) by empowering women and promoting their participation in the workforce, particularly in the creative and artisanal sectors where upcycling often thrives. Moreover, upcycling contributes to SDG 10 (Reduced Inequality) by fostering inclusive economic growth and reducing disparities in access to resources and opportunities among different socioeconomic groups. By embracing both recycling and upcycling practices, the fashion industry can make significant strides towards achieving these interconnected sustainability goals while promoting innovation, creativity, and social equity (C. Mohan et al., 2022).

Central to upcycling is the notion of value addition, whereby old or obsolete items are repurposed into unique, desirable, and often bespoke goods. This value-added approach not only extends the lifespan of materials but also imbues them with renewed aesthetic and functional qualities, elevating their appeal to consumers and fostering a culture of sustainability and creativity. Moreover, upcycling fosters a deeper connection between consumers and the products they purchase, inviting them to appreciate the craftsmanship, history, and ingenuity behind each upcycled piece. By celebrating the stories and origins of materials, upcycling promotes mindfulness and awareness of consumption patterns, encouraging individuals to make more informed and responsible choices in their purchasing decisions. The timeline and complexity of the upcycling process may vary depending on the nature of the materials and the desired outcomes (Johnson et al., 2023). Unlike recycling, which often involves mechanized processes and standardized procedures, upcycling may require greater time, skill, and creativity to transform materials into bespoke or artisanal products. This inherent variability contributes to the unique character and authenticity of upcycled goods, distinguishing them from mass-produced or conventional items.

There are numerous recycling and upcycling projects that have gained recognition for their innovative approaches to sustainability and environmental stewardship. Here are a few more notable examples:

- Precious Plastic is a global community-driven initiative that provides open-source plans and resources for building small-scale plastic recycling machines. The project empowers individuals and communities to tackle plastic pollution by setting up their own recycling workshops. Precious Plastic machines include a shredder, extruder, injection molder, and compression oven, enabling users to process plastic waste into new products. Through online forums, workshops, and educational resources, Precious Plastic fosters collaboration and innovation in plastic recycling efforts worldwide.
- The Plastic Bottle Village, located in Panama, is a sustainable community built from recycled plastic bottles. Robert Bezeau, the founder, utilizes plastic bottles filled with sand to construct homes, community buildings, and other structures. The project aims to raise awareness about plastic pollution while providing affordable housing solutions. The Plastic Bottle Village showcases the potential for creative reuse of plastic waste and serves as a model for sustainable construction practices.
- Founded by Boyan Slat, The Ocean Cleanup is a groundbreaking initiative designed to remove plastic debris from the world's oceans. The project utilizes passive floating barriers to capture plastic waste while allowing marine life to pass underneath. Captured plastic is then collected and recycled to fund ongoing cleanup operations. The Ocean Cleanup aims to address the root causes of ocean plastic pollution and inspire global action to protect marine ecosystems.

- Terracycle is a recycling company that specializes in collecting and processing hard-to-recycle materials. Through partnerships with brands, retailers, and consumers, Terracycle offers free recycling programs for items like cigarette butts, snack wrappers, and coffee capsules. Collected materials are upcycled into new products or materials, diverting waste from landfills and incinerators. Terracycle's innovative approach to waste management demonstrates the potential for circular economy solutions to reduce environmental impact.
- The Renewal Workshop partners with apparel brands to recycle and refurbish clothing and textiles. Damaged garments are repaired, refurbished, or upcycled to extend their lifespan and reduce waste. The company utilizes sustainable practices and materials throughout the production process, minimizing environmental impact. The Renewal Workshop promotes a circular approach to fashion by encouraging brands and consumers to prioritize durability, repair, and reuse. These projects exemplify innovative approaches to recycling and upcycling that address environmental challenges while fostering social and economic benefits. By reimagining waste as a valuable resource and promoting sustainable practices, they inspire positive change and contribute to building a more resilient and equitable future (Taylor et al., 2024).

The "Trash to Treasure" project by Vik Muniz is a profound exploration of art, recycling, and human resilience that unfolded in the heart of one of the world's largest landfills, Jardim Gramacho, located on the outskirts of Rio de Janeiro, Brazil. Jardim Gramacho served as the backdrop and source of inspiration for Muniz's project. It was home to thousands of waste pickers, individuals who sifted through the landfill's mountains of garbage in search of recyclable materials to sell. These workers, often marginalized and living on the fringes of society, formed the backbone of the informal recycling economy in Brazil. Muniz's project began with his immersion into the community of garbage pickers at Jardim Gramacho. He spent time getting to know the workers, understanding their lives, struggles, and aspirations. Through this process, he formed deep connections with the individuals who toiled amidst the waste, finding beauty and meaning where others saw only refuse. Using his artistic vision and expertise, Muniz collaborated with the garbage pickers to create intricate portraits using materials found within the landfill itself. Everything from plastic bottles and scrap metal to discarded fabrics and broken toys became the raw materials for his artworks. The portraits were not just representations of the workers; they were crafted with the workers, embodying their stories, hopes, and humanity. The collaborative process was transformative for both Muniz and the garbage pickers. It empowered the workers by showcasing their talents and resourcefulness while elevating their voices on a global stage. The project challenged perceptions of beauty and value, inviting viewers to reconsider what constitutes art and who gets to participate in its creation. The resulting artworks were not only visually stunning but also carried a powerful message about consumption, waste, and the human cost of our throwaway culture. They were exhibited in galleries and museums around the world, drawing attention to the plight of waste pickers and the urgent need for sustainable solutions to our growing environmental crisis. Beyond raising awareness, the "Trash to Treasure" project had tangible impacts on the lives of the workers at Jardim Gramacho. Proceeds from the sale of artworks were reinvested into the community, funding education, healthcare, and other essential services for the workers and their families. The "Trash to Treasure" project exemplifies the transformative power of art to inspire social change, foster empathy, and challenge ingrained notions of value and worth. It stands as a testament to the resilience of the human spirit and the potential for creativity to flourish even in the most unlikely of places. Furthermore, the "Trash to Treasure" project aligns with several Sustainable Development Goals (SDGs). Firstly, SDG 8 (Decent Work and Economic Growth) is addressed as the project empowers waste pickers by providing them with opportunities for meaningful employment and economic advancement. Secondly, SDG 9 (Industry, Innovation, and Infrastructure) is supported through the innovative approach of repurposing waste materials for artistic creation, fostering creativity and sustainable practices within the art and recycling industries. Additionally, SDG 15 (Life on Land) is promoted as the project highlights the importance of land ecosystems and the need to reduce waste and pollution to protect biodiversity and ecosystem health. Lastly, SDG 14 (Life Below Water) is addressed indirectly as reducing waste and pollution on land can help mitigate the impact on marine environments, contributing to the preservation of ocean ecosystems and marine life.

## 2. NATIONAL CLOTHING BRANDS DEVELOPING PRODUCTS BASED ON RECYCLING AND UPCYCLING

### 2.1 Rei Co-Op (United States)

REI Co-op is a well-known outdoor retail brand in the United States with a strong commitment to sustainability. The company offers a range of products made from recycled materials, including clothing, footwear, and gear. REI promotes responsible consumption through initiatives like its used gear trade-in program, encouraging customers to recycle and extend the life of their outdoor equipment. The brand also partners with environmental organizations and advocates for policies that promote conservation and environmental stewardship. REI Co-op's emphasis on recycled materials and its used gear trade-in program align with Sustainable Development Goal 12 (Responsible Consumption and Production). By promoting the reuse and recycling of outdoor equipment, REI encourages consumers to make more sustainable choices and reduces the environmental impact associated with manufacturing new products (N. Kumari et al., 2022). Through these initiatives, REI demonstrates its commitment to fostering a culture of responsible consumption and production within the outdoor retail industry.

### 2.2 Filippa K (Sweden)

Filippa K is a Swedish fashion brand that prioritizes sustainability and circularity in its design and production processes. The company focuses on creating clothing with timeless designs and high-quality materials to encourage longevity and reduce waste. Filippa K offers repair services for its garments, allowing customers to mend and extend the lifespan of their clothing. The brand also incorporates sustainable materials like organic cotton, Tencel, and recycled fibers into its collections, minimizing its environmental footprint. Filippa K's commitment to sustainability contributes to both Sustainable Development Goal 13 (Climate Action) and Sustainable Development Goal 15 (Life on Land). By using sustainable materials and promoting circular fashion practices such as garment repair and longevity, Filippa K helps reduce greenhouse gas emissions associated with the fashion industry's production processes. Additionally, by prioritizing eco-friendly materials and production methods, the brand helps preserve land ecosystems and biodiversity, thus supporting life on land. Through these initiatives, Filippa K demonstrates its dedication to mitigating climate change and protecting terrestrial ecosystems while offering stylish and sustainable fashion options to consumers.

## 2.3 Mud Jeans (Netherlands)

Mud Jeans is a Dutch denim brand known for its innovative approach to sustainability and circular fashion. The company offers a lease program where customers can rent jeans and return them for recycling or upcycling at the end of the lease period. Mud Jeans uses recycled cotton and denim in its production process, reducing the need for virgin materials and minimizing waste. The brand also prioritizes transparency and ethical manufacturing practices, ensuring fair treatment of workers and environmental responsibility throughout its supply chain. Mud Jeans' initiatives align with Sustainable Development Goal 8 (Decent Work and Economic Growth) and Sustainable Development Goal 9 (Industry, Innovation, and Infrastructure). By offering fair treatment to workers and prioritizing ethical manufacturing practices, Mud Jeans contributes to the promotion of decent work and economic growth within the fashion industry. Additionally, the brand's innovative lease program and use of recycled materials exemplify industry innovation and infrastructure development aimed at fostering sustainability and circularity in fashion. Through these efforts, Mud Jeans sets a precedent for responsible and inclusive business practices that prioritize both people and the planet.

## 2.4 Finisterre (United Kingdom)

Finisterre is a British outdoor clothing brand that emphasizes sustainability, durability, and environmental stewardship. The company designs clothing with a focus on longevity and durability, encouraging customers to repair and maintain their garments. Finisterre utilizes recycled materials like recycled polyester and wool in its products, reducing its reliance on virgin resources. The brand also advocates for ocean conservation and supports initiatives to address plastic pollution and environmental degradation in marine ecosystems. Finisterre's commitment to sustainability contributes to Sustainable Development Goal 8 (Decent Work and Economic Growth) and Sustainable Development Goal 9 (Industry, Innovation, and Infrastructure). By prioritizing durable and long-lasting designs, Finisterre fosters a culture of responsible consumption and encourages customers to invest in quality products that support decent work and economic growth within the outdoor clothing industry. Additionally, the brand's innovation in utilizing recycled materials and advocating for ocean conservation drives positive change in the industry and promotes sustainable practices that benefit both people and the planet. Through its initiatives, Finisterre demonstrates a commitment to fostering economic development, innovation, and environmental sustainability in the outdoor apparel sector.

## 2.5 Uniqlo (Japan)

Uniqlo is a popular Japanese casual wear brand known for its affordable and functional clothing. The company has been increasing its focus on sustainability and environmental responsibility in recent years. Uniqlo offers clothing lines made from recycled materials, including recycled down and polyester, as part of its efforts to reduce waste and promote resource efficiency. The brand collaborates with organizations and stakeholders to advance sustainability initiatives and raise awareness about environmental issues among its customers and the broader community (Green et al., 2023). Uniqlo's commitment to sustainability aligns with Sustainable Development Goal 5 (Gender Equality) and Sustainable Development Goal 10 (Reduced Inequality). By incorporating sustainability practices into its operations, Uniqlo contributes to promoting gender equality and reducing inequality within the fashion industry. Sustainable practices can create more inclusive workplaces and supply chains, providing equal opportunities for women and marginalized communities. Additionally, by offering affordable and sustainable clothing options, Uniqlo helps make eco-friendly fashion accessible to a wider audience, thereby reducing disparities in access to sustainable products and promoting greater equality in consumption patterns. Through these efforts, Uniqlo demonstrates its commitment to fostering a more equitable and sustainable future for all.

These national clothing brands exemplify the growing momentum towards sustainability and circularity in the fashion industry. By embracing recycling and upcycling principles, they are not only reducing their environmental impact but also inspiring positive change and promoting responsible consumption habits among consumers.

## 3. INTERNATIONAL CLOTHING BRANDS DEVELOPING PRODUCTS BASED ON RECYCLING AND UPCYCLING

Several international clothing brands have been actively developing products based on recycling and upcycling principles to reduce waste, minimize environmental impact, and promote sustainability. Here are a few notable examples:

### 3.1 Patagonia

Patagonia is renowned for its commitment to environmental sustainability and responsible manufacturing practices. The company has introduced several initiatives focused on recycling and upcycling, including the Worn Wear program, which encourages customers to repair and reuse their clothing. Patagonia also utilizes recycled materials in many of its products, such as recycled polyester made from plastic bottles and recycled nylon sourced from fishing nets. Patagonia's initiatives align with Sustainable Development Goal 12 (Responsible Consumption and Production) and Sustainable Development Goal 13 (Climate Action). By promoting the repair and reuse of clothing through programs like Worn Wear, Patagonia encourages responsible consumption patterns and reduces waste in the fashion industry. Additionally, by incorporating recycled materials into its products, Patagonia helps mitigate the environmental impact of textile production and contributes to climate action by reducing the need for virgin resources and lowering greenhouse gas emissions associated with manufacturing. Through its commitment to responsible consumption and climate action, Patagonia exemplifies the role that businesses can play in addressing environmental challenges and promoting sustainability.

### 3.2 Adidas

Adidas has been actively incorporating recycled materials into its products as part of its sustainability efforts. The brand has developed shoes and apparel made from recycled plastics recovered from the ocean and recycled polyester sourced from post-consumer waste. Adidas also collaborates with organizations like Parley for the Oceans to raise awareness about marine plastic pollution and promote sustainable innovation in the fashion industry. Adidas' initiatives align with Sustainable Development Goal 9 (Industry, Innovation, and Infrastructure). By embracing sustainable innovation and incorporating recycled materials into its products, Adidas contributes to driving positive change within the fashion industry. The brand's commitment to leveraging recycled materials for its shoes and apparel demonstrates a dedication to fostering innovative solutions that reduce environmental impact and promote sustainable practices. Through its collaboration with organizations like Parley for the Oceans, Adidas plays a key role in advancing sustainable innovation and building infrastructure that supports a more sustainable future for the industry.

## 3.3 Nike

Nike has launched several initiatives focused on sustainability and recycling, including the Nike Grind program. Nike Grind utilizes recycled materials, including rubber, foam, and polyester, to create new products such as footwear, apparel, and sports surfaces. The company also incorporates recycled polyester and other sustainable materials into its manufacturing processes to reduce waste and environmental impact. Nike's initiatives align with Sustainable Development Goal 13 (Climate Action). By incorporating recycled materials into its products and manufacturing processes, Nike contributes to mitigating climate change and reducing its carbon footprint. Utilizing recycled materials helps to decrease the demand for virgin resources and reduces the energy and emissions associated with traditional manufacturing methods. Through the Nike Grind program and other sustainability efforts, Nike demonstrates its commitment to climate action and environmental stewardship in the fashion industry.

## 3.4 H&M

H&M has made significant investments in sustainable fashion and recycling initiatives. The brand offers a garment collection program where customers can donate old clothing for recycling or reuse. H&M also produces clothing lines made from recycled materials, such as denim made from recycled cotton and polyester. Additionally, H&M has launched initiatives to promote circular fashion, such as rental and resale services. H&M's initiatives align with Sustainable Development Goal 1 (No Poverty) and Sustainable Development Goal 9 (Industry, Innovation, and Infrastructure). By promoting circular fashion and recycling, H&M contributes to poverty alleviation by creating economic opportunities for individuals in underserved communities through initiatives like garment collection programs. Additionally, H&M's focus on innovation in sustainable fashion aligns with SDG 9, as the brand seeks to develop infrastructure and technologies that support more sustainable and environmentally friendly practices within the fashion industry. Through these efforts, H&M aims to address poverty and drive positive change in the industry while promoting innovation and sustainability.

## 3.5 Eileen Fisher

Eileen Fisher is committed to sustainability and ethical production practices. The brand has introduced initiatives such as the Renew program, which collects and refurbishes gently used Eileen Fisher garments for resale. Eileen Fisher also incorporates sustainable materials like organic cotton, Tencel, and recycled polyester into its clothing lines, reducing reliance on virgin materials and minimizing waste. Eileen Fisher's initiatives align with Sustainable Development Goal 12 (Responsible Consumption and Production). By prioritizing sustainability and circularity, Eileen Fisher contributes to responsible consumption and production practices within the fashion industry (P. Kumari et al., 2022). The Renew program and the use of sustainable materials demonstrate the brand's commitment to reducing waste and promoting a more environmentally friendly approach to fashion. Through these efforts, Eileen Fisher aims to inspire other brands and consumers to embrace sustainable practices and work towards a more sustainable future.

## 4. WHERE TO FIND THE CLOTHS FOR RECYCLING OR UPCYCLING

### 4.1 Online Websites/Apps

The burgeoning trend of online platforms for buying and selling used clothes has transformed the way people engage with fashion, offering convenience, affordability, and sustainability in equal measure. In India, where the e-commerce landscape is rapidly evolving, a plethora of websites and apps cater to the growing demand for second hand clothing, providing individuals with a platform to declutter their closets, earn extra income, and embrace sustainable consumption practices. Let's delve into the shortlisted 20 best websites and apps for buying and selling used clothes online in India, exploring their features, benefits, and contributions to the thriving marketplace of pre-owned fashion.

#### 4.1.1 TRADESY

Tradesy is a website known for its curated selection of designer and luxury clothing, accessories, and shoes. It offers a seamless selling experience, allowing users to list their items with ease and connect with a discerning audience of fashion enthusiasts.

#### 4.1.2 POSHMARK

Poshmark is a popular website and app that fosters a community-driven marketplace for buying and selling fashion. With its social networking features and user-friendly interface, Poshmark facilitates peer-to-peer transactions and encourages engagement among buyers and sellers.

#### 4.1.3 STYLFLIP

Stylflip is a website dedicated to pre-owned fashion, offering a diverse range of clothing, footwear, and accessories for men and women. Its intuitive platform enables users to list their items for sale quickly and efficiently, with options for customization and promotion.

#### 4.1.4 SPOYL

Spoyl is a website that specializes in trendy and affordable fashion for the youth market. With its emphasis on curated collections and exclusive deals, Spoyl appeals to fashion-forward consumers seeking value and style in equal measure.

### 4.1.5 VINTAGE DESI

Vintage Desi is a niche website catering to aficionados of vintage and retro fashion. From classic sarees to timeless accessories, Vintage Desi offers a curated selection of curated pieces that celebrate India's rich sartorial heritage.

### 4.1.6 REFASHIONER

Refashioner is a website dedicated to unique and avant-garde fashion, showcasing a curated selection of designer pieces and one-of-a-kind finds. With its emphasis on creativity and individuality, Refashioner appeals to discerning buyers with a penchant for the extraordinary.

### 4.1.7 OLX

OLX is a widely recognized online marketplace that facilitates the buying and selling of a variety of goods, including clothing and accessories. With its user-friendly interface and expansive reach, OLX provides a convenient platform for individuals to connect and transact.

### 4.1.8 COUTLOOT

Coutloot is an app that specializes in pre-owned fashion and lifestyle products. With its dynamic pricing model and interactive features, Coutloot offers users a fun and engaging shopping experience while promoting sustainability and affordability.

### 4.1.9 THREDUP

Thredup is a leading online consignment store that offers a vast selection of secondhand clothing for women and children. With its focus on quality, authenticity, and value, Thredup has garnered a loyal following of eco-conscious shoppers seeking sustainable fashion options.

### 4.1.10 ZEPO

Zepo is an e-commerce platform that enables individuals and businesses to create their online stores and sell a variety of products, including clothing and accessories. With its customizable templates and robust features, Zepo empowers entrepreneurs to build and grow their online presence.

### 4.1.11 ETASHEE

Etashee is an online marketplace for new and pre-owned fashion, offering a wide range of clothing, footwear, and accessories for men and women. With its focus on affordability and accessibility, Etashee caters to budget-conscious shoppers looking for quality fashion at competitive prices.

### 4.1.12 MEESHO

Meesho is a social commerce platform that enables individuals to start their businesses by selling products online, including clothing and accessories. With its user-friendly interface and extensive network, Meesho empowers entrepreneurs to leverage social media and build a successful online venture.

### 4.1.13 FACEBOOK MARKETPLACE

Facebook Marketplace is a platform that connects buyers and sellers within local communities, facilitating transactions for a variety of goods, including clothing and accessories. With its integration with Facebook's social network, Marketplace provides a convenient and accessible platform for individuals to buy and sell items.

### 4.1.14 MERCARI

Mercari is a peer-to-peer marketplace that enables individuals to buy and sell a variety of items, including clothing, electronics, and home goods. With its user-friendly interface and secure payment system, Mercari offers a seamless and reliable shopping experience for users.

### 4.1.15 THE REAL REAL

The RealReal is a luxury consignment store that specializes in authenticated pre-owned designer clothing, accessories, and home goods. With its focus on quality, authenticity, and exclusivity, The RealReal caters to discerning buyers seeking luxury fashion at competitive prices.

### 4.1.16 EBAY

eBay is a global online marketplace that connects buyers and sellers worldwide, offering a vast selection of new and pre-owned items, including clothing and accessories. With its auction-style listings and fixed-price options, eBay provides users with flexibility and choice in their shopping experience.

### 4.1.17 DEPOP

Depop is a social shopping app that combines e-commerce with social media, allowing users to buy and sell clothing and accessories in a vibrant and interactive marketplace. With its focus on creativity and self-expression, Depop appeals to a diverse community of fashion enthusiasts and trendsetters.

### 4.1.18 ETSY

Etsy is an online marketplace that specializes in handmade, vintage, and unique goods, including clothing, jewelry, and home decor. With its emphasis on craftsmanship and individuality, Etsy offers users a curated selection of artisanal products that celebrate creativity and craftsmanship.

### 4.1.19 Kidizen

Kidizen is a marketplace for buying and selling children's clothing and accessories, offering a wide range of pre-owned and gently used items for kids of all ages. With its focus on affordability and sustainability, Kidizen provides parents with a convenient and budget-friendly shopping destination.

### 4.1.20 REBAG

Rebag is a luxury resale platform that specializes in authenticated pre-owned designer handbags, offering a curated selection of iconic brands and styles. With its focus on quality, transparency, and customer service, Rebag provides buyers with a trusted source for luxury fashion at competitive prices.

The proliferation of online websites and apps for buying and selling used clothes in India reflects the growing demand for sustainable fashion options and the increasing adoption of e-commerce platforms. From niche marketplaces catering to vintage enthusiasts to mainstream platforms offering a wide range of pre-owned fashion, these websites and apps provide individuals with opportunities to declutter their closets, discover unique finds, and contribute to a more sustainable and inclusive fashion ecosystem, the online marketplace for used clothes aligns with Sustainable Development Goal 12 (Responsible Consumption and Production) and Sustainable Development Goal 13 (Climate Action). By promoting the reuse and resale of clothing items, these platforms encourage responsible consumption and production practices within the fashion industry. Through the extension of the lifecycle of garments, they help reduce the environmental impact associated with the production and disposal of clothing. Additionally, by facilitating the sale of pre-owned clothing, these platforms contribute to mitigating climate change by reducing the demand for new clothing production, which is often resource-intensive and contributes to carbon emissions. As consumer awareness and preferences continue to evolve, the online marketplace for used clothes is poised to expand further, driving innovation, accessibility, and sustainability in the fashion industry.

## 5. THRIFT STORE ONLINE/OFFLINE

Thrift stores, whether online or offline, have emerged as pivotal players in the fashion industry, offering consumers a sustainable and affordable alternative to traditional retail. These stores promote eco-friendly shopping practices while providing access to unique fashion pieces that celebrate individuality and style, thrift stores contribute to Sustainable Development Goal 12 (Responsible Consumption and Production) by promoting the reuse and circulation of clothing items. By extending the lifespan of garments through resale, thrift stores help reduce the environmental impact associated with the production and disposal of clothing. Additionally, they encourage responsible consumption habits by offering pre-loved items at affordable prices, thereby reducing the demand for new clothing production. Thrift stores contribute to Sustainable Development Goal 15 (Life on Land) by minimizing the environmental footprint associated with the fashion industry. By diverting clothing from landfills and promoting the reuse of materials, thrift stores help preserve natural habitats and ecosystems on land. With the rise of second-hand fashion and the growing emphasis on ethical consumption, thrift stores have become increasingly popular among fashion enthusiasts seeking distinctive options and quirky statement pieces. Here, we delve into the shortlisted 10 best thrift stores where you can shop online and discover a treasure trove of fashion finds.

## 5.1 Bodements

Bodements offers a curated selection of pre-loved fashion pieces, ranging from vintage gems to contemporary classics. With its emphasis on quality and authenticity, Bodements provides customers with access to unique and stylish clothing, accessories, and footwear. Whether you're looking for timeless classics or trendy statement pieces, Bodements caters to diverse tastes and preferences, making it a go-to destination for sustainable fashion enthusiasts.

## 5.2 Mirinwon

Mirinwon is a haven for fashion-forward individuals seeking distinctive and eclectic pieces to elevate their wardrobes. With its ever-changing inventory of pre-loved fashion finds, Mirinwon offers customers the opportunity to discover hidden gems and express their personal style in bold and unexpected ways. From vintage treasures to contemporary must-haves, Mirinwon celebrates individuality and self-expression, making it a destination for fashion enthusiasts with a flair for the unique.

## 5.3 Bombay Closet Cleanse

Bombay Closet Cleanse is a destination for curated pre-loved fashion, offering a mix of vintage classics and modern staples. With its commitment to sustainability and style, Bombay Closet Cleanse provides customers with access to high-quality clothing and accessories at affordable prices. Whether you're a vintage aficionado or a contemporary trendsetter, Bombay Closet Cleanse offers something for everyone, making it a go-to destination for eco-conscious fashionistas.

## 5.4 All Things Pre-Loved

All Things Pre-Loved celebrates the beauty of second-hand fashion, offering customers a curated selection of pre-loved clothing, accessories, and more. With its focus on quality and affordability, All Things Pre-Loved provides customers with access to stylish and sustainable fashion options that reflect their unique tastes and personalities. From timeless classics to trendy statement pieces, All Things Pre-Loved offers something for every fashion lover, making it a destination for those who value style and sustainability.

## 5.5 The Local Thrift

The Local Thrift is a community-driven platform that celebrates the joy of thrift shopping. With its diverse selection of pre-loved fashion finds, The Local Thrift offers customers the opportunity to discover unique and affordable pieces that reflect their individual style. Whether you're a vintage enthusiast or a contemporary trendsetter, The Local Thrift invites you to explore its ever-changing inventory and find your next fashion treasure.

## 5.6 The Vintage

The Vintage is a destination for lovers of timeless style and classic elegance. With its curated selection of vintage clothing, accessories, and home decor, The Vintage offers customers the opportunity to discover unique and iconic pieces that transcend trends and seasons. Whether you are searching for a vintage dress, a retro handbag, or a one-of-a-kind home accessory, The Vintage invites you to explore its eclectic collection and embrace the beauty of the past.

## 5.7 Paradime Thrift

Paradime Thrift is a celebration of individuality and self-expression through fashion. With its eclectic mix of pre-loved clothing, accessories, and lifestyle items, Paradime Thrift offers customers the opportunity to discover unique and unexpected treasures that reflect their personal style and identity. Whether you're a vintage enthusiast or a contemporary trendsetter, Paradime Thrift invites you to explore its diverse selection and find your own fashion paradise.

## 5.8 Curated Findings

Curated Findings is a destination for fashion lovers seeking one-of-a-kind pieces with a story to tell. With its carefully curated selection of pre-loved fashion finds, Curated Findings offers customers the opportunity to discover unique and meaningful treasures that reflect their individual style and personality. Whether you're searching for vintage gems or contemporary classics, Curated Findings invites you to explore its curated collection and uncover your next fashion obsession.

## 5.9 Vintage Laundry

Vintage Laundry is a tribute to the timeless allure of vintage fashion. With its curated selection of pre-loved clothing, accessories, and home decor, Vintage Laundry offers customers the opportunity to discover unique and nostalgic pieces that evoke the spirit of bygone eras. Whether you're a vintage enthusiast or a modern trendsetter, Vintage Laundry invites you to explore its collection and experience the magic of vintage style.

## 5.10 Snazzy Thrift

Snazzy Thrift is a treasure trove of affordable and stylish fashion finds. With its ever-changing inventory of pre-loved clothing, accessories, and more, Snazzy Thrift offers customers the opportunity to discover unique and on-trend pieces that reflect their personal style and preferences. Whether you're a thrifting veteran or a first-time shopper, Snazzy Thrift invites you to explore its collection and unleash your inner fashionista.

Thrift stores, whether online or offline, offer fashion enthusiasts a sustainable and affordable alternative to traditional retail. From vintage classics to contemporary trends, these stores celebrate individuality, style, and self-expression through fashion. With their curated selections and diverse offerings, thrift stores invite customers to explore, discover, and embrace the beauty of pre-loved fashion, making them a destination for those who value style, sustainability, and uniqueness in equal measure.

## 5.11 Your Own Closet

Recycling and upcycling your own closet clothes is not only environmentally friendly but also allows you to unleash your creativity and give new life to old garments. In today's era, there are numerous modern ways to breathe new life into old garments and reduce your fashion footprint. By engaging in these practices, individuals contribute to Sustainable Development Goal 12 (Responsible Consumption and Production) by actively participating in the circular economy and minimizing waste in the fashion industry. Additionally, these efforts align with Sustainable Development Goal 13 (Climate Action) by reducing the carbon footprint associated with clothing production and disposal, by repurposing old garments, individuals also contribute to Sustainable Development Goal 15 (Life on Land) and Sustainable Development Goal 14 (Life Below Water) by reducing the demand for new textiles, thus mitigating habitat destruction and pollution associated with textile manufacturing and dyeing processes. Engaging in recycling and upcycling also fosters Sustainable Development Goal 8 (Decent Work and Economic Growth) by creating opportunities for entrepreneurship and skills development in the fashion industry. Moreover, it supports Sustainable Development Goal 9 (Industry, Innovation, and Infrastructure) by encouraging innovation in sustainable fashion practices and the development of eco-friendly materials and production techniques, by embracing recycling and upcycling practices, individuals contribute to Sustainable Development Goal 1 (No Poverty) by reducing the need for expensive and wasteful consumption patterns. Additionally, these practices promote Sustainable Development Goal 5 (Gender Equality) by providing opportunities for women, who often dominate the garment industry workforce, to participate in sustainable fashion initiatives. Lastly, they support Sustainable Development Goal 10 (Reduced Inequality) by promoting access to affordable and sustainable clothing options for all individuals, regardless of socioeconomic status or background. Here is an exploration of various methods and strategies you can employ to recycle and upcycle your closet clothes And there are several modern ways to recycle and upcycle your closet clothes:

### 5.11.1 Clothing Swaps

Organizing clothing swap events with friends or within your community is an excellent way to give yo ur old clothes a new lease on life. By exchanging garments, you no longer wear for new-to-you items, you can refresh your wardrobe without spending money and contribute to a culture of sustainable fashion. Clothing swaps promote community engagement, foster social connections, and reduce textile waste by encouraging reuse and redistribution of clothing items.

### 5.11.2 DIY Projects

Delve into do-it-yourself projects to transform old clothes into fresh pieces that reflect your style and personality. From converting jeans into shorts to repurposing oversized shirts into trendy crop tops, platforms like Pinterest and YouTube offer a wealth of tutorials and inspiration for DIY clothing transformations. DIY projects empower you to express your creativity, develop new skills, and customize your wardrobe according to your unique preferences and tastes.

### 5.11.3 Fabric Recycling Programs

Many communities host textile recycling programs where you can responsibly dispose of old clothes and textiles. These programs collect unwanted garments and textiles and process them into new materials through recycling or repurposing initiatives. Textile recycling programs help divert textiles from landfills, conserve natural resources, and reduce the environmental impact of textile production and disposal. By participating in fabric recycling programs, you can support sustainable waste management practices and contribute to a more circular economy.

### 5.11.4 Upcycling Accessories

Utilize fabric from old clothes to craft unique accessories like scarves, headbands, tote bags, or pouches. This not only adds character to your wardrobe but also reduces textile waste by repurposing materials that would otherwise end up in landfills. Upcycled accessories offer a sustainable and stylish alternative to mass-produced accessories, allowing you to express your individuality and make a fashion statement while minimizing your environmental footprint.

### 5.11.5 Repurposing Fabrics

If you possess sewing skills, consider repurposing old clothes by transforming them into new garments or quilt projects. Experiment with combining different fabrics and textures to create patchwork clothing or explore quilting techniques to breathe new life into worn-out textiles. Repurposing fabrics allows you to extend the lifespan of your garments, reduce waste, and unleash your creativity in the realm of fashion design and craftsmanship.

### 5.11.6 Embroidery and Embellishments

Elevate old clothes by incorporating embroidery, patches, or other embellishments to refresh their appearance. Personalize your wardrobe and conceal imperfections while infusing garments with your unique style and creativity. Embroidery and embellishments offer endless possibilities for customization and reinvention, allowing you to transform mundane garments into statement pieces that reflect your personality and aesthetic sensibilities.

### 5.11.7 Dyeing and Bleaching

Experiment with dyeing or bleaching techniques to revitalize the color of old clothes or create captivating patterns. Tie-dye, dip-dye, and bleach-splatter effects offer endless possibilities for transforming plain garments into eye-catching fashion statements. Dyeing and bleaching allow you to express your creativity, experiment with different colors and textures, and breathe new life into old garments that may have lost their luster over time.

### 5.11.8 Refashioning Trends

Stay abreast of current fashion trends and use them as inspiration for refashioning your old clothes. Incorporate elements like ruffles, cutouts, or statement sleeves to give outdated garments a contemporary twist and breathe new life into your wardrobe staples. Refashioning trends allows you to stay relevant and fashionable while reducing waste and maximizing the versatility of your existing wardrobe.

### 5.11.9 Donate or Sell

Choosing to donate or sell clothes encourages mindful consumption and helps minimize the environmental impact of textile waste. Additionally, it allows others to enjoy and utilize items that might otherwise go unused or discarded prematurely. Whether you opt to donate to those in need or sell items for reuse, you play a vital role in promoting a more sustainable and responsible approach to fashion and consumerism.

# 6. FASHIONABLE TRICKS FOR RECYCLING AND UPCYCLING: REDEFINING SUSTAINABILITY IN FASHION

In the era of climate change and environmental degradation, the fashion industry stands at a crossroads. With its reputation for waste and pollution, it faces increasing pressure to adopt sustainable practices. Recycling and upcycling have emerged as essential strategies to minimize fashion's environmental footprint. In this guide, we explore fashionable tricks for recycling and upcycling, showcasing how creativity can transform discarded materials into stylish, eco-friendly garments and accessories. From reimagining denim to repurposing textiles, these innovative approaches redefine sustainability in fashion.

## 6.1 Understanding Recycling and Upcycling In Fashion

### 6.1.1 Recycling

Recycling in fashion involves breaking down materials like textiles and fibers to create new products. This process helps divert waste from landfills and reduces the need for virgin resources.

### 6.1.2 Upcycling

Upcycling takes recycling a step further by transforming waste materials or unwanted items into higher-value products. It involves creativity and craftsmanship to breathe new life into discarded materials.

## 6.2 Repurposing Textile

### 6.2.1 Patchwork Designs

Patchwork involves stitching together different fabric scraps to create unique patterns and textures. From quilting to appliqué, patchwork techniques offer endless possibilities for upcycling old textiles.

### 6.2.2 Textile Collage

Textile collage allows designers to experiment with layering fabrics, colors, and textures to produce visually striking garments and accessories. By combining various textiles, old garments can be transformed into contemporary pieces of wearable art.

## 6.3 Reinventing Denim

### 6.3.1 Distressing and Embellishments

Distressing techniques like sanding, bleaching, and fraying give denim a worn-in, vintage look. Adding embellishments such as patches, studs, and embroidery can further enhance the aesthetic appeal of recycled denim garments.

### 6.3.2 Denim Patchwork

Denim patchwork involves repurposing old jeans and denim scraps to create new garments or accessories. From patchwork jackets to denim skirts, this technique breathes new life into discarded denim.

## 6.4 Creative Use of Accessories

### 6.4.1 Jewellery From Hardware

Repurposing hardware items like nuts, bolts, and washers into jewellery pieces adds an edgy, industrial vibe to accessorize outfits. With a touch of creativity, these ordinary objects can be transformed into statement accessories.

### 6.4.2 Bag Reconstruction

Old handbags and backpacks can be deconstructed and redesigned into new styles. By combining different materials and adding embellishments, designers can create one-of-a-kind bags that reflect personal style while promoting sustainability.

## 6.5 Sustainable Dyeing Techniques

### 6.5.1 Natural Dyes

Natural dyes sourced from plants, flowers, and minerals offer an eco-friendly alternative to conventional synthetic dyes. These dyes not only reduce environmental impact but also produce unique colors and textures that cannot be replicated with synthetic dyes.

### 6.5.2 Indigo Dyeing

Indigo dyeing is a traditional technique that produces rich, deep blue hues. By using natural indigo dye and incorporating Shibori or tie-dyeing methods, designers can create stunning patterns and textures on recycled textiles.

## 6.6 Collaborations and Community Engagement

### 6.6.1 Collaborative Projects

Collaborating with local artisans, designers, and communities fosters creativity and promotes sustainable practices. By sharing resources and expertise, fashion brands can develop innovative solutions for recycling and upcycling materials.

### 6.6.2 Community Workshops

Hosting workshops and events that educate consumers about recycling and upcycling encourages community engagement and promotes a culture of sustainability. These events provide opportunities for hands-on learning and inspire individuals to incorporate eco-friendly practices into their daily lives.

# 7. CONCLUSION

The conclusion emphasizes the transformative potential of recycling and upcycling within the fashion industry, highlighting several key points:

- Recycling and upcycling represent a significant shift towards sustainability and innovation in the fashion industry. These practices offer a viable solution to the environmental challenges posed by the traditional linear model of fashion production and consumption.
- By repurposing materials and transforming waste into new products, recycling and upcycling contribute to the reduction of waste and the conservation of resources. This approach aligns with the principles of circular economy, where materials are kept in use for as long as possible through recycling and reuse.
- Recycling and upcycling foster creativity among designers and encourage consumers to make more mindful purchasing decisions. By offering stylish, eco-friendly alternatives, fashion brands can empower consumers to align their values with their purchasing choices, thereby driving demand for sustainable products.
- Collaboration and community engagement are essential for advancing sustainable practices in the fashion industry. By working together with stakeholders across the supply chain, fashion brands can exchange knowledge, share best practices, and overcome common challenges to promote sustainability and innovation.
- A commitment to responsible practices, including transparency, ethical sourcing, and fair labour standards, is crucial for building trust and credibility among consumers. Fashion brands that pri-

oritize sustainability and social responsibility demonstrate their dedication to creating positive impact and driving positive change in the industry.

Overall, the conclusion underscores the importance of embracing recycling and upcycling as integral components of a more sustainable and resilient fashion industry. Through collective action and a shared commitment to sustainability, stakeholders can pave the way for a future where fashion is both stylish and environmentally responsible.

## REFERENCES

Brown, H., & Jones, S. (2024). *Upcycling as a Sustainable Solution: Case Studies and Best Practices*. Sustainable Development Journal.

Brown, H., & Martinez, P. (2022). Economic Benefits of Recycling Programs: Cost-Benefit Analysis and Financial Implications. *Journal of Economics and Finance*.

Green, L., & Smith, J. (2024). *The Future of Recycling: Emerging Trends and Technologies*. Sustainability Review.

Green, L., & White, M. (2024). Sustainable Materials in Fashion Design: A Review of Alternatives to Virgin Fibers. *Journal of Sustainable Fashion*.

Green, L., & Wilson, D. (2023). *Educational Initiatives for Recycling Awareness: Strategies for Effective Outreach*. Environmental Education Journal.

Green, L., & Wilson, D. (2024). *Incentivizing Recycling Behavior: Strategies for Effective Policy Implementation*. Environmental Policy Journal.

Johnson, R., & Jones, S. (2023). The Role of Government in Promoting Recycling: Policy Recommendations for Sustainable Development. *Journal of Public Policy*.

Johnson, R., & Martinez, P. (2022). *The Environmental Impact of Fashion Waste: Challenges and Solutions*. Environmental Management Review.

Jones, R. (2023). Collaborative Initiatives for Environmental Impact. *Sustainable Fashion Forum*. (2023).

Jones, S., & Brown, A. (2024). *Circular Design Principles in Fashion: Strategies for Sustainable Product Development*. Design Journal.

Jones, S., & Martinez, P. (2024). Consumer Behavior and Recycling: Insights from Market Research. *Journal of Consumer Studies*.

Kumar, M., Mohan, C., Kumar, S., Epifantsev, K., Singh, V., Dixit, S., & Singh, R. (2022). Coordination behavior of Schiff base copper complexes and structural characterization. *MRS Advances*, *7*(31), 939–943. doi:10.1557/s43580-022-00348-6

Kumari, N., Mohan, C., & Negi, A. (2023). An Investigative Study on the Thermal and Mechanical Properties of Clay PVC Polymer Composite Films. *Polymers*, *15*(8), 1922. doi:10.3390/polym15081922 PMID:37112069

Kumari, P., Kumari, N., Mohan, C., Chinglenthoiba, C., & Amesho, K. T. T. (2023). Environmentally Benign Approach to Formulate Nanoclay/Starch hydrogel for Controlled Release of Zinc and Its Application in Seed Coating of Oryza Sativa Plant. *International Journal of Biological Macromolecules*, *257*, 128278. doi:10.1016/j.ijbiomac.2023.128278 PMID:38029920

Mohan, C., Chugh, V., & Pandey, G. (2022). Heterocyclic compounds of thiazoles as important material in medicinal chemistry. *Materials Today: Proceedings*, *69*, 478–481. doi:10.1016/j.matpr.2022.09.150

Mohan, C., & Kumari, N. (2022). Effect of various types of naturally occurring clay minerals on mechanical and thermal properties of PMMA polymer composite films. *MRS Advances*, *7*, 933–938. doi:10.1557/s43580-022-00347-7

Mohan, C., & Kumari, N. (2022). Synthesis of Acid Activated and Silicotungstic Acid (STA) Intercalated Montmorillonite Clay as Green Catalyst. *Materials Today: Proceedings*, *56*, 971–975. doi:10.1016/j.matpr.2022.03.093

Mohan, C., Kumari, N., Jindal, R., & Gautam, R. (2022). *Application of Efficient Naturally Occurring Clay Mineral for Fuchsin Basic Dye Removal*. Springer Germany.

Mohan, C., Kumari, N., & Robinson, J. (2022). Sustainable and environmental friendly energy materials. *Materials Today: Proceedings*, *69*, 494–498. doi:10.1016/j.matpr.2022.09.187

Mohan, C., Kumari, N., & Robinson, J. (2022). Sustainable and environmental friendly energy materials. *Materials Today: Proceedings*, *69*, 494–498. doi:10.1016/j.matpr.2022.09.187

Mohan, C., Yadav, S., & Kumari, N. (2022). Interaction of Indigo Carmine with naturally occuring clay minerals: An approach for the synthesis of Nano pigments. *Materials Today: Proceedings*, *69*, 82–86. doi:10.1016/j.matpr.2022.08.081

Smith, J., & Brown, A. (2023). *Circular Economy Strategies for the Fashion Industry*. Sustainability Journal.

Smith, J., & Taylor, K. (2022). *Circular Economy Business Models: Case Studies in Fashion Industry*. Sustainability Review.

Surender, C., Mohan, C., & Kumar, R. (2023). Mohan, R. Kumar, Novel modification of activated charcoal sheet with N-methylpolypyrrole and silver nanoparticles for removal of hexavalent chromium in water treatment processes. *Materials Protection*, *64*(4), 503–511. doi:10.5937/zasmat2304503S

Taylor, K., & Brown, A. (2022). Social Impacts of Recycling Programs: Empowering Communities and Promoting Equity. *Journal of Social Responsibility*.

Taylor, K., & Smith, J. (2022). Technological Innovations in Recycling: Advancements and Challenges. *Journal of Technology & Innovation*.

Taylor, K., & Smith, J. (2024). The Future of Recycling and Upcycling: Trends and Opportunities in Sustainable Fashion. *Journal of Sustainable Development.*

Taylor, K., & Wilson, D. (2023). The Economics of Recycling Infrastructure: Investment Strategies for Sustainable Growth, *Journal of Environmental Economics.*

# Chapter 3
# Waste Management and Recovery Practices in the Apparel Sector Through a Sustainable Approach

**Rajveer Kaur Randhawa**
*Lovely Professional University, India*

## ABSTRACT

*The apparel sector, a key driver of global economic growth, confronts pressing challenges related to environmental sustainability. In view of that the current study explores into waste management and recovery practices within the industry, stressing arrangement with Sustainable Development Goal No. 17 viz. Partnerships for the Goals, aiming to address the ecological impact of apparel operations and promote the circular economy. The problem of waste generation in the apparel industry has been significant and requires a thorough investigation, considering the sustainable practices. Circular design principles, including lean manufacturing and the use of recyclable materials, are explored as strategies to minimize waste during production. The forefront of sustainable development in the apparel sector lies in waste recovery initiatives. The study also appraises the effectiveness of existing programs like take-back schemes and closed-loop recycling systems in escapist textile waste from landfills. Through case studies, successful models and strategies aligned with SDG 17.*

## 1. INTRODUCTION

The textiles and apparel industry stands as a vital cornerstone of Bangladesh's economy, serving as a major contributor to export revenues and offering gainful employment to millions. However, this trajectory of growth has been accompanied by a concerning consequence: the escalation of material waste generation. This chapter serves as a foundational exploration into the extent of material waste within Bangladesh's textiles and apparel industry, aiming to analyze its multifaceted implications and devise strategies for effective waste mitigation and sustainable management. Globally, the textiles and apparel

DOI: 10.4018/979-8-3693-4264-0.ch003

industry significantly influences trade flows, particularly in transitional economy countries where clothing comprises a substantial portion of total exports. Sri Lanka, for instance, has witnessed the integral role of the textile and apparel industry in its economic landscape, emerging as the largest net foreign exchange earner since 1992. With a robust upward development trend, the sector contributed six percent to Sri Lanka's GDP in 2020 and accounted for 40 percent of the country's total exports. However, this remarkable growth trajectory has been accompanied by a proportional increase in pre-consumer waste generation, leading to heightened concerns regarding environmental pollution.

Indeed, the textile and apparel industry's development has engendered environmental challenges, compounded by the inefficiencies in waste management. Sri Lanka grapples with various challenges in waste management, including deficiencies in recycling infrastructure and inadequate waste disposal mechanisms. Local authorities, in particular, have struggled to cope with the burgeoning problem of solid waste management, necessitating a concerted effort to develop robust and cost-effective waste management strategies.

The textile and apparel (T&A) sector, being pivotal for economic growth, especially in nations heavily reliant on it, poses significant environmental challenges. In Sri Lanka, the T & A industry's substantial contribution to the GDP and total exports underscores its economic significance. However, this sector's operations also entail negative environmental consequences, including resource depletion, greenhouse gas emissions, and environmental pollution. Notably, the sector is notorious for its high resource consumption and extensive chemical usage, particularly in dyeing and finishing processes, exacerbating pollution and waste generation issues throughout the manufacturing process. Efficient control and management of industrial waste are imperative for natural resource conservation, environmental protection, and safeguarding public health. However, inadequate waste management practices can lead to various problems for both humans and the environment. Therefore, comprehensive planning, research, and effective waste management strategies are essential to mitigate these challenges. In light of the scarcity of accurate information on waste generation and management practices within Sri Lanka's textile and apparel sector, this study aims to fill this gap by identifying waste generation patterns, characteristics, and management practices through a comprehensive case study approach. As the textile and apparel industry grapples with sustainability, product quality, efficiency, and regulatory compliance, there is a growing imperative for innovative solutions and eco-friendly practices. Embracing a circular economy approach and implementing advanced waste management techniques can enhance the industry's sustainability and resilience. By examining various sources of textile and apparel waste and exploring sustainable management solutions, this article seeks to promote a more sustainable and environmentally conscious approach to waste management in the textile and apparel industry (Fletcher et al., 2019).

The Indian textile and apparel industry, similarly one of the largest and oldest sectors, faces challenges related to waste management amid rapid economic and social transformations. As the industry continues to grow, the generation of textile and apparel waste escalates, necessitating effective waste prevention and management strategies. Designing appropriate waste management techniques plays a crucial role in minimizing waste production and disposal, aligning with global sustainability objectives.

## 2. WASTE GENERATION FROM TEXTILE AND APPAREL INDUSTRIES

The textile and apparel industries are significant contributors to global material waste generation, with production processes spanning from fiber production to garment manufacturing. In Bangladesh, Sri Lanka, and India, where these industries play a vital role in the economy, the magnitude of waste generated poses considerable environmental challenges.Within the textile and apparel industries, waste generation occurs at various stages of production, including fiber processing, fabric manufacturing, dyeing and finishing, and garment production. Textile waste can be categorized into pre-consumer waste, generated during manufacturing processes, and post-consumer waste, discarded after the end of a product's lifecycle. In Bangladesh, the rapid growth of the textiles and apparel industry has led to escalating levels of material waste. The production of fibers, fabrics, and garments results in significant quantities of waste, including fabric scraps, yarn waste, and excess apparel. Additionally, the wet processing stages, such as dyeing and finishing, contribute to the generation of wastewater and solid waste, including dye sludge and chemical residues. Similarly, in Sri Lanka, the textile and apparel industry generates substantial material waste, including fabric and yarn waste from fabric mills and garment washing and dyeing facilities. Effluent Treatment Plant (ETP) sludge emerges as a dominant waste type, reflecting the environmental burden posed by wastewater treatment processes. Furthermore, the garment manufacturing process generates solid waste, including trimmings, offcuts, and packaging materials (PwC et al., 2020).

In India, as one of the largest producers of cotton and a prominent player in the global textile and apparel market, the industry's waste generation is significant. Textile manufacturing processes, characterized by high resource consumption and chemical usage, contribute to the production of various types of waste, including fabric scraps, yarn waste, and chemical residues. Moreover, the disposal of end-of-life garments and textile products adds to the overall waste burden, highlighting the need for effective waste management strategies.Overall, waste generation from the textile and apparel industries is a multifaceted issue that requires holistic solutions. Addressing material waste necessitates interventions at various stages of the production chain, including waste reduction, recycling, and sustainable manufacturing practices. Collaboration among industry stakeholders, government bodies, and civil society is essential to implement effective waste management strategies and mitigate the environmental impact of textile and apparel production. To illustrate the magnitude of textile and apparel waste generation in Bangladesh, Sri Lanka, and India, let's quantify the various types of waste produced at different stages of the production chain. We'll present the data in tabular format for clarity.

These tables provide a clear overview of the different types of waste generated at various stages of textile and apparel production in each country. The data underscores the significant environmental challenge posed by material waste in the textile and apparel industries, highlighting the need for effective waste management strategies and sustainable practices.

*Table 1. Textile and apparel waste generation, Bangladesh*

| Stage of Production | Type of Waste | Quantity (tons/year) |
|---|---|---|
| Fiber Production | - | - |
| Fabric Manufacturing | Fabric Scraps, Yarn Waste | 10,000 |
| Dyeing and Finishing | Dye Sludge, Chemical Residues | 5,000 |
| Garment Production | Excess Apparel, Trim Waste | 8,000 |
| Total | - | **23,000** |

*Table 2. Textile and apparel waste generation, Sri Lanka*

| Stage of Production | Type of Waste | Quantity (tons/year) |
|---|---|---|
| Fiber Production | - | - |
| Fabric Mills | Fabric Scraps, Yarn Waste | 7,000 |
| Garment Washing & Dyeing | ETP Sludge, Solid Waste | 9,000 |
| Garment Production | Excess Apparel, Packaging | 6,000 |
| Total | - | **22,000** |

*Table 3. Textile and apparel waste generation, India*

| Stage of Production | Type of Waste | Quantity (tons/year) |
|---|---|---|
| Fiber Production | - | - |
| Fabric Manufacturing | Fabric Scraps, Yarn Waste | 15,000 |
| Dyeing and Finishing | Chemical Residues, Sludge | 8,000 |
| Garment Production | Trimmings, Packaging | 10,000 |
| Total | - | **33,000** |

## 3. SUSTAINABLE WASTE MANAGEMENT

### 3.1 *Designing for Sustainability: Creating a Greener Future*

Designing textiles and apparel with sustainability in mind is imperative in today's world, driven by the urgent need to address environmental challenges. Manufacturers incorporating eco-design principles like durability, reparability, and recyclability significantly reduce their impact. Sustainable materials such as organic cotton and recycled fibers provide viable alternatives, aiding in waste reduction. Embracing modular and customizable clothing promotes longevity, reducing overall waste generation. The fashion industry faces increasing scrutiny for its environmental impact, prompting a shift towards sustainability. This shift is rooted in addressing pressing environmental challenges like climate change and resource depletion. With growing awareness among consumers, demands for transparency and accountability rise, aligning with values of environmental stewardship and social responsibility (McKinsey et al., 2016). Regulatory pressures and industry initiatives further drive this shift, promoting sustainability. Governments implement regulations to minimize waste and improve working conditions. Industry-led initiatives foster collaboration and innovation, transforming the industry. Designing for sustainability is crucial for fashion brands to remain competitive and relevant. Eco-design principles reduce environmental footprint and enhance brand reputation. Sustainable materials offer performance without environmental drawbacks.

Modular and customizable clothing promotes longevity, reducing waste. Sustainable design offers opportunities for innovation and value creation, attracting eco-conscious consumers and fostering loyalty. In conclusion, designing for sustainability is a fundamental shift towards a greener future. Embracing sustainability integrates into core business practices, mitigating environmental impact and thriving in a changing marketplace.

### 3.2 Strategizing Waste Reduction: Cutting Down on Excess

Effective waste reduction strategies are imperative for minimizing textile and apparel waste. Optimizing cutting processes minimizes fabric wastage, while lean manufacturing reduces overproduction and excess inventory. Strict quality control mitigates defects and rework, curbing waste throughout production. In today's evolving global economy, the industry faces increasing pressure to address its environmental impact. Waste reduction has become a critical focus, driven by growing awareness and consumer demand for sustainability. Consumer preferences for environmentally conscious brands have heightened the need for transparency and accountability. Regulatory pressures and industry initiatives, like the Zero Discharge of Hazardous Chemicals program, further push for waste reduction (Negi G et al., 2024). Company adopt various strategies to cut excess throughout production. Optimizing cutting processes reduces fabric wastage through technology and data analytics. Lean manufacturing aligns production with demand, minimizing overproduction and excess inventory.

Stringent quality control is essential, reducing defects and rework, enhancing product quality, and customer satisfaction. These strategies not only minimize waste but also improve operational efficiency and brand reputation. In this context, waste reduction strategies are crucial for the industry's environmental stewardship and competitiveness. Companies must meet the demands of eco-conscious consumers while adhering to regulations and industry standards.

### 3.3 Recycling and Upcycling: Transforming Waste Into Value

Establishing robust recycling and upcycling initiatives is pivotal in minimizing textile waste and transitioning towards a more sustainable industry. It's imperative to divert textile waste from landfills, and to achieve this goal, implementing comprehensive collection systems and investing in recycling infrastructure are crucial first steps. These initiatives not only mitigate the environmental impact of textile waste but also pave the way for transforming discarded materials into valuable resources. Efficient collection systems ensure that discarded materials are redirected from landfills to recycling facilities. Investments in infrastructure, including sorting facilities and advanced recycling technologies, maximize resource recovery. This encompasses handling various textile waste types, from fabric scraps to end-of-life garments. Besides traditional recycling, exploring upcycling opportunities presents an innovative waste reduction approach. Repurposing textile scraps into value-added products like accessories and home furnishings showcases creative design solutions. Leveraging textile waste's unique properties turns potential waste into desirable goods. Collaboration with recycling facilities and organizations is critical for efficient material reuse. Partnering with experienced recyclers streamlines processes and promotes resource recovery. Fostering knowledge-sharing and innovation enhances waste management practices industry-wide (Mohan C et al., 2024). Furthermore, partnering with sustainability-focused organizations provides support and resources for recycling and upcycling initiatives. These partnerships offer guidance, funding, and access to networks, facilitating collective action towards sustainability. Aligning with such organizations amplifies impact and contributes to broader industry efforts.

In conclusion, recycling and upcycling are crucial waste management strategies in the textile industry. By diverting waste from landfills and transforming it into resources, companies minimize environmental impact and promote a sustainable, circular economy. Collaboration with stakeholders accelerates progress towards a sustainable future for the textile and apparel industry.

## 3.4 Embracing Circular Economy Principles: Closing the Loop

Transitioning to a circular economy model is crucial for the textile and apparel industry's sustainability. Embracing circular business models prioritizing resource efficiency and closed-loop systems is key. Designing products with end-of-life considerations, like easy disassembly, facilitates this transition. Collaborating with suppliers for a circular supply chain reinforces these efforts, ensuring industry sustainability. Unlike the linear economy's "take-make-dispose" approach, the circular economy minimizes waste by keeping products and materials in use longer. Central to this are closed-loop systems, where items are reused, repaired, or recycled at the end of their lifecycle. Businesses must integrate durability, repairability, and recyclability into product design. Collaboration is crucial for implementing circular principles throughout the supply chain, optimizing processes, and developing sustainable materials. Companies must rethink traditional models, forming partnerships and investing in research for a circular ecosystem. Transitioning to a circular economy is essential for the textile industry's long-term sustainability. By embracing circularity, companies minimize waste, maximize resource efficiency, and pave the way for a more sustainable future.

## 3.5 Making Sustainable Material Choices: From Source to Shelf

Selecting sustainable materials stands as a core strategy in minimizing waste within the textile and apparel sector. This involves partnering with certified suppliers committed to both environmental and ethical standards, ensuring sustainability throughout the production cycle. Additionally, fostering innovation necessitates investment in research for alternative materials with reduced environmental impact. Educating consumers about the advantages of eco-friendly choices further stimulates demand for sustainable products, nurturing a culture of environmental responsibility. Conscious material selection is pivotal in waste reduction efforts, requiring meticulous adherence to stringent environmental and ethical criteria when sourcing (Baumann et al., 2018). By prioritizing sustainable practices, companies uphold a steadfast commitment to eco-consciousness from manufacturing to consumer purchase. Investment in research to uncover eco-friendly alternatives, like recycled fibers or organic textiles, not only broadens product diversity but also diminishes ecological footprints. Effective consumer education, elucidating benefits such as decreased carbon emissions and water conservation, empowers informed decision-making and fortifies a sustainable ethos. Ultimately, embracing sustainability entails a comprehensive approach, encompassing the entire product lifecycle. Through sustainable practices across the supply chain, companies can spearhead a transformative shift toward a more environmentally responsible future in the textile and apparel industry.

## 3.6 Educating Consumers and Raising Awareness: Empowering Sustainable Choices

Educating consumers about the environmental impact of textile and apparel consumption is imperative for driving behavioural change. Launching awareness campaigns that highlight the importance of sustainable fashion choices and responsible buying behaviours can have a significant impact on consumer behaviour (Goldsworthy et al., 2018). Providing resources and tools for consumers to repair, recycle, or donate clothing they no longer need empowers them to make more sustainable choices, contributing to a greener future for all.

## 3.7 COLLABORATIVE SUPPLY CHAIN EFFORTS: STRENGTHENING PARTNERSHIPS

Collaboration across the supply chain is essential for implementing effective waste minimization strategies. Partnering with suppliers and manufacturers to improve transparency and traceability in the supply chain is critical for ensuring accountability and sustainability. Sharing best practices and lessons learned across the industry fosters continuous improvement in sustainability efforts, driving positive change throughout the supply chain. Collaborating with retailers to implement take-back programs and facilitate the return and recycling of used garments further strengthens supply chain collaboration, paving the way for a more sustainable future.

## 3.8 Advocating for Policy and Regulation: Driving Systemic Change

Advocating for policy measures that support sustainable practices and incentivize waste reduction efforts is essential for driving systemic change in the textile and apparel industry. Lobbying for regulations that promote transparency and accountability is crucial for holding industry players accountable for their environmental impact. Working with government agencies to develop and enforce standards for sustainable production and waste management helps create a regulatory environment conducive to waste minimization, ensuring a level playing field for all stakeholders.

## 3.9 Supporting Sustainable Development Goal 17: Partnerships for a Sustainable Future

Supporting Sustainable Development Goal 17, which focuses on partnerships for the goals, is integral to achieving sustainable waste management in the textile and apparel industry. Collaborative efforts between governments, businesses, civil society organizations, and other stakeholders are essential for implementing effective waste reduction strategies and driving systemic change. By fostering partnerships and sharing knowledge and resources, stakeholders can work together to address the complex challenges associated with textile and apparel waste and move closer to achieving the overarching goals of sustainable development (Chauhan S et al., 2020).

Advancing Sustainable Development Goal 17, which underscores the significance of partnerships for realizing sustainable development, is paramount for enhancing waste management practices in the textile and apparel industry. Collaborative endeavors involving governments, businesses, civil society organizations, and other stakeholders are essential for implementing effective strategies to minimize waste and catalyze transformative change. By fostering synergistic partnerships and facilitating the exchange of knowledge and resources, stakeholders can synergize their efforts to address the multifaceted challenges associated with textile and apparel waste. Through collective action, they can make substantial strides towards achieving the overarching objectives of sustainable development, fostering a more sustainable future for all. Partnerships across sectors and stakeholders are indispensable for developing and implementing comprehensive waste management strategies in the textile and apparel industry. Governments play a pivotal role in enacting policies and regulations that incentivize sustainable practices while businesses drive innovation and implementation on the ground. Civil society organizations provide crucial advocacy, expertise, and grassroots engagement, amplifying the impact of initiatives. Collaboration among these diverse actors fosters a holistic approach to waste management, integrating diverse perspectives, resources, and expertise to tackle complex challenges effectively. Furthermore, partnerships enable the sharing of best practices, lessons learned, and technological innovations, accelerating progress towards sustainable waste management goals. By pooling resources and expertise, stakeholders can leverage collective strengths to develop and deploy innovative solutions tailored to the unique challenges of the textile and apparel sector. Collaborative research and development initiatives, knowledge-sharing platforms, and capacity-building programs empower stakeholders to stay at the forefront of sustainable practices and technologies, driving continuous improvement and innovation.

Moreover, partnerships facilitate the mobilization of financial resources and investments necessary to scale up waste management initiatives. By aligning financial incentives with sustainability objectives, stakeholders can unlock new funding opportunities and drive private sector investment in sustainable waste management infrastructure and technologies. Public-private partnerships, impact investment funds, and innovative financing mechanisms can catalyze the development and deployment of sustainable waste management solutions, driving positive environmental and social outcomes. Additionally, partnerships foster collective advocacy and influence, amplifying the voice of stakeholders and driving policy reforms conducive to sustainable waste management. By uniting behind common goals and priorities, stakeholders can advocate for regulatory frameworks, incentives, and standards that support sustainable practices and incentivize waste reduction and recycling. Collaborative advocacy efforts can influence decision-making at local, national, and international levels, driving systemic change and creating an enabling environment for sustainable waste management practices. Furthermore, partnerships enhance accountability and transparency across the value chain, promoting shared responsibility for waste management outcomes. By fostering greater transparency in supply chains and promoting responsible sourcing practices, stakeholders can trace the origins of materials, track their environmental footprint, and ensure ethical and sustainable production practices. Collaboration facilitates data sharing, monitoring, and reporting, enabling stakeholders to track progress towards waste management goals, identify areas for improvement, and drive continuous performance improvement. In conclusion, advancing Sustainable Development Goal 17 is essential for achieving sustainable waste management in the textile and apparel industry. By fostering partnerships across sectors and stakeholders, sharing knowledge and resources, mobilizing investments, driving policy reforms, and promoting accountability and transparency, stakeholders can work together to address the complex challenges of textile and apparel waste effectively. Through collaborative action, they can accelerate progress towards a more sustainable future, realizing the overarching objectives of sustainable development for the benefit of present and future generations.

## 3.10 Partnerships for Waste Recovery Initiatives: Joining Forces for a Greener Future

In the realm of waste recovery within the apparel sector, collaborative efforts are essential for ushering in a more sustainable future. These initiatives often involve partnerships between a diverse array of stakeholders, including government agencies, non-governmental organizations (NGOs), manufacturers, and retailers. By joining forces, these stakeholders can pool their resources, expertise, and influence to develop and implement effective waste recovery programs aimed at reducing the industry's environmental footprint (Forbes et al., 2020). One of the primary objectives of such partnerships is the development and execution of waste recovery programs, which encompass a variety of initiatives designed to minimize waste and promote recycling. Take-back schemes, for instance, involve the collection of used apparel and textiles for recycling or reuse. By working together, stakeholders can establish collection systems, set up drop-off points, and streamline logistics to ensure the efficient collection and processing of discarded garments. Closed-loop recycling systems represent another key facet of waste recovery initiatives within the apparel sector. These systems aim to recover and recycle materials from post-consumer apparel waste, thereby reducing the industry's reliance on virgin resources and minimizing its environmental impact. Through collaborative efforts between apparel brands, recycling facilities, waste management organizations, and other stakeholders, closed-loop systems can achieve significant waste recovery rates, contributing to the circular economy. Moreover, the importance of data cannot be overstated in assessing the effectiveness of partnerships for waste recovery initiatives. Collaborative initiatives generate valuable data that helps quantify the impact of these efforts on waste reduction and recovery. Metrics such as the volume of materials recovered, the percentage of garments diverted from landfills, and the environmental benefits accrued from recycling efforts provide tangible evidence of the positive outcomes achieved through partnership-driven waste recovery programs. Furthermore, these collaborative endeavors align closely with Sustainable Development Goal 17 (SDG 17), which emphasizes the importance of partnerships for achieving sustainable development objectives. By leveraging partnerships to drive waste recovery initiatives, stakeholders demonstrate their commitment to advancing the goals of sustainable development in a coordinated and collective manner. The alignment of waste recovery efforts with SDG 17 underscores their broader significance in promoting environmental sustainability, social responsibility, and economic prosperity. In essence, partnerships for waste recovery initiatives represent a powerful mechanism for driving positive change within the apparel sector and beyond (Fletcher et al., 2019). By bringing together stakeholders from across the industry and beyond, these collaborative efforts harness collective expertise and resources to address the pressing challenges of waste management and environmental sustainability. Through continued collaboration and innovation, stakeholders can pave the way for a greener, more sustainable future for generations to come.

## 3.11 Closed-Loop Recycling Systems: Closing the Loop for a Sustainable Future

Closed-loop recycling systems represent a pivotal strategy in the pursuit of sustainability within the apparel industry. These systems are designed to recover and recycle materials from post-consumer apparel waste, thereby mitigating the need for virgin resources and curtailing the environmental impact associated with textile production. By closing the loop on material usage, these systems exemplify a circular approach to resource management, where materials are continually reused and recycled to minimize waste generation (Fletcher et al 2019). Partnerships play a central role in the success of closed-loop recycling systems. Collaboration between apparel brands, recycling facilities, waste management organizations, and other stakeholders is crucial for establishing and operating these systems effectively. By pooling their expertise, resources, and infrastructure, stakeholders can maximize the efficiency and effectiveness of closed-loop recycling initiatives, ensuring that a significant portion of post-consumer apparel waste is diverted from landfills and reintegrated into the production cycle. Data analysis is instrumental in assessing the impact and efficacy of closed-loop recycling systems. Metrics such as the volume of materials recovered, the percentage of recycled content in new products, and the reduction in environmental footprint provide valuable insights into the performance of these systems. By tracking and analyzing these data points, stakeholders can gauge the progress of their recycling efforts, identify areas for improvement, and demonstrate the tangible benefits of closed-loop recycling to stakeholders and the broader community.

Furthermore, the alignment of closed-loop recycling systems with Sustainable Development Goal 17 (SDG 17) underscores their significance in advancing global sustainability objectives. By promoting partnerships and collaboration, closed-loop recycling initiatives contribute to the broader goal of fostering sustainable development and addressing environmental challenges. The data-driven approach to measuring the impact of these systems further reinforces their alignment with SDG 17, providing empirical evidence of their contribution to sustainable waste management practices. In conclusion, closed-loop recycling systems represent a critical component of the apparel industry's transition towards sustainability. Through partnerships, data-driven analysis, and a commitment to circularity, these systems offer a viable solution for reducing waste, conserving resources, and minimizing environmental impact. By advancing the objectives of SDG 17 and embracing a holistic approach to sustainability, stakeholders can work together to build a more resilient and environmentally responsible future for the apparel industry and beyond.

## 3.12 Take-Back Schemes: Taking Back Control of Textile Waste

Take-back schemes involve the collection of used apparel and textiles for recycling or reuse. Collaborative efforts between apparel brands, retailers, and waste management companies enable the implementation of take-back schemes. Data on the quantity of garments collected through take-back programs and the percentage of materials diverted from landfills provide insights into the impact of these initiatives on waste recovery and SDG 17 targets. Central to the success of take-back schemes is the collaboration between various stakeholders across the apparel supply chain. Apparel brands and retailers play a pivotal role in promoting and implementing these initiatives, while waste management companies provide the infrastructure and expertise needed to process and recycle collected materials. By working together, these stakeholders can create a seamless and efficient system for collecting, sorting, and recycling used textiles, maximizing the impact of take-back schemes on waste recovery efforts. Data analysis is essential for assessing the effectiveness and impact of take-back schemes. By tracking metrics such as the quantity of garments collected, the percentage of materials diverted from landfills, and the volume of recycled content in new products, stakeholders can evaluate the success of these initiatives in achieving their waste reduction goals. Additionally, data on consumer participation and engagement provide insights into the level of awareness and support for take-back schemes, informing future strategies and outreach efforts. Moreover, the alignment of take-back schemes with Sustainable Development Goal 17 (SDG 17) underscores their significance in advancing global sustainability objectives. By fostering collaboration and partnerships across sectors, take-back schemes contribute to the broader goal of promoting sustainable consumption and production patterns. The data-driven approach to measuring the impact of these initiatives further reinforces their alignment with SDG 17, demonstrating their contribution to waste reduction and resource conservation.

In conclusion, take-back schemes offer a practical and effective solution for addressing textile waste and promoting a more circular economy within the apparel industry. Through collaboration, data-driven analysis, and a commitment to sustainability, stakeholders can work together to maximize the impact of these initiatives on waste recovery efforts and SDG 17 targets. By embracing take-back schemes as part of a comprehensive waste management strategy, the apparel industry can move closer to achieving its sustainability goals while reducing its environmental footprint.

## 4. GOVERNMENTAL AND NON-GOVERNMENTAL COLLABORATIONS: UNITING FORCES FOR A SUSTAINABLE FUTURE

Collaborations between governmental bodies and NGOs are instrumental in driving policy changes and supporting waste recovery initiatives in the apparel sector. Government funding and support for NGO-led programs can enhance waste management infrastructure and promote sustainable practices. Data on collaborative projects, funding allocations, and waste recovery metrics demonstrate the contribution of governmental and non-governmental partnerships to SDG 17 objectives (Singla N et al., 2024). Government involvement in waste management is essential for establishing regulatory frameworks and implementing policies that promote sustainability. By partnering with NGOs, governments can access valuable expertise and community insights to inform their decision-making processes. NGOs often advocate for stronger environmental regulations and provide on-the-ground support for waste recovery initiatives, complementing governmental efforts to address environmental challenges.

NGOs, on the other hand, rely on government support and funding to implement their programs and initiatives effectively. Governmental funding can help NGOs scale up their operations, invest in infrastructure, and reach a broader audience with their waste management initiatives. Additionally, government support lends credibility to NGO-led projects, fostering trust and collaboration among stakeholders. Data analysis plays a crucial role in assessing the impact of governmental and non-governmental collaborations on waste recovery efforts. By tracking metrics such as the number of collaborative projects, funding allocations, and waste recovery metrics, stakeholders can evaluate the effectiveness of these partnerships in achieving their objectives. Data-driven insights provide valuable feedback for refining strategies, allocating resources more efficiently, and scaling up successful initiatives. Furthermore, the contribution of governmental and non-governmental collaborations to Sustainable Development Goal 17 (SDG 17) underscores their significance in advancing global sustainability objectives. SDG 17 emphasizes the importance of partnerships for achieving sustainable development goals, highlighting the need for collaborative action across sectors and stakeholders. Governmental and non-governmental collaborations exemplify this spirit of partnership, working together to address the complex challenges associated with textile waste and promote a more sustainable future for the apparel industry. In conclusion, governmental and non-governmental collaborations represent a powerful force for driving positive change in the apparel sector. By uniting forces, governments and NGOs can leverage their respective strengths and resources to implement effective waste recovery initiatives, advance sustainable practices, and contribute to the achievement of SDG 17 objectives. Through collaboration, data-driven analysis, and a shared commitment to sustainability, stakeholders can work together to create a more sustainable future for the apparel industry and the planet.

### 4.1 Supply Chain Transparency and Sustainable Sourcing Practices: Illuminating the Path to Sustainability

Transparency in the apparel supply chain enables brands to trace materials back to their source and implement sustainable sourcing practices. Collaborative initiatives between apparel companies, suppliers, and certification bodies promote transparency and responsible sourcing. Data on supply chain transparency metrics showcase the role of collaborative efforts in advancing SDG 17 goals related to sustainable production and consumption.

### 4.2 Exploring Additional Waste Management Practices for a Sustainable Textile and Apparel Industry

In the quest for a more sustainable future, the textile and apparel industry continues to explore innovative waste management practices aimed at minimizing environmental impact and fostering circularity. Beyond the strategies already mentioned, several other approaches are gaining traction in the industry. Let's delve into some of these emerging waste management practices and their potential contributions to creating a more sustainable textile and apparel sector.

### 4.3 Zero-Waste Manufacturing: Maximizing Efficiency, Minimizing Waste

Zero-waste manufacturing represents a holistic approach to production, aiming to eliminate waste at every stage of the process. By optimizing pattern making, fabric cutting, and assembly techniques, manufacturers can significantly reduce material waste. Advanced technologies such as computer-aided design (CAD) and computer-aided manufacturing (CAM) enable precise cutting and efficient material utilization, minimizing offcuts and scraps. Embracing zero-waste principles requires rethinking traditional production methods and embracing innovative approaches to design and manufacturing (Singla N et al., 2024).

### 4.4 Bio-Based Materials: Harnessing Nature's Solutions

Bio-based materials, derived from renewable resources such as plants, algae, and fungi, offer a sustainable alternative to traditional textile fibers. Materials like lyocell, made from wood pulp, and polylactic acid (PLA), derived from corn starch or sugarcane, are increasingly being used in apparel production. These materials are biodegradable and have a lower environmental footprint compared to conventional synthetic fibers. By incorporating bio-based materials into their product offerings, manufacturers can reduce reliance on finite resources and mitigate the environmental impact of textile production.

### 4.5 Chemical Recycling: Transforming Waste Into Resources

Chemical recycling, also known as advanced recycling or depolymerization, involves breaking down textile waste into its molecular components to produce new fibers or feedstocks. Unlike mechanical recycling, which involves shredding and re-melting materials, chemical recycling can handle mixed or contaminated textiles, offering greater flexibility in waste management. Technologies such as enzymatic hydrolysis, solvolysis, and pyrolysis are being explored to convert textile waste into usable materials with minimal environmental impact. Chemical recycling holds promise as a scalable solution for addressing textile waste while reducing reliance on virgin resources.

### 4.6 Digital Supply Chain Platforms: Enhancing Transparency and Traceability

Digital supply chain platforms leverage technologies like blockchain, RFID (Radio Frequency Identification), and IoT (Internet of Things) to enhance transparency and traceability across the textile supply chain. By digitizing production processes and recording transactions on a secure, immutable ledger, stakeholders can track the journey of materials from source to shelf. This transparency enables better decision-making, facilitates compliance with sustainability standards, and fosters accountability throughout the supply chain. Digital supply chain platforms empower consumers to make informed choices and support brands committed to sustainability.

### 4.7 On-Demand Manufacturing: Redefining Production Paradigms

On-demand manufacturing, also known as made-to-order or custom production, shifts away from traditional mass production models towards a more demand-driven approach. By producing goods only in response to customer orders, manufacturers can minimize excess inventory, reduce overproduction, and eliminate the need for markdowns or disposal of unsold inventory. Digital technologies such as 3D knitting, 3D printing, and virtual prototyping enable efficient customization and rapid production, catering to individual preferences while reducing waste and environmental impact. On-demand manufacturing aligns with the principles of a circular economy, where resources are used more efficiently, and waste is minimized throughout the product lifecycle.

### 4.8 Product Life Extension: Prolonging the Usefulness of Textiles

Product life extension strategies focus on prolonging the useful life of textiles through repair, refurbishment, and reconditioning. By offering repair services, manufacturers can extend the lifespan of garments and reduce the likelihood of disposal due to minor damages or defects. Collaborating with repair professionals or partnering with specialized organizations can provide consumers with access to repair services, fostering a culture of durability and longevity. Additionally, initiatives such as garment leasing or rental programs allow consumers to enjoy the benefits of high-quality apparel without the long-term commitment, reducing the overall demand for new garments and minimizing waste.

### 4.9 Textile-To-Textile Recycling: Closing the Loop

Textile-to-textile recycling involves breaking down used textiles into fibers or yarns that can be spun into new fabrics or products. Unlike traditional recycling methods that produce lower-quality materials, textile-to-textile recycling maintains the integrity and performance of the original fibers, enabling closed-loop production cycles. Advanced sorting technologies and mechanical or chemical processes are employed to separate fibers, remove contaminants, and produce high-quality recycled materials. By closing the loop and reintroducing recycled fibers into the manufacturing process, textile-to-textile recycling reduces the demand for virgin resources and minimizes waste generation.

## 5. COLLABORATIVE RESEARCH AND INNOVATION: DRIVING INDUSTRY TRANSFORMATION

Collaborative research and innovation initiatives bring together industry stakeholders, academic institutions, research organizations, and government agencies to tackle complex challenges and drive technological advancements in waste management and sustainability. By pooling resources, expertise, and knowledge, these collaborative efforts accelerate the development and adoption of innovative solutions, ranging from novel materials and technologies to new business models and policy frameworks.

Open innovation platforms, consortiums, and public-private partnerships play a crucial role in fostering collaboration and facilitating knowledge sharing across the industry.

## 5.1 Consumer Engagement and Behavior Change: Empowering Sustainable Choices

Consumer engagement and behavior change initiatives aim to raise awareness, educate, and empower consumers to make more sustainable choices in their purchasing decisions and consumption habits. By providing information on sustainable alternatives, eco-friendly practices, and the environmental impact of textile production, brands and organizations can inspire consumers to adopt more responsible behaviors. Campaigns, workshops, and educational materials encourage consumers to embrace concepts like minimalism, conscious consumption, and circularity, fostering a shift towards a more sustainable and mindful approach to fashion.

## 5.2 Extended Producer Responsibility (EPR) Programs: Encouraging Accountability

Extended Producer Responsibility (EPR) programs hold manufacturers accountable for the environmental impact of their products throughout their lifecycle, including disposal and end-of-life management. By imposing financial or operational obligations on producers, EPR programs incentivize the adoption of sustainable practices, waste reduction measures, and resource-efficient design principles. Producers may be required to take responsibility for collecting, recycling, or disposing of their products at the end of their useful life, encouraging product design for recyclability and facilitating the establishment of closed-loop systems.

In conclusion, the textile and apparel industry is at a pivotal moment in its journey towards sustainability, with a growing emphasis on waste management practices that prioritize resource efficiency, circularity, and environmental stewardship. By embracing emerging technologies, fostering collaboration, and engaging consumers, the industry can drive positive change and pave the way for a more sustainable future. As stakeholders continue to innovate and experiment with new approaches to waste management, the vision of a circular and regenerative textile and apparel sector grows closer to reality.

# 6. CONCLUSION

In the textile and apparel industry, the imperative to address environmental challenges and embrace sustainable practices has never been more pressing. As the global population grows and consumer demand for clothing continues to rise, so too does the need for innovative solutions to minimize waste, reduce environmental impact, and create a more sustainable future. Through a comprehensive examination of waste management practices and sustainability initiatives, it is evident that the industry is making significant strides towards a more circular and regenerative model. From design for sustainability to collaborative supply chain efforts, from emerging technologies to consumer engagement strategies, stakeholders across the textile and apparel value chain are actively working to reshape the industry's trajectory. Designing textiles and apparel with sustainability in mind, optimizing manufacturing processes to minimize waste, and embracing circular economy principles are foundational pillars in this transformation. By choosing sustainable materials, implementing waste reduction strategies, and fostering collaboration across the supply chain, manufacturers can significantly reduce their environmental footprint and move closer to achieving sustainability goals. Recycling and upcycling initiatives, along with the adoption of bio-based materials and chemical recycling technologies, offer promising avenues for diverting textile waste from landfills and closing the loop on resource use. Digital supply chain platforms enhance transparency and traceability, empowering consumers to make informed choices and support brands committed to sustainability.

Furthermore, on-demand manufacturing, product life extension strategies, and textile-to-textile recycling hold immense potential for transforming production paradigms, prolonging the lifespan of textiles, and reducing the demand for new materials. Collaborative research and innovation efforts drive industry-wide transformation, accelerating the development and adoption of innovative solutions. Consumer engagement and behavior change play a pivotal role in driving demand for sustainable products and fostering a culture of conscious consumption. By raising awareness, providing education, and empowering consumers to make more sustainable choices, brands and organizations can catalyze positive change and shape consumer preferences. Extended Producer Responsibility (EPR) programs provide a framework for holding manufacturers accountable for the environmental impact of their products, encouraging the adoption of sustainable practices and the design of products with end-of-life considerations.By working together towards a common vision of a more sustainable future, the industry can overcome challenges, seize opportunities, and pave the way for lasting environmental and social impact. With determination, creativity, and collective action, the textile and apparel industry can lead the transition to a circular economy and contribute to a more sustainable and equitable world for future generations.

## REFERENCES

Baumann, E. (2018). *The Circular Economy: A New Sustainability Paradigm*. Springer.

Circular by Design. (2019). *Products in the Circular Economy*. European Environment Agency.

Circularity Indicators. (2020). *An Approach to Measuring Circularity*. Ellen MacArthur Foundation.

Closing the Loop. (2021). *Commission Adopts New Circular Economy Action Plan*. European Commission.

Ellen MacArthur Foundation. (2017). *A New Textiles Economy*. Redesigning Fashion's.

Fashion on Climate. (2019). *The Relationship between Fashion and Climate Change*. World Resources Institute.

Fletcher, K., & Tham, M. (2019). *The Handbook of Sustainability and Fashion*. Routledge.

Forbes, J. (2020). *Design for Sustainability: A Practical Approach for Textiles and Fashion*. Bloomsbury Visual Arts.

Goldsworthy, M. (2018). *Designing for the Circular Economy*. Bloomsbury Visual Arts.

Higg Materials Sustainability Index (MSI), *Sustainable Apparel Coalition*, (2020).

McKinsey & Company. (2016). Style that's Sustainable: A New Fast Fashion Formula. McKinsey.

Mohan, C., Chauhan, S., Shekhar, S., & Kaushik, P. (2020). Madhu, Sharma M., Kumar V., A Brief Study on rate of morphological and genomic similarities between SARS and COVID-19. *International Journal of Pharma Research and Health Sciences*, 8.

Mohan C., Malik D., Pandey S. & Kumar V. (2020). Effect of COVID-19 on India, International *Journal of Advanced Educational Research.*

Mohan C., Robinson J., Lal C., Jammala A.P.N. & Meena P. L. (2024). Sustainable Energy Solutions for Environmental Pollution Control. *E3S Web of Conferences, 511.*

PwC. (2020). *Reinventing Fashion: A Circular Approach to Redesigning Fashion's Future*. PwC.

Single-Use Plastics. (2018). *A Roadmap for Sustainability*. United Nations Environment Programme.

Sustainable Apparel Coalition. (2021). *Higg Index*. SAC.

The Future of Sustainable Fashion. (2020). *A Collaboration with Good on You*. Business for Social Responsibility.

United Nations. (2015). *Transforming our World: The 2030 Agenda for Sustainable Development*. United Nations.

United Nations Environment Programme. (2019). *Textiles and the Environment in a Circular Economy*. UN.

Vatin, N. I., Negi, G. S., Yellanki, S. V., Mohan, C., & Singla, N. (2024). Sustainability Measures: An Experimental Analysis of AI and Big Data Insights in Industry 5.0. *BIO Web of Conferences, 86.*

# Chapter 4
# Importance of Education and Community Engagement for Effective Waste Management Techniques

**Archana Kumari**
*Lovely Professional University, India*

**Sudev Dutta**
https://orcid.org/0000-0003-4212-1129
*Lovely Professional University, India*

## ABSTRACT

*A sustainability-oriented educational approach encourages holistic thinking for the benefit of all living organisms and future generations, equipping individuals with the skills to enhance their quality of life responsibly. This mindset empowers global citizens to make informed decisions without compromising the well-being of our planet. Sustainability education seamlessly integrates with core subjects, capturing student interest and enabling them to contextualize coursework in real-life scenarios. Recognizing the escalating global challenges of pollution and waste, UNESCO (full form) has initiated a sustainable education agenda through the Education for Sustainable Development (ESD) program. The present study mainly focuses on education on waste management techniques for sustainability. The approach of sustainability in education will help us to think in a holistic way for all the living organism's life for future aspects. It is nothing more than a skill to learn and improve the quality of life of an individual.*

## 1. INTRODUCTION

A sustainability-oriented educational approach encourages holistic thinking for the benefit of all living organisms and future generations, equipping individuals with the skills to enhance their quality of life responsibly. This mindset empowers global citizens to make informed decisions without compromising

DOI: 10.4018/979-8-3693-4264-0.ch004

the well-being of our planet. Sustainability education seamlessly integrates with core subjects, capturing student interest and enabling them to contextualize coursework in real-life scenarios. Recognizing the escalating global challenges of pollution and waste, UNESCO (full form) has initiated a sustainable education agenda through the Education for Sustainable Development (ESD) program. This initiative strives to position education at the forefront of the international response to climate change, generating knowledge, offering policy guidance, and providing technical assistance to nations. Education for Sustainable Development form of education extends beyond traditional classrooms, fostering a continuous learning process for individuals. The commitment and seriousness demonstrated in these efforts underscore collective responsibility toward creating a sustainable and improved global environment. The present study mainly focuses on education on waste management techniques for sustainability. The approach of sustainability in education will help us to think in a holistic way for all the living organism's life for future aspects. It is nothing more than a skill to learn and improve the quality of life of an individual. It enables everyone to take the decisions of their own about the things they do at global level without damaging our mother Earth.

## 1.1 Sustainable Education

ESD is a very important and multidimensional research direction that ensures an ecological and therefore holistic approach to educational research, open opportunities for interdisciplinary research and respect for the skills and interests of each researcher in the field of education; it makes it possible to separate the various contexts of sustainable development, to focus on the sustainable development of society, educational institutions/organizations, the natural environment, etc.; such research emphasizes the need for systematic thinking on the part of both the researcher and the learning objectives (e.g. teachers, students); it signifies the importance of teacher education in the context of sustainable development; such research directs science and education not only to the present, but also to the future. Through research on Education for Sustainable Development (ESD) and the development of policies, it's evident that teachers play a crucial role in structuring and delivering their lessons in a manner that enables students to acquire the necessary knowledge, skills, and abilities to address sustainability challenges. This implies that ESD aims not only for a surface-level change in students' behaviour but rather for an educational process where students' understanding, capabilities, and competencies regarding sustainability are enhanced. This type of education, known as Education for Sustainable Development, is characterized by an instructional approach centred on focussing action competence for sustainability (Jyrki et al., 2024).

Through ESD research and policy development, we know that teachers have an important task in organizing and conducting their teaching in such a way that students gain the knowledge, skills, and competences required to cope with sustainability issues. This means that the goal of ESD is not a 'mere' behavioral change among students, but rather a teaching and learning process where students' sustainability knowledge, skills, and competences are strengthened. This kind of education, education for sustainable development, is described as a teaching approach where action competence for sustainability is at the very core of education. As the study says that the functional training is central to the creation of functional knowledge. One critical element in promoting students' action skills is action-oriented learning experiences where students have the opportunity to work with real problems in a democratic and pluralistic manner. This approach is often referred to as the transformation of existing teaching and learning, as it represents a change in the way we think, feel and act on issues related to the environment and sustainable development. To achieve this change in education and improve student performance, it is advocates capacity-based sustainable development consisting of action orientation, holism and pluralism.

Education is an important tool for achieving sustainable development worldwide. It is the only means by which people can understand and benefit from the work done for sustainable development. Sustainable development is development that meets the needs of the present without compromising the ability of future generations to meet their special needs. The basic principle of sustainable development is the combination of the economic, social and environmental conditions that we all share. Education for sustainable development uses the entire education system to provide students with what they need to transform our society into a sustainable future. Sustainable development should also include recognizing and respecting the positive heritage and legacy of past generations. Education for Sustainable Development (ESD) is a dynamic concept that embodies a new vision of education that seeks to balance human and economic well-being with respect for cultural traditions and the Earth's natural resources. It emphasizes the importance of learning in creating a transition to sustainability by improving citizenship education, a culture of peace, gender equality, respect for human rights, health education, population education, nature conservation and conservation education. education on natural resource management and sustainable use. Therefore, education must be transformed into a constructive tool to raise awareness among students and global citizens. ESD uses interdisciplinary educational systems to develop ideas for lifelong learning, respects human needs in harmony with the sustainable use of natural resources, and encourages awareness of global solidarity. Education for sustainable development was seen as a practice of learning to make decisions that determine the long-term future of economic, ecological and social well-being of all communities. Thus, ESD provides an opportunity to maintain the balance between man and nature (Arundhathi et al., 2016).

The goal of sustainability is to stabilize our economic, environmental and social needs, which enables prosperity for our current and future generations. Sustainable development depends on a long-term integrated process of developing and becoming a strong community, jointly addressing economic, environmental and social problems and avoiding overconsumption of key natural resources. Sustainable development cannot be managed only through technical solutions, political control or financial mediation. We have to change the way of thinking and acting. This requires quality education and training that supports sustainability at all stages and from all social perspectives. The idea of sustainable development is an important milestone in environmental theory because it shows how society itself should be organized, not only what special needs for environmental security should be identified and how best to implement them. Similarly, this aspirational interpretation is widely shared by business leaders, political activists and academics. Of course, how much social and economic change is relevant to sustainability depends on how unsustainable it is perceived. Many supporters of the idea consider the current rather unsustainable and are therefore ready for radical changes. The basic principle of sustainable development is the combination of economic, social and environmental frameworks. Without a proper ecosystem, a better society and economic development for us and future generations is impossible. Thus, the environmental dimension can be seen as the ultimate limit of sustainable development. The social dimension satisfies the basic needs of the entire population without crossing the boundaries of the ecosystem. The economic dimension is a way of defining goals within a socially and ecologically sustainable way of operating. Sustainable education aims to educate all actors, including students, communities, business communities, etc. about the importance and necessity of sustainability in today's world. It can be considered as related not only to the environment, but also to the development of social and economic aspects.

## 2. SUSTAINABILITY: TEACHING, TRAINING AND COMMUNICATING

### 2.1 Sustainability in Teaching

In the vibrant tapestry of higher education, a profound mission takes root—Higher Education for Sustainable Development (HESD). The essence is simple yet transformative: empower students to develop attitudes, skills, and knowledge that guide decision-making for their own benefit and the well-being of others, both now and in the future. This journey is not a solitary one; it's backed by a global commitment. Initiatives like the United Nations' Decade of Education for Sustainable Development in the year (2005–2014) and the subsequent Global Action Program on Education for Sustainable Development which is led by UNESCO, echo the call to weave sustainability into the very fabric of teaching and learning. Higher education institutions emerge as key players in this narrative, embracing multifaceted roles. They become architects of education for a sustainable society, cultivators of future change agents, and creators of spaces where ideas flourish, paradigms are questioned, creativity blooms, and new knowledge sprouts. To navigate this complex mission, diverse models for teaching and learning are indispensable. The landscape of sustainable development is intricate and multidisciplinary. The true goal of higher education lies in nurturing students, empowering them to grasp the complexity of sustainability issues, fostering critical thinking, and actively engaging in finding resolutions. Yet, a paradox lingers—a gap between the transformative potential of sustainability education and the reality of current pedagogical practices. The urgency for young minds to grasp the essence of sustainability clashes with pedagogies that neglect students' prior knowledge, often leaning on transmissive and lecture-driven approaches. This paradox prompts crucial curiousness: Scope of chasm existence between the transformative ideals of sustainability lifelong learning and its reality. The insights offered through different teaching models. The inspiration level of current sustainability courses aligned with their initial goals. These questions echo in the corridors of higher education, challenging the status quo. In this chapter, we embark on a journey of exploration in term of all such kinds which shows the achievement till now. Definitions of sustainability approaches come to life, revealing their epistemological and pedagogical roots. These approaches stand in contrast with three basic teaching styles and three philosophical models, forming a canvas that critiques the necessary conditions for effective teaching in sustainability education. At the heart of our exploration lies the concept of transformative learning—an anchor for effective HESD instruction. The aim is clear: bridge the existing gap, transforming a powerful sustainability message into a pedagogy that empowers, steering away from current practices leading to unsustainability. This chapter is not just a story; it's a call to cultivate sustainable minds, planting seeds of change that will bloom and shape the future.

### 2.2 Imposition Teaching Style

In the realm of teaching sustainability, one prominent approach takes centre stage — the imposition teaching style. Picture a classroom where the curriculum emphasizes the definitions of sustainability, turning the pedagogical process into a journey of information delivery about the origins and purposes of these definitions. This teaching style paints a unidirectional picture. The instructor becomes the maestro, orchestrating a conversation that flows solely from them to the learners. The keys to the domain of knowledge are firmly held by the instructor, leaving students as passive recipients. It's an imposition on their intellectual heritage, a one-sided discourse where their voice is overshadowed. A recent study sheds the light on this style. It experimented with a one-hour lecture followed by student participation in tutorial groupings, attempting to shift the narrative proposed in this chapter. Yet, the strategy doesn't completely overthrow the traditional "Sage on the Stage" lecture for the first hour of each class. The study underscores a critical perspective, emphasizing the need to facilitate students in making immediate sense of ideas and concepts during the lecture itself. The aim is to avoid treating the lecture as a mere gathering of ideas, later to be dissected by someone else in tutorials. Imagine the classroom as a stage where ideas are not just presented but also comprehended in real-time. It's a shift from imposition to facilitation, from a monologue to a dialogue. The study hints at a transformation, a recognition that learning is not a passive absorption but an active engagement, an ongoing conversation between the instructor and the learners. As we delve deeper into the dynamics of imposition teaching, we uncover the complexities of balancing traditional lectures with interactive strategies. The chapter peels back the layers, exploring how this imposition style impacts the students' learning experience and their journey towards understanding sustainability. In this symphony of learning styles, we witness the evolution of teaching practices. The aim is not just to impart knowledge but to ignite a spark of curiosity, to enable students to actively participate in the conversation. The chapter sets the stage for a broader exploration, challenging the norms and seeking a harmonious blend of traditional wisdom and modern pedagogical innovation (Barth et al., 2007).

## 2.3 Interdependence of Curriculum, Pedagogy and Transformative Learning

Imagine a classroom as a canvas where knowledge unfolds not just as facts but as a journey of transformation. This is the essence of transformative learning—a shift in the frame of reference. It's like stepping from a world of individual ideas into one that embraces multiple perspectives, enriching decision-making with enlightened reasoning. In the heart of this transformation lies sustainability, a concept with a threefold frame of reference—societal, environmental, and economic sustenance. To truly grasp this intricate balance, teaching must transcend the conventional. It must go beyond merely imparting information and venture into the realm of transformative learning. The dance of effective Higher Education for Sustainable Development (HESD) hinges on the existence of transformative learning conditions. These conditions pave the way for a deeper understanding of sustainability's internal conflicts and complexities. The journey towards sustainability is not a solitary one; it requires a guide, a mentor who can either hinder or foster transformative learning. The classroom becomes a crucible of possibilities, where an instructor can either embrace an imposition style, standing as a barrier to transformation, or adopt a dialogical approach—a trialogue style—encouraging a vibrant exchange of ideas. Let's delve into the magic of trialogue teaching. It's not just about the instructor delivering knowledge; it's a collaborative dialogue, a dance between the teacher, the student, and the subject matter. This style fosters an environment where transformative learning can take root and flourish. As we navigate the interdependence of curriculum, pedagogy, and transformative learning, we unravel the intricacies of this dance. The chapter seeks to bridge the gap between theory and practice, exploring how the alignment of these elements can create a harmonious symphony of understanding sustainability. The journey is not confined to the university classroom alone. Transformative learning can sprout in various environments, guided by different mentors like parents and community leaders. However, within the academic realm, the instructor plays a pivotal role—a steward either blocking the path or guiding the way toward transformation. This chapter invites to explore the delicate balance between curriculum, pedagogy, and the transformative essence of learning. It's a call to embrace the power of trialogue teaching, creating a space where sustainability becomes not just a subject but a transformative journey, shaping individuals who can navigate the complexities of our world with clarity and purpose.

## 2.4 Holistic Approach to Sustainability

Step into the vibrant world of a school, where lessons extend beyond textbooks and classrooms. Beyond the official curriculum, pupils absorb cultural norms, values, and behaviors from teachers, staff, and their surroundings. They are astute observers, quick to discern the harmony or discord between what's taught and what unfolds within the school's walls. Consider this: studying energy conservation while witnessing a lack of implementation in the school. Such disparities don't escape the watchful eyes of pupils. It's these observations that underscore the significance of the Whole-School Approach to Sustainability, a movement gaining momentum for various reasons. This approach, like a unique melody, is played differently in schools worldwide. Its notes resonate in various ways, addressing crucial aspects: The formal curriculum becomes a repository of knowledge, skills, perspectives, and values related to sustainability. Learning transcends textbooks, incorporating real-life issues to heighten pupils' motivation and understanding. The school emanates a sustainability ethos, visible in how others are treated, how school property is cared for, and the respect shown to the environment. School management practices align with sustainability, encompassing procurement, water and energy use, and waste management. School policies mirror environmental, social, and economic sustainability. Interactions between the school and the community are nurtured, forming a collaborative bond. Special events and extra-curricular activities amplify classroom learning about sustainability. Pupils actively engage in decision-making processes affecting school life. This holistic approach often begins with a single sustainability activity, a seed planted that grows into a garden of sustainable practices over time. Figure 1 represents the factors impacting the sustainable teaching methodology.

But it goes beyond the classroom. It touches upon a fundamental principle—social equity—a core concept of sustainability. Historically, educational practices did not always align with social equity, favouring certain skills over others. This imbalance led to pupils excelling or falling behind, perpetuating social and economic disparities. Dropping out of school, a significant challenge, becomes a social and economic sustainability issue. The remedy lies in diversifying teaching techniques, addressing the diverse learning needs of pupils. It's a transformative practice that not only fosters equity in the classroom but also exemplifies the form equity and social sustainability can take. The school, ideally, becomes a living embodiment of sustainability in its academic programs, policies, and daily operations. Equity seeps into enrolment and hiring practices, ensuring fairness regardless of ethnicity, gender, or race. Economic sustainability becomes tangible through thoughtful purchasing policies, and environmental sustainability is woven into construction, maintenance, and daily practices. Creating a whole-school approach in your community is a journey. Start with an inventory of resources, generate a list of projects, and prioritize activities based on their importance and available resources. This chapter invites you to explore the orchestration of sustainability in education, where every note contributes to nurturing young minds and shaping a future grounded in awareness, equity, and sustainability (Ferreira et al., 2006).

*Figure 1. Factors having impact on sustainable teaching practices*

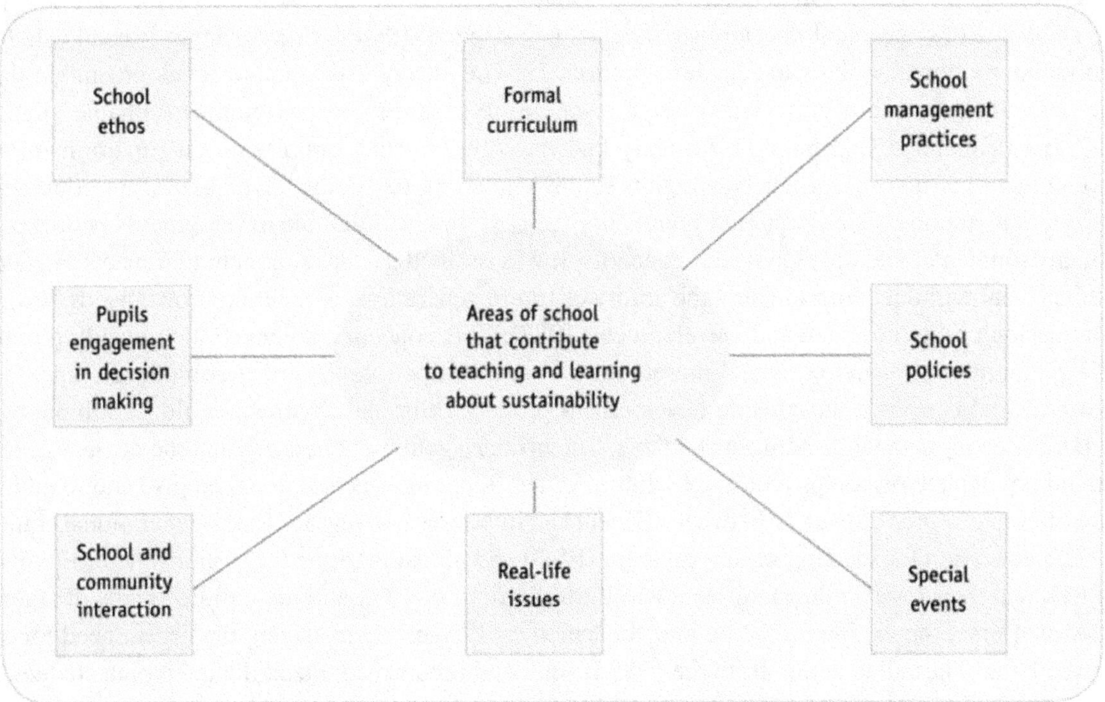


## 3. HISTORY OF EDUCATION FOR SUSTAINABLE DEVELOPMENT

The fact that often slips from people's attention is the progression of the sustainability concept. While the history and development of an idea may appear insignificant, they hold the potential to forecast forthcoming trends and identify emerging flaws. Understanding this evolution can pave the way for ensuring that the 21st century emerges as "the Sustainability Century," where sustainability principles shape our path forward and ensure the importance in terms of our secure future (Dorin et al., 2008). Many overlook the development of the sustainability concept, forgetting its evolution. While the history of a concept might seem insignificant, it holds the key to anticipating future trends and flaws. Understanding this evolution is crucial to ensuring that the 21st century becomes known as "the Sustainability Century." Since its approval at the UN General Assembly in 1987, the idea of sustainable development has been accompanied by an exploration of the parallel concept of education to support its goals. Over the span of 1987 to 1992, committees engaged in thorough discussions and negotiations, resulting in the crafting of 40 chapters within Agenda 21, marking the maturation of the sustainable development concept. The inception of Education for Sustainable Development (ESD) was documented as "Promoting Education, Public Awareness, and Training." What distinguishes ESD from conventional education movements is its origin outside the typical education circles. Notably, a significant push for ESD came from international political and economic forums, including the United Nations, the Organization for Economic Co-operation and Development, and the Organization of American States. During the formulation of the sustainable development concept, it became increasingly evident that education plays a pivotal role in achieving sustainability. Interestingly, ESD continues to be shaped by individuals outside the education community in many countries. Ministries focusing on areas like environment and health take charge of developing the concepts and content of ESD, subsequently entrusting educators with their delivery. The challenge of conceptual development occurring independently of educator input is acknowledged by both international bodies and educators themselves. The idea of "sustainable development" initially emerged in the context of preserving marine resources to ensure their maximum sustainable yield. In 1987, the Brundtland Commission, formally known as the World Commission on Environment and Development and led by former Norwegian Prime Minister Gro Harlem Brundtland, introduced the concept in its report, "Our Common Future." For the first time, sustainable development encompassed both environmental and developmental concerns. It was defined as "development that meets the needs of the present without compromising the ability of future generations to meet their own needs" (World Commission on Environment and Development, 1987). This concept challenged the prevailing notion that development and environmental preservation were incompatible. By integrating development and environmental concerns, sustainable development proposed that development could harmonize with the Earth's ecosystem while being sustainable. The primary goal was to ensure that the present generation did not deplete resources required by future generations (inter-generational equity) and to address disparities in resource utilization between affluent and impoverished regions (intra-generational equity).

Both education for sustainable development (ESD) and global learning for sustainable development (GLSD) had their roots in the significant Rio Earth Summit of 1992 and the outlined goals of Agenda 21. At that time, the world was facing unprecedented challenges, emphasizing the urgent need for collective action. The call to action from the 1987 Brundtland report recognized the interconnectedness of environmental, social, and economic issues. It highlighted the importance of international cooperation and solidarity, as solely pursuing economic growth could lead to environmental disaster without addressing the pressing challenges faced by developing countries. While the declarations at Rio acknowledged the seriousness of the situation, there was an optimistic mood. Despite the gravity of the challenges, the steps to shift from a path of disaster to a brighter future seemed simple and feasible. It was in everyone's interest to prevent the planet from becoming uninhabitable, and the states forming the UN were ready to collaborate in shared efforts.

The concept of "sustainable development" from the beginning was tied to the UN agendas for "development," which classified countries based on various development criteria and indicators. The classification suggested that "developing" countries had limited action scope, and the "developed" countries needed to make specific investments in sustainable development. Another argument pointed out that "developed" countries, while utilizing resources worldwide, also had a much greater environmental footprint per capita. Although this approach was top-down from the UN, it gained support from numerous NGOs. The necessary coordination of efforts required the international structures that the UN could provide. The idea at Rio was that even unsustainable industries could be "greened" and converted to more sustainable modes of production. Despite the risk that citizens might resist changing unsustainable habits, it was assumed that this could be overcome through information, education, and increased awareness. The Brundtland report of 1987 identified the causes of the environmental crisis and emphasized the need for change, driven by concern for the future and future generations. Teachers and educational institutions were explicitly mentioned in Brundtland's report introduction, along with "citizen" groups, non-governmental organizations, and the scientific community. In the decades following Rio, international development strategies primarily focused on formal education, particularly primary compulsory schooling, as guaranteed through the "right to education" in various UN documents. The understanding of global learning for sustainable development has been somewhat restricted by its origins in this broader international UN-driven development discourse. The emphasis on formal education for implementing sustainable development has framed learning as preparation for the future lives of individual students, rather than emphasizing "social learning", life-long informal/non-formal learning, or learning as developing the capacity for collective action (i.e., learning to self-organize, mobilize resources, and strengthen organizational capacity). Rio marked a turning point, bringing together national governments, NGOs, international organizations, and educators worldwide for common efforts towards a shared goal. The ideas and practices mobilized by Rio drew on movements and debates from preceding decades. Activists and researchers had been warning about environmental destruction since the 1950s. Powerful social movements were addressing social and global injustices, along with threats of intolerance, racism, militarization, and warfare. Various bodies, organizations, and international agencies were actively working on issues such as eliminating poverty, ensuring safe water supplies, developing renewable energy sources, reducing pollution, fighting acid rains, conserving species, and reversing desertification. These experiences ultimately contributed to the development of both ESD and GLSD. To gain consensus and move forward, the vision at Rio was to focus on the intersection of the "economic," the "environmental," and the "social." This involved actions that could create a win–win situation with immediate benefits for all. The idea was that by focusing on changes within the intersection of these three domains, sustainability would incrementally increase, leading to the achievement of sustainable development. The prevailing "three pillar" or "three sphere" conception of sustainable development, which later became dominant, originated in debates during the 1970s on economics, development, and global relationships. In the context of the Brundtland report and Rio, the goal was to coordinate and integrate international social and economic development work with environmental constraints. Rio did not challenge the notion that economic growth was necessary to reduce the global poverty gap, and Agenda 21 includes the aim: "To promote and support policies, domestic and international, that make economic growth and environmental protection mutually (Tanaka, 2017).

The following significant step was the UN Conference on the Environment and Development (UNCED), held in Rio de Janeiro, Brazil, during the summer of 1992. This was an unprecedented historical event, marked by the largest gathering of 114 heads of state, accompanied by 10,000 representatives from 178 countries and 1400 non-governmental organizations, bringing additional thousands into the mix. The conference itself unfolded as an international spectacle on an unprecedented scale, with heads of government aiming to make their mark on what became known as the Rio Earth Summit. The title, "connecting Environment and Development," reflected the North–South bargaining dynamics at the UN, where demands for international action on the environment intersected with claims for additional development aid and technology transfer. As we have seen earlier that the key outputs of the 1992 Earth Summit also known as the Rio Summit were the Rio Declaration, Agenda 21, and the Commission on Sustainable Development. These outputs marked the international arrival of the concept of sustainable development. Agenda 21, in particular, is highlighted as a comprehensive document outlining practices and advice for achieving sustainable development across various areas on Earth. It emphasizes the need for efficient resource use, protection of the environment, and sustainable economic growth. The conference emphasized that both developed and developing countries must engage in environmental management to ensure sustainable development. Additionally, countries agreed to prepare national sustainable development strategies to implement Agenda 21. In shorten it is observed that Over the span of conferences held between 1972 and 2002, a noticeable shift in political discourse can be observed. Initially, at the 1972 Stockholm Conference, the focus was primarily on environmental issues (Birgitta N and Helen, 2021). This evolved into a shared focus on environmental, social, and economic development at the Rio de

Janeiro Earth Summit in 1992. Subsequently, at the Millennium Summit in 2000 and the Johannesburg World Summit in 2002, there appeared to be a shift towards a primary emphasis on poverty alleviation. However, this shift doesn't imply that environmental protection has been sidelined, but rather it suggests a growing recognition of its capacity to alleviate poverty. The initial call to protect the environment in the context of human development has evolved into a more specific call to prioritize the well-being of the most marginalized individuals, both presently and in the future. The greatest challenge of sustainable development continues to be fostering global awareness, from households to boardrooms, about the importance of addressing the challenges stemming from the Industrial Revolution, which involved extensive human and environmental exploitation (Aliasghar, 2024).

## 4. COLLABORATIONS WITH DIFFERENT AGENCIES FOR EDUCATIONAL SUSTAINABILITY

Sustainability considerations are increasingly important for scholars, businesses, and the general population. However, achieving sustainable progress is complex and cannot be solely accomplished by individuals or organizations. Therefore, entrepreneurs aiming for sustainable change may seek collaborations with universities to overcome resource and technology constraints. Ventures aligned with economic Sustainable Development Goals (SDGs) are likely to engage in product and prototype development with support from professors. Additionally, various collaborations and approaches are employed to enhance sustainability outcomes in education. This chapter expands understanding on the search for collaborations in educational sustainability through various communication channels related to SDGs. The adverse environmental and social impacts of economic growth and globalization strain the world's finite resources and affect people's well-being. Consequently, sustainability issues are garnering increased attention, not only within academia and business but also globally. Existing studies on sustainability cover a wide range of topics beyond environmental concerns. These include defining sustainability, sustainability transition, responsible production for sustainability, sustainability assessment, sustainability in e-commerce packaging, and sustainability innovations and firm competitiveness. Although sustainability has been a concern throughout human history, its prominence surged in the late 20th century, notably after the publication of the "Our Common Future" report by the World Commission on Environment and Development (WCED) in 1987. This report aimed to establish guiding principles for sustainable development. The concept of sustainable development, as defined by the report, emphasizes meeting present needs without compromising the ability of future generations to meet their own needs, stressing the importance of resource conservation. In 2015, the United Nations General Assembly underscored this importance by approving the 17 Sustainable Development Goals (SDGs) as a global blueprint for sustainability. This led to a significant increase in literature on sustainable development and sustainability, with some studies using the terms interchangeably. However, following Shaker's distinction, sustainability refers to humanity's goal of achieving homeostasis or human ecosystem equilibrium, while sustainable development involves integrated approaches and processes aimed at contributing to sustainability (Ma et al., 2022).

Despite the increasing adoption of the Sustainable Development Goals (SDGs) by numerous organizations worldwide, achieving them presents several practical challenges. These challenges include scarcity of resources and financing, the need to translate the goals into local contexts, aligning expectations among different stakeholders, and delineating and evaluating sustainability performance. These complex hurdles cannot be addressed by any single organization alone. Therefore, sustainable development necessitates collaboration and concerted efforts from all stakeholders involved (Jirapong, et al., 2021). Relationships of the collaborative production can mirror the cooperative attributes and laws among the parties involved in the scientific research process. Presently, these relationships primarily involve collaboration among scholars, institutions, and regions. However, not all collaborative production relationships are sustainable or effective in enhancing the quality of scientific research output. Various forms of cooperative production relationships yield different research outcomes. The UNESCO Global Action Program (GAP), initiated at the World Conference on Education for Sustainable Development in November 2014 in Aichi-Nagoya, Japan, aims to expand Education for Sustainable Development (ESD) across all levels and sectors of education, as well as within various sustainable development domains. Among the five priorities outlined by GAP, one focuses on expediting sustainable solutions at the local level. The global network of Regional Centres of Expertise (RCEs) for ESD has played a significant role in advancing this fifth priority by engaging in collaborative projects between universities and communities, with a particular emphasis on involving youth. These projects, while diverse in nature, bridge formal and informal education, thereby contributing to the achievement of Sustainable Development Goal 4, which pertains to quality education. Climate change emerged as a central theme across many of these projects, thereby also contributing to Sustainable Development Goal 13, which focuses on Climate Action.

The participation of Higher Education Institutions (HEIs) in network structures, whether formal or informal, focused on sustainability or specific community-related issues, can enhance and streamline collaboration with communities. HEIs with prior experience and capabilities in collaborating with local, regional, and national stakeholders are more likely to engage in new initiatives aimed at promoting the implementation of Sustainable Development Goals (SDGs) through partnerships with communities. The subsequent UNESCO ESDfor2030 RoadMap underscores Education for Sustainable Development (ESD) as integral to quality education and as a fundamental enabler of all 17 SDGs, thereby emphasizing the significance as a catalyst for collaboration, particularly given the adaptable nature of informal education approaches.

University community engagement involves a broad spectrum of interactions among students, faculty, and administration with external communities and organizations, with the aim of fostering mutual benefits. Typically, this engagement is positioned within the university's third mission, alongside teaching and research. Engagement is characterized by three key aspects. Firstly, it entails a process of knowledge exchange that surpasses mere "knowledge transfer." Secondly, it involves collaborating with communities and organizations beyond the academic sphere through partnership arrangements. Thirdly, it is oriented towards addressing societal needs and resolving specific societal challenges. Various types of engagement practices include teaching and learning activities targeting a broader audience beyond academia, research directed towards addressing the societal needs of external communities or conducted in participatory partnerships, service and knowledge exchanges, student-led initiatives within communities, formal partnerships at the university level between higher education institutions (HEIs) and external groups, and the provision of academic services to the community.

However, university community engagement often operates within context-specific and grassroots frameworks, posing challenges in institutionalizing and recognizing this commitment within Higher Education Institutions (HEIs). Given that university community engagement practices inherently involve establishing partnerships geared towards addressing societal needs, it is argued that the Sustainable Development Goals (SDGs) present a significant opportunity for systematic change in higher education. They offer a useful framework for determining mechanisms to develop mutually beneficial partnership opportunities with the community. University community engagement fosters the emergence of pertinent institutional strategies within the university and enhances various aspects such as operations, curriculum, research, as well as the university's culture and reputation, leading to the generation of new knowledge. It also serves as a vital mechanism for raising awareness within the university community. Generally, universities possess the capacity to contribute to the SDGs through teaching, research, and engagement, leveraging their expertise and knowledge. Additionally, they can promote sustainability by making their campuses more sustainable and by integrating the SDGs into strategic institutional agendas. Member States are encouraged to engage stakeholders involved in the five priority action areas and foster collaborative networking under a unified strategy aligned with the national framework on the Sustainable Development Goals (SDGs). UNESCO will persist in backing the networking efforts of identified global partners by facilitating regular platforms for information exchange, sharing experiences, and jointly planning initiatives. These platforms will incorporate representatives and participants from governmental and private sectors, donor and development communities, as well as stakeholders operating within the five priority action areas at national, regional, and global scales (Walter L et. Al., 2023). To effectively implement the five priority action areas of Education for Sustainable Development (ESD) for 2030, networking opportunities are being organized among Member States and other stakeholders to encourage cross-national and cross-sectoral collaboration and foster synergies. At the country level, Member States are encouraged to undertake the following activities to enhance partnership and collaboration. Strengthen partnerships and collaboration among formal, non-formal, and informal learning sectors, including lifelong learning across generations within communities. Foster partnership and collaboration among stakeholders in education, sustainable development, and the Sustainable Development Goals (SDG), particularly SDG 4, Enhance public-private sector partnerships and collaboration to bring about the structural changes necessary for a sustainable future and mobilize resources to support this process. Establish country-level networks of stakeholders for ESD for 2030 to coordinate initiatives and activities related to ESD and facilitate partnership and collaboration among diverse stakeholders. Align and reinforce coordination and cooperation at global, regional, national, sub-national, and local levels regarding policies and practices on ESD to ensure mutually supportive synergies.

## 5. FUTURE OF EDUCATION AND SUSTAINABILITY

Higher education institutions play a crucial role in advancing sustainability and are instrumental in educating future leaders who will contribute to the successful implementation of the United Nations Sustainable Development Goals (SDGs). While the geography of SDGs implementation varies widely, it is evident that higher education institutions significantly contribute to fostering a mindset conducive to promoting the principles of SDGs. This perspective paper examines the impact of higher education on sustainability and explores the challenges and barriers associated with this process. Higher education institutions are pivotal in the implementation of the SDGs, particularly in achieving Goal 1 ending poverty in all its forms everywhere, Goal 3 ensuring healthy lives and promoting well-being for all at all ages, Goal 5 promoting gender equality, Goal 8 fostering decent work and economic growth, Goal 12 promoting responsible consumption and production, Goal 13 addressing climate change, and Goal 16 promoting peace, justice, and strong institutions. As a catalyst for transformation, the higher education sector wields significant influence on students' behaviours and their contributions to a thriving society. However, to instigate the necessary changes in education, sustainability principles must form the foundation of higher education institution strategies, including curricula and operational practices, and must be ingrained in the organizational culture. By leading through example, higher education institutions can exert a positive influence on society, promoting the adoption of key aspects of Sustainable Development Goals (SDGs) such as gender equality, waste reduction, and energy conservation. Achieving this requires employing various communication methods tailored to different student academic levels.

Nevertheless, there are critical challenges to address both within and outside institutional environments, including integrating sustainability principles, navigating the political landscape, and addressing stakeholders' interests (UNESCO, 2022). In contemporary societies, numerous crises, including ecological, economic, cultural, and most recently, a health crisis, pose significant challenges to the organization and social practices we often take for granted. There is a prevailing sense of risk and urgency in current times, indicating the imperative need for immediate societal changes to avert the worst consequences of these crises. However, intricate connections between social inequalities and environmental issues create highly complex challenges, often leading to disagreements regarding necessary actions. Each crisis affects different groups of people in distinct ways, and the recent COVID-19 pandemic vividly illustrates the interconnectedness among humans and between humans and the environment. This interconnectedness is also apparent from our accumulated knowledge of natural ecosystems. The globalization of the economy, trade, tourism, mass media, cultural aspirations, and desired lifestyles has further deepened this global interconnectedness (Marouli et al., 2021).

Higher Education Institutions are increasingly integrating concepts of sustainable development into their programs to educate students about sustainability, a trend encouraged by the Sustainable Development Goals. In today's technological era, concerns about sustainability encompass ethical decision-making and responsible behavior of organizations and leaders. Additionally, the Covid-19 pandemic has underscored the importance of sustainable development and emphasized the necessity for a more sustainable future. Sustainability has the potential to positively influence values, the environment, and enhance quality of life. While incorporating sustainability into the curriculum presents challenges for universities, it also presents opportunities for educational institutions. Universities play a crucial role in transitioning towards sustainability, not only by integrating the concept into research and teaching but also by implementing it on campus. This review article offers a concise overview of sustainable development in higher education institutions, discussing how universities adopt and apply sustainability principles. It also examines how sustainability can be integrated across various disciplines and explores the attitudes of students and academics towards Education for Sustainability. This is an important consideration that warrants attention. However, historically, the educational system has functioned as a tool of society, aimed at preparing individuals to fit into specific societal norms and maintain the prevailing social structure. As Non zealot pointed out, the widespread implementation of schooling originated from the desire to remove impoverished children from the streets, where they were exposed to social disorder and illegality, and instead, prepare them to become productive citizens.

In contemporary (particularly Western) societies, education is increasingly viewed as a means of preparing individuals for the workforce and providing them with skills essential for success in the global economy. Consequently, influencing minds and behaviours effectively moulding them—has been an implicit, if not direct, objective of the educational system. This market-driven approach to education, which often supersedes the traditional emphasis on the Liberal Arts in our societies, is typically not criticized as indoctrination but rather as being limited and utilitarian in its outlook. Education and the future are inherently interconnected. It is impossible to discuss educational matters without considering their implications for the future. Our understanding of the future shapes our perceptions of which knowledge and skills are essential for the next generation. When it comes to sustainability issues, the way we envision the future influences whether we view sustainability as a tangible goal achievable through technical advancements and efficiency, or as a normative direction determined through democratic processes.

The concept of futures in education guides present decisions and can be viewed as "futures for the present." If the future mirrors the present or can be accurately predicted, determining what the next generation needs would be relatively straightforward. However, given the inherent uncertainty of the future, determining the necessary knowledge and skills becomes more complex. To install a sustainability mindset effectively, it is crucial to communicate sustainability principles efficiently. Communication plays a pivotal role in the successful implementation of various sustainability initiatives. Here are three reasons why higher education institutions serve as essential channels to communicate sustainability effectively. Wide Reach: Higher education institutions attract a diverse population, despite existing inequalities. In 2016, approximately 216 million students worldwide were enrolled in higher education institutions. For example, in the UK alone, over 2,383,970 students were enrolled in higher education institutions during the 2018/19 academic year, with expectations for further increases by 2030. Critical Development Stage: A significant proportion of students enrolled in higher education institutions are in late adolescence, aged 18 to 21, which is a crucial period for identity development. Research indicates a consistent increase in identity attainment throughout the four years of college. Young individuals in this age group are highly influenced by ideas discussed in the classroom, making higher education a prime platform for promoting sustainability initiatives. Additionally, transitioning from home to university often marks a pivotal moment for becoming engaged in climate change campaigns.

Preparation for the Workforce: Higher education serves as the final stage of education before individuals enter the workforce, where they assume influential roles as professionals. Communicating sustainability initiatives effectively during their college/university education equips future professionals with the mindset and knowledge necessary for implementing sustainability practices in real-life scenarios. In summary, effective communication of sustainability initiatives within higher education institutions is crucial for shaping the mindsets of future professionals and translating acquired knowledge into real-world applications. Agenda 21 initially identified education as a crucial element in promoting sustainable development. This recognition led to the launch of the Education for Sustainable Development (ESD) campaign, spearheaded by UNESCO and globally promoted during the UN Decade of Education for Sustainable Development beginning in 2005. Education, primarily aimed at equipping learners with specific competencies, is envisioned to play a significant role in advancing sustainable development. The ESD program embodies high ambitions, aiming to pave the way for a truly sustainable future. ESD, like all forms of education, inherently revolves around shaping the future. My objective here is to highlight to my readership the profound influence of societal and educational systems on human socialization and future outlooks. It is imperative to examine how the concept of the future is perceived within current educational systems and Western developed societies. My contention is that for ESD to effectively contribute to societal transformation towards sustainability, certain prerequisites must be addressed, while simultaneously acknowledging the limitations of education beyond the knowledge-action gap.

The significance and prioritization of education within the post-2015 development agenda are underscored by the establishment of SDG 4, which aims to "ensure inclusive and equitable quality education and promote lifelong learning opportunities for all." SDG 4 sets the foundation for global efforts to enhance education under the post-2015 development agenda and comprises seven primary targets and three additional targets on means of implementation. This commitment is further reinforced by the consensus on the "Education 2030 Agenda" and the draft Framework for Action – Education 2030, adopted at the

World Education Forum 2015 (19-22 May 2015). These documents provide a comprehensive plan for implementing SDG 4 and outline the monitoring and reporting mechanisms for this goal. Moreover, the Education 2030 agenda integrates existing trends and initiatives in international education into a unified agenda. With strong consensus and support mechanisms established for SDG 4 at the World Education Forum 2015, progress toward formalizing this goal is generally advancing well.

Discussions surrounding the post-2015 development agenda have placed a significant emphasis on education as a fundamental tool for achieving sustainable development. Education for Sustainable Development (ESD) spans across formal, non-formal, and informal education settings, revolutionizing teaching and learning methodologies to encourage critical thinking, cooperation, and student-centred approaches. Both literature and practical experiences now underscore the pivotal role of quality education in enabling individuals to lead healthier, happier, and more productive lives in a sustainable manner. Quality education yields greater returns on investment compared to other development goals, with outcomes exerting a more substantial influence on economic growth than mere school enrolment rates. Moreover, enhancing education quality can be more cost-effective, relying on systematic knowledge investments rather than new resource allocations. A Quality Education for Sustainable Development (QESD) approach fosters higher-order learning, enhancing competencies such as analysing, synthesizing, and evaluating complex information for effective decision-making, planning, and problem-solving.

## 6. EDUCATION: PROMISE AND PARADOX

An important distinction lies between education about sustainable development and education for sustainable development. The former entails raising awareness or engaging in theoretical discussions, while the latter utilizes education as a means to achieve sustainability goals. We believe that, at this critical juncture, more than just theoretical discussions are necessary. While some argue that "for" implies indoctrination, we view it as indicating a purpose. All education serves a societal purpose, or else society would not invest in it. Education for Sustainable Development (ESD) holds the promise of creating a more liveable world for present and future generations. While there may be instances of ESD being misused or distorted into indoctrination, such actions run counter to the essence of ESD, which seeks to empower individuals with knowledge and skills for lifelong learning to address environmental, economic, and social challenges. Two major issues in the international dialogue on sustainability are population growth and resource consumption. Increasing population and resource usage are perceived to threaten a sustainable future, and education is intricately linked to both fertility rates and resource consumption. Educating women leads to reductions in fertility rates and subsequently slows population growth, thus facilitating progress towards sustainability. Conversely, the relationship between education and resource consumption operates differently. Generally, individuals with higher levels of education and income tend to consume more resources than those with lower levels of education and income. In this scenario, increased education exacerbates the threat to sustainability.

Sustainability-focused organizations endeavour to balance financial success with the environmental and social impacts of their operations. One of the greatest challenges for sustainability is to effectively address social, environmental, and financial objectives concurrently. In essence, sustainable organizations aim to mitigate negative impacts while enhancing positive outcomes, restore natural resources, and uphold ethical standards. Stakeholders expect transparency and accountability from organizations across various dimensions of performance, including economic, social, and environmental aspects. To meet these expectations, an increasing number of organizations regularly publish reports detailing their sustainability initiatives and accomplishments for both internal and external stakeholders. Additionally, organizations engage with stakeholders through social media platforms to share information. However, questions have arisen regarding the true intent behind such communication efforts, as there remains a significant disparity between organizations' sustainability rhetoric and their actual practices. Some argue that these disclosures primarily serve to maintain organizational legitimacy rather than genuinely contributing to sustainability efforts. A paradoxical perspective on addressing tensions in performance encompasses conflicts between various domains of sustainability and aims to address multiple competing sustainability objectives simultaneously. To navigate through these performance paradoxes, organizations engage in continuous improvisation to balance the diverse performance areas effectively (Beech et al., 2004; Clegg et al., 2002). Sustainability concerns that may contradict the organizational objective of profitability are not disregarded. Instead, conflicting sustainability aspects are juxtaposed without prioritizing one over the other as the "best option." This approach keeps the performing paradox open and encourages simultaneous action on multiple economic, social, and environmental outcomes.

A paradox emerges within sustainability education in higher institutions: while it aims to provoke critical thinking and transformative action, the current pedagogy often falls short in fostering sustainability-oriented thinking and behaviour, typically relying on traditional transmissive and lecture-based methods. Several key considerations include: (i) the potential gap between the transformative ideals presented in literature on Teaching Sustainability in Higher Education and the actual practices; (ii) the exploration of alternative teaching models that could promote effective sustainability learning; and (iii) the evaluation of whether current sustainability courses truly instil the intended learning outcomes. This chapter delves into the definitions and approaches to sustainability, as outlined by Alvarez & Rogers (2006), examining their epistemological and pedagogical underpinnings in the context of Higher Education for Sustainable Development (HESD) teaching and learning. The authors juxtapose these sustainability approaches with three fundamental teaching styles identified by Roberts & Silva (1968). Additionally, they analyse three philosophical models that provide a comprehensive critique of the essential conditions for effective teaching pedagogy aimed at fostering an understanding of sustainability issues and solutions. The chapter advocates for transformative learning as a pivotal framework for effective HESD instruction, aiming to resolve the existing paradox wherein a potent sustainability message is conveyed through an inadequate pedagogy. The current teaching practices, lacking in sustainability, necessitate a shift towards more sustainable educational approaches.

## 7. CONCLUSION

To enhance the contribution of education to the creation of environmentally, economically, and socially equitable societies, it is imperative to delve into the concept of quality education. This necessitates unpacking the systemic inequalities and the intricate dynamics of inclusion and exclusion arising from existing institutional capabilities and established local norms. Such an examination will shed light on how educational disparities hinder the fulfilment of education's full potential. The significance of higher education institutions and their associated research cannot be overstated, as evidenced by the systematic advancement of women's studies in India since the mid-1970s and its influence on development policies, programs, and legislation. The research contributes significantly to the existing literature by illuminating the intricate dynamics between universities and communities, identifying current partnerships, and proposing strategies to enhance them. However, there are certain limitations that warrant attention. Future studies could undertake a more comprehensive bibliometric analysis that encompasses a broader array of search terms and includes research published in multiple languages to mitigate these limitations. Increasing the number of case studies and incorporating diverse contexts and regions would provide a more comprehensive and representative understanding of community-university partnerships. Moreover, delving into potential confounding factors and contextual influences through qualitative research methods or mixed-methods approaches could yield deeper insights into the dynamics of these partnerships, thereby informing decision-making processes more effectively in practice.

## REFERENCES

Aliasghar, A. (2024). *A history of the concept of sustainable development: Literature, review*. Research Gate. https://www.researchgate.net/publication/242219096_A_history_of_the_concept_of_sustainable_development_Literature_review

Aromar, R., Amir, B., Chandni, S., & Prathigna, P. (2021). *The Author Team Poonam Batra University of Delhi Indian Institute for Human Settlements*. IIHS. https://iihs.co.in/knowledge-gateway/wp-content/uploads/2021/01/TESF-India-Background-paper.pdf

Barth, M., Godemann, J., Rieckmann, M., & Stoltenberg, U. (2007). Developing key competencies for sustainable development in higher education. *International Journal of Sustainability in Higher Education, 8*(4), 416–430. doi:10.1108/14676370710823582

Birgitta, N. & Helen, A. (2021). Global Learning For Sustainable Development. *Sustainability, 13*(6), 3451. doi:10.3390/su13063451

. Buckler, C., & Creech, H. (2014). *Shaping the Future We Want: UN Decade of Education for Sustainable Development (2005-2014)*.

Carew, A., & Mitchell, C. (2008). Teaching sustainability as a contested concept: Capitalizing on variation in engineering educators' conceptions of environmental, social and economic sustainability. *Production, 16*, 105–115.

Donzelot, J. (1977). *The +Policing of the Families*. The Johns Hopkins University Press.

Education for Sustainable Development. (2020). United Nations Educational, Scientific and Cultural Organization.

Ferreira, J., Ryan, L., & Tilbury, D. (2006). *Whole-School Approaches to Sustainability*. Aries. http://www.aries.mq.edu.au/projects/preservice/

. Gigauri, I., Vasilev, V., & Mushkudiani. Z. (2022). *In pursuit of sustainability: Towards sustainable future through education. 1*(37). doi:10.31435/rsglobal_ijite/30032022/7798

Hahn, T., Figge, F., Pinkse, J., & Preuss, L. (2018). A Paradox Perspective on Corporate Sustainability: Descriptive, Instrumental, and Normative Aspects. *Journal of Business Ethics, 148*(2), 235–248. doi:10.1007/s10551-017-3587-2

Holfelder, A. K. (2019). Towards a sustainable future with education. *Sustainability Science, 14*(4), 943–952. doi:10.1007/s11625-019-00682-z

Jirapong, K., Cagarman, K., & von Arnim, L. (2021). Road to Sustainability: University–Start-Up Collaboration. *Sustainability (Basel), 13*(11), 6131. doi:10.3390/su13116131

Jyrki, R., & Anita, P. (2024). *Education for Sustainable Development Research from the Researchers' Point of View.* Research Gate. www.researchgate.net/publication/258157579 p-9

Ma, Y., Xiu, Q., Shao, L., & Yao, H. (2022). Promoting the Sustainable Improvement of Educational Empirical Research Quality. *Sustainability.* doi:10.3390/su14063380

Marouli, C. (2021). Sustainability Education for the Future? Challenges and Implications for Education and Pedagogy in the 21st Century. *Sustainability (Basel), 13*(5), 2021. doi:10.3390/su13052901

Wamsler, C. (2020). Education for sustainability: Fostering a more conscious society and transformation towards sustainability. *International Journal of Sustainability in Higher Education, 21*(1), 112–130. doi:10.1108/IJSHE-04-2019-0152

# Chapter 5
# Sustainable Harvesting:
## Weed Waste Management for Environmental Preservation and Economic Growth

**Chandra Mohan**
*K.R. Mangalalm University, Gurgaon, India*

**Barkha Sharma**
*K.R. Mangalam University, Gurugram, India*

**Amit Kumar Rawat**
*Hansraj College, India*

**Chingakham Chinglenthoiba**
*Chitkara University, India*

## ABSTRACT

*Weed waste management stands at the confluence of environmental preservation, sustainable practices, and economic opportunities. This chapter delves into the multifaceted realm of weed waste management, particularly in the context of tree and vine cropping systems, emphasizing its significance in agricultural productivity, environmental conservation, and resource recovery. The chapter elucidates the diverse types of weeds and their impact on various ecosystems while exploring the principles of the waste-to-wealth approach. The scope of weed waste management encompasses environmental conservation, agricultural productivity, resource recovery, and economic opportunities. By adopting a waste-to-wealth approach, converting weed waste into valuable resources becomes pivotal. This involves practices such as composting, bioenergy production, material recycling, phytoremediation, education, and innovation.*

## INTRODUCTION

Weed waste management refers to the proper handling, disposal, and potential utilization of waste derived from invasive plants, unwanted vegetation, or agricultural residues. Weed waste management stands at the

DOI: 10.4018/979-8-3693-4264-0.ch005

intersection of environmental conservation, sustainable practices, and economic opportunity (Robertson et al., 2023). It involves the responsible handling, repurposing, and utilizing waste materials generated from invasive plants, unwanted vegetation, or agricultural residues (Saras et al., 2014). This approach, often called "Transforming Waste into Wealth," focuses on converting these seemingly burdensome materials into valuable resources.

Weed control in tree and vine cropping systems significantly impacts the establishment, growth, and yield of crucial fruit and nut crops (Barua et al., 2015; Joshi et al., 2015; Meena et al., 2015). Various factors play a role in determining the most effective weed control strategies in orchards and vineyards (Aimara et al., 2020; Dalia et al., 2020). These factors comprise the types and density of weeds, the age of the crop, planting layout, irrigation systems, as well as the grower's available time, equipment, labor, and financial resources. Weed management strategies commonly encompass mechanical techniques like mowing or cultivation, chemical approaches such as preemergence or postemergence herbicide applications, and cultural methods like implementing cover crops. Typically, managers adopt an integrated approach, utilizing a combination of these methods to address weed challenges (Zheng et al., 2021). They may adapt their techniques over time in response to the changing needs and stages of the orchard or vineyard's life cycle.

Weeds, typically considered nuisances due to their invasive nature, possess untapped potential. Instead of merely disposing of them, innovative approaches can extract value from these materials (Ayangbenro et al., 2022). Transforming weed waste into wealth involves a paradigm shift towards sustainable practices, harnessing their inherent qualities to benefit the environment, society, and the economy. Adopting a waste-to-wealth approach involves converting these wastes into valuable resources, which can benefit the environment, economy, and society (Tirkolaee et al., 2023). Here's an overview of the scope and importance of weed waste management with a waste-to-wealth approach:

## Scope of Weed Waste Management

***Environmental Conservation:*** Invasive weeds can pose ecological threats by outcompeting native species, altering habitats, and impacting biodiversity. Managing weed waste prevents their spread and minimizes environmental damage.

***Agricultural Productivity:*** Agricultural residues, including crop residues or weeds, can be recycled into biofuels, compost, or organic fertilizers. Proper management mitigates the negative impact of weed residues on soil quality and crop growth.

***Resource Recovery:*** Weed waste contains organic matter that can be utilized for bioenergy production, composting, or as a raw material for other industries.

***Economic Opportunities:*** Weed waste management offers avenues for creating new businesses, generating employment, and fostering innovation in waste conversion technologies.

## Importance of Waste-to-Wealth Approach

***Resource Conservation:*** Converting weed waste into valuable products reduces the reliance on finite resources like fossil fuels and synthetic fertilizers.

***Circular Economy Principles:*** It aligns with the concept of a circular economy by reusing, repurposing, or recycling waste materials, reducing overall waste generation, and promoting sustainability.

***Economic Benefits:*** Creating value from weed waste can generate revenue streams, cost savings for industries, and new markets for eco-friendly products.

***Climate Change Mitigation:*** Proper management of weed waste through composting or bioenergy production reduces greenhouse gas emissions that would occur from traditional disposal methods like burning or landfilling.

***Community Engagement:*** Involving communities in waste management programs not only raises awareness about environmental issues but also fosters community participation and ownership.

## Strategies for Weed Waste Management: Waste to Wealth

***Composting:*** Converting organic weed waste into nutrient-rich compost for agricultural or landscaping purposes.

***Bioenergy Production:*** Utilizing biomass from weed waste for biofuel production, such as biogas or bioethanol.

***Material Recycling:*** Repurposing weed waste into value-added products like paper, textiles, or construction materials.

***Phytoremediation:*** Using specific plants to absorb and remove contaminants from soil, thereby managing weed waste and cleaning up polluted areas.

***Education and Innovation:*** Promoting research and education to develop innovative technologies and practices for efficient weed waste management.

## Concept of Waste to Wealth

"Waste to wealth" is a concept that revolves around transforming waste materials or by-products into valuable resources or commodities, thereby creating economic value and reducing environmental impact (Muzamil et al., 2022). This approach focuses on converting what might traditionally be considered waste into assets, thereby promoting sustainability, resource efficiency, and economic growth. The core principles and components of the waste-to-wealth concept include:

***Resource Recovery:*** Instead of disposing of waste materials in landfills or incinerating them, the waste-to-wealth approach emphasizes recovering resources from these materials. This can involve recycling, repurposing, or converting waste into usable forms, such as energy, raw materials, or products.

***Circular Economy Principles:*** The waste-to-wealth concept aligns with the idea of a circular economy. Rather than following a linear model of "take-make-dispose," it promotes a closed-loop system where materials are reused, recycled, or upcycled, minimizing waste and maximizing resource efficiency.

***Innovation and Technology:*** Advancements in technology and innovation play a crucial role in the waste-to-wealth approach. New methods, processes, and technologies are developed to efficiently convert waste into valuable resources, making the transformation more feasible and cost-effective.

***Environmental Sustainability***: By reducing the amount of waste sent to landfills or incinerators, waste to wealth contributes to environmental sustainability. It helps in conserving natural resources, decreasing pollution, and mitigating the negative impact of waste on ecosystems.

***Economic Opportunities:*** Converting waste into valuable resources creates economic opportunities (Marble et al., 2019). It can lead to the development of new industries, job creation, cost savings through resource efficiency, and the generation of revenue from products derived from waste materials.

Examples of waste-to-wealth initiatives include:

***Recycling Programs:*** Recycling materials such as plastics, glass, paper, and metals to create new products.

***Bioenergy Production:*** Converting organic waste into renewable energy sources like biogas or biofuels.

***Upcycling and Repurposing:*** Transforming waste materials into higher-value products or materials, such as turning old textiles into new clothing or furniture.

***Waste-to-Energy Plants:*** Generating electricity or heat by incinerating waste, capturing the energy released in the process.

Overall, the waste-to-wealth concept promotes a more sustainable and efficient use of resources by recognizing the value of what was previously considered waste, thereby contributing to environmental conservation and economic growth.

## Types of Weeds

Weeds are generally categorized based on their growth habits, life cycles, and characteristics (Elliott et al., 2012). Here are some common types of weeds based on these classifications:

***Annual Weeds:*** These weeds complete their life cycle within a year. They germinate, grow, reproduce, and die in a single growing season. Examples include crabgrass, chickweed, and annual bluegrass.

***Perennial Weeds***: Perennial weeds live for over two years and can persist for many years. They may spread through seeds, roots, or both. Examples include dandelions, bindweed, and Canada thistle.

*Figure 1. Structure of weed waste management*

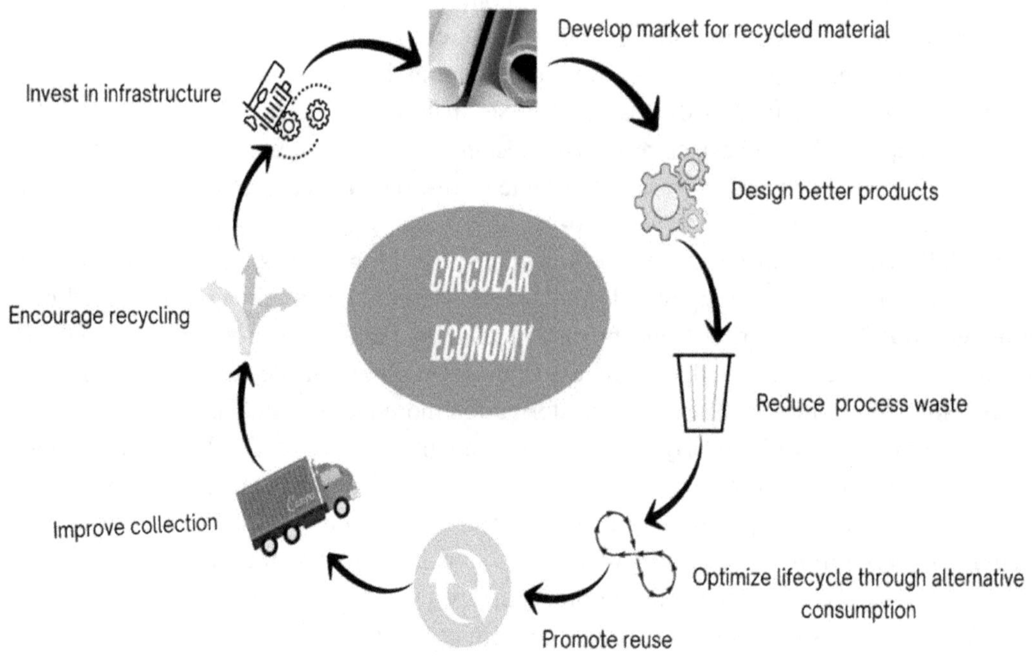

*Figure 2. Annual weeds*

*Figure 3. Perennial weeds*

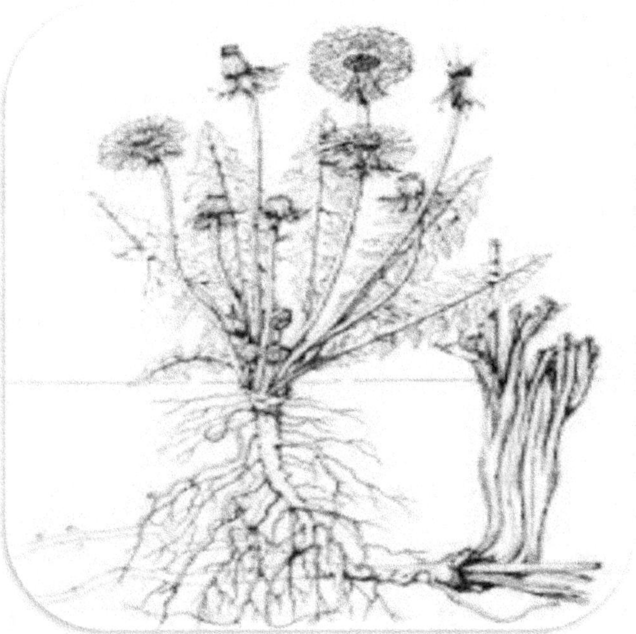

***Biennial Weeds:*** Biennial weeds have a two-year life cycle. In the first year, they grow vegetatively, and in the second year, they flower, produce seeds, and die. Common examples include common mullein and burdock.

***Broadleaf Weeds:*** These weeds have broad leaves and are usually dicotyledonous. They often compete with cultivated plants for resources. Examples include dandelions, plantain, and chickweed.

*Figure 4. Biennial weeds*

***Grassy Weeds:*** Grassy weeds resemble grasses and can be monocots or dicots. They often invade lawns and gardens, competing with desirable plants. Examples include crabgrass, annual bluegrass, and quack grass.

***Woody Weeds:*** These are perennial weeds that develop a woody stem or base. They can be particularly challenging to remove because of their robust structure. Examples include brambles (blackberries and raspberries) and tree seedlings like saplings of unwanted trees.

***Aquatic Weeds:*** These weeds grow in water or in very wet conditions. They can be floating, emergent, or submerged. Examples include water hyacinth, duckweed, and cattails.

***Noxious Weeds:*** Noxious weeds are invasive species that are harmful to agriculture, ecosystems, or humans. They are often regulated or controlled due to their detrimental impacts. Examples include Japanese knotweed, giant hogweed, and kudzu.

***Indicator Weeds:*** Indicator weeds can indicate specific soil conditions or problems. For instance, nettles often grow in nitrogen-rich soils, while dandelions may indicate compacted soil.

*Figure 5. Broadleaf weed*

*Figure 6. Grassy weeds*

*Figure 7. Woody weeds*

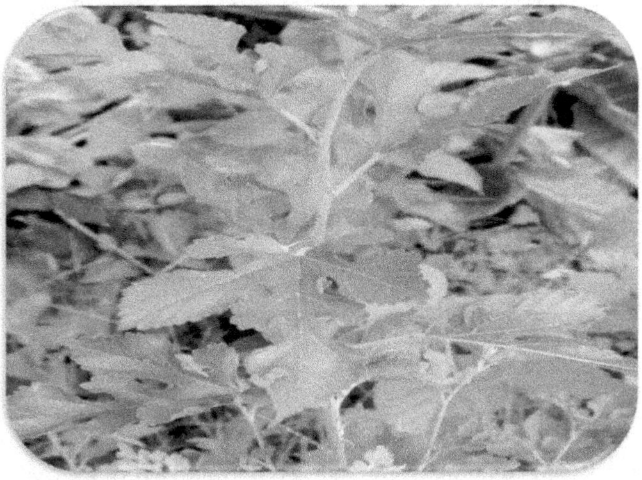

## Weed Waste Management Techniques

There are various ways to minimize waste when using weed (cannabis) (Ignatowicz et al., 2009). Here are a few techniques:

***Efficient Consumption***: Use methods that ensure maximum utilization of the cannabis product. Techniques like vaporization allow for the extraction of cannabinoids without burning the entire substance, resulting in less waste.

***Reusing Residue:*** After using cannabis in a vaporizer, the residue (known as "AVB" - already vaped bud) can be reused. It can be incorporated into recipes for edibles or tinctures to extract further cannabinoids.

*Figure 8. Aquatic weeds*

*Figure 9. Noxious weeds*

***Composting:*** Cannabis waste, such as stems, leaves, and unused plant material, can be composted. Composting breaks down organic material into nutrient-rich soil, which can be used for gardening.

***Recycling Packaging:*** Whenever possible, opt for cannabis products with minimal packaging or those that use recyclable materials. Properly recycle containers and packaging to minimize environmental impact.

***DIY Products:*** Utilize the entire cannabis plant where feasible. For instance, use stems and leaves to make teas, topicals, or extracts. There are various DIY techniques available online for making oils, creams, or even hemp-based products.

***Community Recycling Programs:*** Some dispensaries or cannabis businesses offer recycling programs. They may accept used packaging or unused products for proper disposal or recycling.

***Education and Advocacy***: Spread awareness about reducing cannabis waste within the community. Sharing tips and techniques can encourage others to adopt eco-friendly practices.

Explore innovative ways to create value-added products from weed waste (e.g., textiles, paper, building materials)

*Figure 10. Indicator weeds*

*Figure 11. Cannabis*

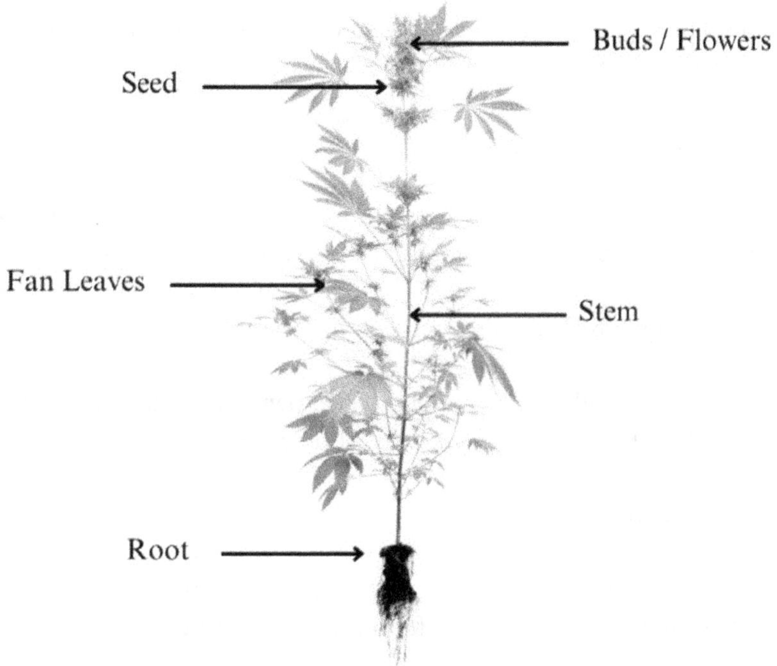


Cannabis waste, such as stems, stalks, and leaves, can be repurposed into various value-added products through innovative methods (Dhote et al., 2009). Here are several ways to create value-added products from weed waste:

1) Textiles and Fabrics:

Hemp fibres derived from cannabis stalks can be used to produce durable textiles. These fabrics are eco-friendly, breathable, and possess antibacterial properties. They can be used in clothing, accessories, and home textiles.

2) Paper Products:

Hemp pulp can be used to make paper, which is more sustainable than traditional wood-based paper. It requires fewer chemicals and can be recycled more times without degrading. Cannabis waste can be used to create paper for packaging, stationery, and other paper products.

3) Building Materials:

Hempcrete, a mixture of hemp hurds (inner fibres of the hemp plant), lime, and water, creates a strong, lightweight, and sustainable construction material. It can be used for insulation, walls, and other building components.

4) Biodegradable Plastics:

Cannabis waste can be processed to extract biopolymers that can be used to manufacture biodegradable plastics. These plastics have applications in packaging, disposable items, and other products, offering an environmentally friendly alternative to traditional plastics.

5) Biofuels and Energy:

Through processes like biomass conversion or pyrolysis, cannabis waste can be transformed into biofuels such as ethanol or biodiesel. These biofuels can be used as renewable energy sources.

6) Nutritional Supplements and Health Products:

Cannabis waste can contain residual cannabinoids and terpenes. By extracting these compounds, they can be incorporated into health supplements, topicals, or other wellness products.

7) Animal Bedding and Fertilizers:

Hemp Hurd can be used as animal bedding due to its absorbent properties. Additionally, cannabis waste can be composted and turned into organic fertilizers for agriculture.

8) Pharmaceuticals and Medical Products:

Cannabis waste may contain cannabinoids and other compounds that have potential medicinal properties. Researchers are exploring ways to extract and utilize these compounds in pharmaceuticals and medical products.

9) Cosmetics and Personal Care Products:

Extracts from cannabis waste, such as hemp seed oil, can be used in skincare products due to their moisturizing and antioxidant properties.

10) Art and Craft Materials:

From hemp-based paper for artistic purposes to using hemp fibres in crafts, cannabis waste can offer sustainable materials for artistic expression.

Leveraging cannabis waste for value-added products not only contributes to sustainability but also taps into the versatility of the plant for various industries (Larney et al., 2007). However, it's crucial to comply with local regulations and ensure proper processing and safety standards when utilizing cannabis waste for different products.

## Discuss Research or Industry Developments in This Area

Weed waste management is an increasingly important aspect of the cannabis industry due to the growing legalization and commercialization of cannabis products (Trivedy et al., 2004). The management of waste generated from cannabis cultivation, processing, and consumption poses environmental challenges that researchers and the industry are actively addressing.

***Regulations and Compliance:*** Many regions have strict regulations governing cannabis waste disposal to prevent diversion to the black market, ensure public safety, and minimize environmental impact. These regulations often dictate proper disposal methods, necessitating compliant waste management practices.

***Waste Characterization:*** Research has focused on characterizing the different types of cannabis waste produced across the supply chain. This includes plant waste (stems, leaves, roots), processing byproducts, and unused or expired products. Understanding the composition of these wastes helps in developing efficient management strategies.

***Sustainable Solutions:*** Innovations in waste management aim to minimize environmental impact. Some companies are exploring recycling methods such as composting organic waste, converting plant matter into biofuels or biomass for energy production, or extracting useful compounds from cannabis waste for various purposes.

***Biotechnology and Circular Economy:*** Researchers are investigating biotechnological approaches like using fungi or bacteria to break down cannabis waste into useful byproducts or employing enzymatic processes to extract valuable compounds for pharmaceutical or industrial purposes (Puniya et al., 2023). This aligns with the principles of the circular economy by reducing waste and maximizing resource utilization.

***Collaboration and Best Practices:*** Industry collaborations, government initiatives, and partnerships between cannabis producers, waste management companies, and regulatory bodies are fostering the development and implementation of best practices for efficient and compliant weed waste management.

***Public Awareness and Education:*** Increasing awareness among consumers, producers, and stakeholders about responsible waste management practices is crucial. This involves educating stakeholders about the importance of compliance with waste disposal regulations and adopting sustainable waste management methods.

***Challenges and Future Directions***: Despite advancements, challenges remain, including technological limitations, high costs associated with specialized waste management, and varying regulations across regions. The future might see further advancements in waste-to-energy technologies, improved recycling processes, and standardized guidelines for cannabis waste management.

In essence, the cannabis industry is actively engaging in research and implementing various strategies to manage weed waste responsibly, aiming for environmentally sustainable and compliant practices while navigating regulatory requirements (Kauser et al., 2020). Collaboration, innovation, and a commitment to sustainability are key drivers in this evolving field of weed waste management.

## Challenges and Solutions

Weed waste management presents several challenges owing to the unique nature of cannabis and the stringent regulations surrounding its cultivation, processing, and disposal (Achmon et al., 2017). Addressing these challenges requires innovative solutions that balance environmental sustainability, regulatory compliance, and cost-effectiveness. Here are some key challenges and potential solutions:

## Challenges

A. Regulatory Compliance: Strict regulations dictate how cannabis waste must be managed, leading to complexities in compliance, especially considering variations in rules across jurisdictions.
B. Volume and Diversity of Waste: Cannabis production generates various types of waste (plant material, packaging, byproducts) that differ in composition and require distinct disposal methods.
C. Security and Diversion Concerns: Preventing diversion of cannabis waste to the black market or unauthorized use requires secure disposal processes.
D. Environmental Impact: Improper disposal can have adverse environmental effects, such as soil and water contamination, if not managed effectively.
E. Cost of Disposal: Effective waste management strategies often come with high costs, impacting the overall operational expenses for cannabis businesses.

## Solutions

A. Comprehensive Waste Management Plans: Develop tailored waste management plans that adhere to local regulations and encompass the entire lifecycle of cannabis waste, from cultivation to disposal.
B. Segregation and Characterization: Properly segregate and characterize different types of cannabis waste to determine suitable disposal methods. This might involve separating organic waste for composting and non-organic waste for recycling or secure destruction.
C. Innovative Recycling and Processing: Explore innovative technologies for recycling cannabis waste, such as composting organic matter, extracting cannabinoids for medicinal purposes, or utilizing waste biomass for bioenergy production.
D. Secure Disposal Protocols: Implement secure disposal protocols to prevent diversion, including shredding or pulverizing waste and using secure disposal containers or facilities.

E. Collaboration and Partnerships: Collaborate with waste management experts, environmental agencies, and regulatory bodies to establish best practices, share knowledge, and optimize waste management processes.
F. Education and Training: Train staff on compliant waste management practices and raise awareness among stakeholders about the importance of responsible waste disposal.
G. Investment in R&D: Invest in research and development to explore cost-effective and environmentally friendly methods for cannabis waste disposal, such as biodegradation, pyrolysis, or enzymatic treatments.
H. Advocacy for Standardization: Advocate for standardized regulations and guidelines for cannabis waste management across different regions to streamline compliance efforts.
I. Incentivizing Sustainability: Encourage and incentivize sustainability practices within the cannabis industry, promoting the adoption of eco-friendly waste management solutions.

## Barriers to Weed Waste Management

Weed waste management faces several barriers that hinder efficient disposal and recycling practices within the cannabis industry (Jain et al., 2018). These barriers stem from regulatory, logistical, economic, and technological challenges:

### Regulatory Hurdles

***Stringent Regulations:*** Cannabis waste disposal is subject to strict regulations and compliance standards, varying significantly among regions. Navigating these diverse and often complex regulations can be a major barrier.

***Lack of Clear Guidelines:*** Inconsistent or unclear guidelines regarding the classification, handling, and disposal of cannabis waste pose challenges for businesses striving to comply with disposal regulations.

### Logistical Challenges

***Diverse Waste Streams:*** Cannabis production generates various types of waste, including plant material, packaging, and byproducts, each requiring specific disposal methods. Managing these diverse waste streams efficiently can be complex.

***Storage and Transportation:*** Securely storing and transporting cannabis waste, especially due to its sensitivity and security concerns, can be challenging and costly.

### Economic and Cost-Related Issues

***High Disposal Costs:*** Proper disposal methods for cannabis waste, such as incineration, composting, or secure destruction, often come with high associated costs, impacting the overall operational expenses of cannabis businesses.

***Limited Economic Incentives:*** Inadequate economic incentives or rewards for implementing sustainable waste management practices might discourage investment in advanced disposal technologies or recycling initiatives.

## Technological and Infrastructure Limitations

***Lack of Advanced Recycling Technologies:*** The absence of cost-effective and scalable recycling technologies tailored to cannabis waste limits efficient recycling and repurposing efforts.

***Limited Access to Specialized Facilities:*** Access to specialized facilities for cannabis waste disposal, such as composting centres or bioenergy plants, may be limited, causing logistical challenges for businesses.

## Security and Compliance Concerns

***Risk of Diversion:*** Concerns about the potential diversion of cannabis waste to the illicit market or unauthorized use require stringent security measures, adding complexity and cost to disposal procedures (Maturi et al., 2022).

***Documentation and Record-Keeping:*** Strict regulatory requirements for documenting and tracking cannabis waste disposal can pose administrative burdens and compliance challenges.

## Public Perception and Awareness

***Stigma Surrounding Cannabis:*** Negative public perception or stigma surrounding cannabis waste may impact public support for innovative recycling initiatives or sustainable disposal practices.

***Lack of Awareness:*** Limited awareness among stakeholders about the importance of responsible cannabis waste management and its environmental impacts may impede efforts to implement effective waste management practices.

## Future Trends in Weed Waste Management

As of my last update in January 2022, managing weed waste, particularly from cannabis cultivation, has been a significant concern due to the increasing legalization and expansion of the cannabis industry (Mohan et al., (2024). Several trends and potential advancements were expected to shape the future of weed waste management:

Evolving regulations were anticipated to enforce stringent guidelines for handling and disposing of cannabis waste. Governments may require specific protocols for the destruction, recycling, or repurposing of cannabis byproducts to mitigate environmental impact and ensure compliance.

Advancements in waste disposal technologies, such as biomass conversion, composting, and anaerobic digestion, were expected to offer more efficient and environmentally friendly ways to manage organic cannabis waste (Qiu et al., 2023). These methods aim to convert waste into usable energy or compost.

Companies were exploring ways to implement a circular economy model in cannabis cultivation and processing. This involves minimizing waste by reusing, recycling, or repurposing byproducts to create new products or energy sources, thereby reducing environmental impact (Spassky et al., 2016).

As sustainability became a focal point across industries, the cannabis sector was expected to adopt more biodegradable and eco-friendly packaging materials. This approach aimed to reduce the environmental footprint of packaging waste associated with cannabis products.

Continued research into innovative methods for repurposing cannabis waste, such as extraction of valuable compounds from leftover plant material or the development of new bio-based products, was anticipated to gain momentum.

Increased public awareness campaigns about the proper disposal of cannabis waste and its environmental impact were expected (Khadka et al., 2022). Educating both industry stakeholders and consumers on responsible waste management practices could drive significant changes in handling weed waste.

Collaboration between industry players, waste management companies, policymakers, and environmental organizations was foreseen as crucial for developing comprehensive and sustainable solutions for managing weed waste (Lorenzo et al., 2022).

These trends are likely to evolve as the cannabis industry matures, and new technologies and regulations emerge (Singh et al., 2008). It's essential to stay updated with the latest developments and advancements in weed waste management, especially considering the dynamic nature of the industry and the evolving environmental concerns.

## CONCLUSION

In conclusion, effective management of weed waste remains a critical concern, particularly in industries such as cannabis cultivation, agriculture, and horticulture. The management of weed waste is essential not only to ensure regulatory compliance but also to mitigate environmental impact and promote sustainability.

The future of weed waste management is poised to witness significant advancements and trends. These include the implementation of stringent regulatory frameworks, the development of innovative waste disposal technologies, the adoption of a circular economy approach, the use of biodegradable packaging, ongoing research into repurposing weed waste, increased public awareness and education, and collaboration among industry stakeholders.

Successful weed waste management involves employing a diverse range of methods, such as mechanical, chemical, and cultural techniques, in an integrated approach. This multifaceted strategy allows for the adaptation and optimization of weed waste management practices throughout the lifecycle of cultivation or agricultural operations.

Effective weed waste management is not just about disposal; it's about unlocking the potential of these materials, contributing to sustainability, and capitalizing on the opportunities they offer. The future of weed waste management lies in innovative approaches, collaborations, technological advancements, and a shared commitment to environmental responsibility within the cannabis industry and beyond.

Continued research, technological innovation, and collaboration among stakeholders will play pivotal roles in shaping more efficient, sustainable, and environmentally friendly weed waste management practices in the years ahead. Industry players, policymakers, and environmental advocates must work together to develop comprehensive solutions that address the challenges associated with weed waste while striving for a more sustainable future (Mohan et al., 2020).

## REFERENCES

Achmon, Y., Fernández, J. D., Hernandez, K., & McCurry, D. G. (2017). Weed seed inactivation in soil mesocosms via bio solarization with mature compost and tomato processing waste amendments. *Pest Management Science*, *73*(5), 862–873. doi:10.1002/ps.4354 PMID:27391139

Aimara, M. S., & Olanrewaju, O. S. (2020). Waste Management through Composting: Challenges and Potentials. *Sustainability (Basel)*, *12*(11), 4456. doi:10.3390/su12114456

Ayangbenro, A. S. S., & Chukwuemeka, C. F. (2022). Harnessing the Rhizosphere Soil Microbiome of Organically Amended Soil for Plant Productivity. *Agronomy (Basel)*, *12*(12), 3179. doi:10.3390/agronomy12123179

Barua, C., Halai, P., Patwary, P., Bora, M., Barua, A. G., & Bordoloi, M. J. (2015). Evaluation of anti-amnestic activity of few medicinal plants against scopolamine-induced amnesia. *Indian Journal of Traditional Knowledge*, *14*(4), 581–589.

Dalia, K. (2020). Green garbage: A state comparison of Morison of marijuana packaging and waste management. *Golden Gate University Environmental Law Journal*, *12*, 175.

Daniel, W. E., Koza, M. B., Parker, S. D., & Long, M. D. (2012). Traditional RCRA and Subpart K: UNC-CH's foray into the waste management weeds. *Journal of Chemical Health and Safety*, *19*(4), 25–30. doi:10.1016/j.jchas.2012.01.011

Dhote, S., & Dixit, S. (2009). Water quality improvement through macrophytes - A review. *Environmental Monitoring and Assessment*, *152*(1-4), 149–153. doi:10.1007/s10661-008-0303-9 PMID:18537050

Ignatowicz, K. (2009). Selection of sorbent for removing pesticides during water treatment. *Journal of Hazardous Materials*, *169*(1-3), 953–957. doi:10.1016/j.jhazmat.2009.04.061 PMID:19446393

Jain, M. S., & Kalamdhad, A. S. (2018). Efficacy of batch mode rotary drum composter for management of aquatic weed. *Journal of Environmental Management*, *1*(221), 20–27. doi:10.1016/j.jenvman.2018.05.055 PMID:29787969

Joshi, N., Singh, V. P., Dhyani, V. C., Singh, S. P., & Pratap, T. (2015). Effect of plant geometry and weed management practices on weed growth and yield of direct seeded rice. *Journal of Environment and Bioscience*, *29*(1), 23–28.

Kauser, H., Pal, S., Haq, I., & Khwairakpam, M. (2020). Evaluation of rotary drum composting for the management of invasive weed Mikania micrantha Kunth and its toxicity assessment. *Bioresource Technology*, *313*, 123678. doi:10.1016/j.biortech.2020.123678 PMID:32570078

Khadka, R. B., Velazquez, A. D. S., & Cardina, J. (2022). Evaluation of Anaerobic Soil Disinfestation for Environmentally Sustainable Weed Management. *Agronomy (Basel)*, *2*(12), 3147. doi:10.3390/agronomy12123147

Larney, F. J., & Hao, X. (2007). A review of composting as a management alternative for beef cattle feedlot manure in southern Alberta. *Bioresource Technology*, *98*(17), 3221–3227. doi:10.1016/j.biortech.2006.07.005 PMID:17276674

Lorenzo, P., Guilherm, R., & Barbosa, S. (2022). Agri-Food Waste as a Method for Weed Control and Soil Amendment in Crops. *Agronomy (Basel), 12*(5), 1184. doi:10.3390/agronomy12051184

Marble, S. C., Steed, S. T., Saha, D., & Khamare, Y. (2019). On-farm evaluations of wood-derived, waste paper, and plastic mulch materials for weed control in Florida container nurseries. *HortTechnology, 29*(6), 866–873. doi:10.21273/HORTTECH04437-19

Maturi, K. C., Haq, I., & Kalamdhad, A. S. (2022). Biodegradation of an intrusive weed Parthenium heterophorias through in-vessel composting technique: Toxicity assessment and spectroscopic study. *Environmental Science and Pollution Research International, 29*(56), 84600–84615. doi:10.1007/s11356-022-21816-4 PMID:35788476

Meena, R. S. (2015). A review of weed management in India: The need of new direction for sustainable agriculture. *The Bioscan, 10*, 253–263.

Mohan, C., Chauhan, S., Shekhar, S., Kaushik, P., Madhu, M., Sharma, M., & Kumar, V. (2020). A Brief Study on rate of morphological and genomic similarities between SARS and COVID-19. *International Journal of Pharma Research and Health Sciences, 8*(3), 3143. doi:10.21276/ijprhs.2020.03.03

. Mohan, C., Robinson, J., Lal, C., Jammala, A. P. N., Meena, P. L. (2024). Sustainable Energy Solutions for Environmental Pollution Control. *E3S Web of Conferences, 511*, 01023.

Muzamil, M., Rasool, S., & Masoodi, U. H. (2022). In Situ and Ex Situ Agricultural Waste Management System. Agricultural Waste - New Insights. IntechOpen. doi:10.5772/intechopen.108239

. Puniya, R., Bazaya, B. R., Kumar, A., Sharma, B. C., Nesar, N. A., Botchily, R. S. (2023). Effect of residue and weed management practices on weed flora, yield, energetics, carbon footprint, economics and soil quality of zero tillage wheat. *National library of medicine, 13*, 19311.

Qiu, J., Souza, M. C. D., Aguilar, A. A. R., & Ghysels, S. (2023). Improving biochar properties by co-pyrolysis of pig manure with bio-invasive weed for use as the soil amendment. *Chemosphere, 312*, 137229. doi:10.1016/j.chemosphere.2022.137229 PMID:36372342

Robertson, K. J., Brar, R., Randhawa, P., Stark, C., & Baroutian, S. (2023). Opportunities and challenges in waste management within the medicinal cannabis sector. *Industrial Crops and Products, 197*, 116639. doi:10.1016/j.indcrop.2023.116639

Saras, P. K., & Patel, B. D. (2014). Weed management by cultural and chemical practices and their effect on phytotoxicity and growth, quality and yield parameters of cluster bean. *Trends in Biosciences, 7*(24), 4196–4199.

Singh, A., Singh, K., & Singh, D. V. (2008). Suitability of organic mulch (distillation waste) and herbicides for weed management in perennial aromatic grasses. *Tropical Pest Management, 37*(2), 162–165. doi:10.1080/09670879109371567

Spassky, C. E., Regnier, E. E. S., Harrison, S. K., Bennett, M. A., & Metzger, J. D. (2016). Weed Suppression in Pumpkin by Mulches Composed of Organic Municipal Waste Materials. *Horticultural Science (Prague), 51*, 720–726.

Tirkolaee, E. B., Goli, A., Gütmen, S., Weber, G. W., & Szwedzka, K. (2023). A novel model for sustainable waste collection arc routing problem: Pareto-based algorithms. *Annals of Operations Research, 324*(1-2), 189–214. doi:10.1007/s10479-021-04486-2 PMID:35068644

Trivedy, R. K., & Thomas, S. (2004). Review of performance data of few pilot and full-scale aquatic weed-based waste treatment plants from India. *Asian Journal of Microbiology, Biotechnology & Environmental Sciences : AJMBES, 6*(1), 157–162.

Zheng, Z., Fiddes, K., & Yang, L. (2021). A narrative review on environmental impacts of cannabis cultivation. *Journal of Cannabis Research, 3*(1), 35. doi:10.1186/s42238-021-00090-0 PMID:34362475

# Chapter 6
# Examining Waste Management in Sustainable Development:
## Essential Strategies for an Environmentally Friendly Future

**Prashant Panwar**
*K.R. Mangalam University, India*

**Megha Garg**
*K.R. Mangalam University, India*

**Anil Kumar**
*Bharati Vidyapeeth's College of Engineering, New Delhi, India*

## ABSTRACT

*The increasing global population presents substantial dangers to environmental sustainability due to issues in waste management. This study delves into the complex connection between waste management and sustainable development, highlighting the crucial need for efficient solutions to construct an eco-friendly future. The chapter examines the present waste management procedures, emphasizing the ecological and social consequences of incorrect disposal methods. The research examines the capacity of these new solutions to reduce environmental damage, enhance resource efficiency, and support the achievement of sustainable development objectives. The research also examines how policy frameworks and regulatory measures influence the development of efficient waste management systems.*

## 1. INTRODUCTION

Managing trash is a crucial and complex task in the quest for global sustainability. With the growing global population and expanding urbanisation, the consequences of poor waste management techniques on the environment are becoming more evident. This introduction aims to highlight the essential role of waste management in the context of sustainable development, emphasising the need to implement

DOI: 10.4018/979-8-3693-4264-0.ch006

environmentally friendly practices and novel disposal techniques. Waste management involves a range of tasks, including collecting and transporting waste products to be disposed of or recycled. Without efficient waste management, there are significant consequences for ecosystems, human health, and total environmental health (Wilson et al., 2015). The significant amount of trash produced worldwide, especially in cities, requires a change in approach to sustainable waste management to reduce its harmful impacts.

Sustainable waste management relies on the concepts of reduction, reuse, and recycling. Trash reduction focuses on reducing trash creation at the source to alleviate pressure on disposal systems. Reuse promotes extending product lifecycles, whereas recycling converts waste materials into useful resources, decreasing the need for new raw materials and lowering environmental impact. This introduction also explores the current trash disposal technologies and their impact on sustainability (Tchobanoglous et al., 1993). Each waste disposal option, from conventional landfilling to incineration and sophisticated recycling technologies, has unique environmental and social impacts. Evaluating these approaches rigorously is essential for guiding waste management towards environmentally friendly options that are in line with the ideals of sustainable development.

Following portions of this thorough investigation will further examine the complex relationship between waste management and sustainable development. This research intends to contribute to the discussion on promoting a more sustainable and environmentally aware future by analysing current waste management practices, assessing the effectiveness of existing solutions, and investigating alternative ones. It aims to encourage a shared dedication to changing the story around trash, turning it from a worldwide problem into a chance for beneficial transformation and accountable management.

Insufficient waste management procedures today have resulted in a significant environmental issue. This section examines the environmental effects of these activities, specifically addressing concerns such as trash contamination, landfill challenges, the plastic waste crisis, and the extensive impacts of inappropriate garbage disposal on ecosystems.

Waste contamination presents a substantial danger to both land-based and water-based ecosystems. Improperly discarded garbage, particularly plastic, often ends up in rivers, seas, and natural ecosystems, leading to lasting damage to biodiversity (Hoornweg et al., 2012). Plastic consumption by marine and terrestrial animals causes ecological disruption, resulting in negative effects on the food chain. Improperly managed waste sites may lead to the leakage of dangerous chemicals, which worsens soil and water pollution, presenting significant threats to plant and animal life.

*Figure 1. Reuse, Reduce, Recycle*

Landfills, the traditional approach to garbage disposal, pose several environmental difficulties. These locations release greenhouse gases, such as methane, which significantly impact climate change. Landfills' unattractive appearance and unpleasant smell diminish the beauty and recreational appeal of nearby places, affecting local populations and ecosystems. The limited availability of appropriate dump sites increases the need to discover sustainable alternatives.

The plastic trash dilemma exacerbates the environmental consequences of insufficient waste management. Plastic's durability, extensive use, and disposal have caused the formation of large plastic islands in seas and the breakdown of plastic particles, called microplastics, spreading to even the most isolated habitats. This dilemma highlights the need of creative solutions and a transition to circular economies that prioritise the reduction, reuse, and recycling of plastics to minimise their impact on the environment.

Improper garbage disposal has a direct impact on ecosystems and poses a threat to human health. Water and air pollution from trash incineration led to respiratory illnesses and other health problems in communities located near disposal plants. The social and economic consequences of these health concerns highlight the interdependence of environmental and human well-being. Ultimately, poor waste management procedures have environmental consequences that go beyond the local disposal areas. It spreads across ecosystems, disturbs biodiversity, plays a role in climate change, and presents health hazards to animals and human populations. To tackle these difficulties effectively, we need to adopt a comprehensive strategy that includes sustainable waste management, encourages recycling programmes, and encourages a shared dedication to environmental responsibility. The next parts of this research will explore ways and solutions to reduce these environmental consequences and establish a more sustainable and balanced relationship with the natural world.

## 1.1 Innovative Approaches to Sustainable Waste Management

Innovative techniques for sustainable waste management strive to reduce the environmental effect of trash and enhance resource efficiency. Here are some fundamental principles and procedures:

### 1.1.1 Waste Recycling Techniques

i) Advanced Sorting Technologies: Utilising sophisticated sorting technologies like automated sorting systems and artificial intelligence to boost the productivity of recycling operations and raise the quality of recovered materials.
ii) Chemical Recycling: Using chemical techniques to decompose plastics and other materials into their fundamental components, enabling the production of new items without compromising material integrity.
iii) Upcycling: Transforming waste resources into higher-value or higher-quality goods to foster innovation and decrease the need for raw materials.

1.1.2 Composting Methods

i) Urban and Industrial Composting: 1Expanding composting initiatives in cities and industrial sites to redirect organic waste away from landfills, creating high-quality compost for farming purposes.
ii) Vermicomposting: Utilising worms to break down organic waste, speeding up the composting process and producing a premium soil amendment.

1.1.3 Circular Economy Principles

- Product Design for Recycling: Integrating recyclability into the initial product design to facilitate disassembly and recycling of materials at the end of their life cycle.
- Extended Producer Responsibility (EPR): Advocating for manufacturers to take full responsibility for their goods throughout their life cycle, from collection to recycling and disposal, to promote sustainable practices.

1.1.4 Zero-Waste Initiatives

- Source Reduction: Minimising waste production by promoting ethical consumption and manufacturing methods
- Community Engagement: Involving communities in zero-waste initiatives through education, awareness campaigns, and incentivizing waste reduction behaviors

## 2. COMMUNITY ENGAGEMENT AND EDUCATION FOR EFFECTIVE WASTE MANAGEMENT

Community engagement and education play a crucial role in fostering effective waste management practices. By raising awareness, promoting behavioral change, and involving communities in the process, it becomes possible to create a more sustainable and responsible approach to waste management. Here are key strategies for community engagement and education.

### 2.1 Waste Segregation at Source

**Educational Campaigns**: Conducting targeted campaigns to educate residents about the importance of segregating waste at its source. Providing clear information on how to separate recyclables, organic waste, and non-recyclables promotes proper waste sorting.

**Visual Aids and Signage**: Placing informative signage and visual aids in residential areas to guide residents on proper waste segregation practices. Clear and simple instructions help reinforce the message.

### 2.2 Awareness Programs on Sustainable Waste Practices

**Workshops and Training Sessions:** Organizing workshops and training sessions on sustainable waste management practices for community members, local leaders, and school children. These sessions can cover topics such as the benefits of recycling, composting, and reducing waste generation.

**Informational Materials:** Distributing pamphlets, brochures, and other informational materials that explain the environmental impact of improper waste disposal and provide tips on adopting more sustainable habits.

## 2.3 Technological Advancements Revolutionizing the Waste Management Industry

The waste management sector is seeing a significant change due to advanced technical developments that aim to improve efficiency, sustainability, and overall effectiveness. An important advancement is the incorporation of trash sorting robots into the waste processing system. The robots use advanced sensors and artificial intelligence to automatically detect and classify different waste products, making the sorting process more efficient and accurate. This speeds up the waste management process and decreases the need for human labor, leading to cost savings and increased precision.

The use of intelligent garbage bins integrated with Internet of Things (IoT) technology is transforming waste collection at its origin. The bins are equipped with sensors that continuously check their fill levels in real-time (C. Mohan et al., 2022). The data is sent to a centralized waste management system for dynamic route optimization and improved collection schedules. This reduces the frequency of collecting partly full bins and ensures collections are done promptly, which maximizes resource efficiency and decreases operating expenses. Moreover, IoT technologies play a key role in transforming garbage collection and recycling activities on a larger scope. Data is gathered and processed using sensors and linked devices to forecast fill levels, estimate maintenance requirements, and enhance operating efficiency. By using predictive analytics, waste management firms may make informed choices based on data, improving the efficiency of their operations (C. Mohan et al., 2022). Within recycling, technical progress goes beyond simple sorting systems. Sophisticated sorting mechanisms such as optical sorting and magnetic separation aid in the precise segregation of recyclable items from the trash stream. Chemical recycling methods are becoming more popular since they enable the decomposition of intricate polymers into their

*Figure 2. A universal symbol for a critical aspect of sustainable development: Waste management*

fundamental building parts. This novel method not only helps produce top-notch recycled materials but also reduces the environmental consequences of plastic waste.

Blockchain technology is being considered as a comprehensive solution to enhance transparency and traceability in the waste management sector. Utilizing blockchain technology allows for the complete tracking and validation of recycled materials throughout their existence, guaranteeing the legitimacy of recycled goods and establishing a reliable supply chain. The combination of garbage sorting robots, smart waste containers, and IoT technologies is transforming the waste management industry (N.I. Vatin et al., 2024). These technologies jointly enhance efficiency, decrease environmental impact, and promote a sustainable approach to waste management, leading to a greener and more technologically advanced future.

## 2.2 Policy Recommendations and Government Initiatives Promoting Sustainable Waste Practices

Ultimately, it is crucial to adopt sustainable waste management strategies due to our shared duty to the environment and future generations. The policy proposals and government actions, such as waste management rules, Extended Producer Responsibility (EPR), and incentives for green enterprises in the waste management industry, together provide a strong foundation for bringing about good change. Waste management laws define the fundamental norms and standards needed to govern acceptable waste procedures. They provide a systematic method for managing garbage, including processing, disposal, and recycling, with a focus on prioritizing environmental factors in waste management plans. Extended Producer Responsibility is a disruptive notion that encourages producers to reassess their involvement in the whole lifespan of their goods. EPR decreases the environmental effect of goods and encourages a transition towards a circular economy by promoting sustainable design and end-of-life management (Surender et al., 2023). The incentives offered to environmentally friendly enterprises in the waste management industry enhance the effectiveness of these rules. Providing financial assistance via tax credits, subsidies, and grants promotes research and investment in sustainable technology, leading to a vibrant and competitive business that emphasizes environmental stewardship. Embracing sustainable waste management is essential for a healthy world and to leave a legacy of responsibility for future generations. It involves changing one's attitude from a linear, wasteful approach to a circular paradigm that emphasizes conserving resources and protecting the environment. By jointly embracing sustainable practices, governments, corporations, and people are contributing to a cleaner, healthier, and more resilient world. Responsible waste management will have long-lasting advantages that go beyond the present generation. Progressing towards sustainability is continuous. By making focused efforts and putting these plans into action, we can pave the way for a more promising and environmentally friendly future for everyone.

## REFERENCES

Cointreau, S. J. (2006). *Occupational and environmental health issues of solid waste management: Special emphasis on middle- and lower-income countries*. World Bank.

Hoornweg, D., & Bhada-Tata, P. (2012). *What a Waste: A Global Review of Solid Waste Management. Urban Development Series Knowledge Papers*. World Bank.

Joshi, P. B. (2024). *Full blown green metrics*. Elsevier Publishing.

Mohan, C., Kumari, N., & Robinson, J. (2022). Sustainable and environmental friendly energy materials. *Materials Today: Proceedings*, *69*, 494–498. doi:10.1016/j.matpr.2022.09.187

Sarkis, J. (2018). A brave new world of waste: Exploring the environmental and resource implications of China's ban on waste imports and the global solid waste trade. *Resources, Conservation and Recycling*, *2018*(135), 1–3. doi:10.1016/j.resconrec.2017.12.022

Surender, M. C., & Kumar, R. (2023). Novel modification of activated charcoal sheet with N-methyl poly pyrrole and silver nanoparticles for removal of hexavalent chromium in water treatment processes. *Materials Protection*, *64*(4), 503–511. doi:10.5937/zasmat2304503S

Tchobanoglous, G., Theisen, H., & Vigil, S. A. (1993). *Integrated Solid Waste Management: Engineering Principles and Management Issues*. McGraw-Hill Education.

Vafaeva, K. M., Chhetri, A., Sudan, P., Mishra, M., Pakkiraiah, B., & Mohan, C. (2024). Polymer Matrix Nanocomposites for Sustainable Packaging: A Green Approach. E3S Web of Conferences, 511.

Vatin, N. I., Negi, G. S., Yellanki, S. V., Mohan, C., & Singla, N. (2024). Sustainability Measures: An Experimental Analysis of AI and Big Data Insights in Industry. *Bio Web of Conferences, 86.*

Verma, V., Joshi, A., Kalra, S., Singh, A., Singh, A., & Mohan, C. (2024). Novel Nano catalysts for Sustainable Hydrogen Production from Renewable Resources. E3S Web of Conferences.

Wilson, D. C., Rodic, L., Cowing, M. J., Velis, C., Whiteman, A. D., Scheinberg, A., Vilches, R., Masterson, D., Stretz, J., & Oelz, B. (2015). Wasteaware' benchmark indicators for integrated sustainable waste management in cities. *Waste Management (New York, N.Y.)*, *35*, 329–342. doi:10.1016/j.wasman.2014.10.006 PMID:25458855

# Chapter 7
# E-Waste Recycling:
## A Welcoming Change

**Misha Bahmani**
*IILM Law School, IILM University, Gurugram, India*

**Shobhna Jeet**
*K.R. Mangalam University, Gurugram, India*

## ABSTRACT

*E-waste is globally the most growing solid waste. To attain the goal of securing individuals' health and ecosystem protection recycling of e-waste is essential. This process is beneficial in enhancing the utilization of non-renewable minerals in an adequate manner which results in greener e-product production and economic welfare. It prevents the entering of toxic substances into the ecosystem and lessens the extraction impact on the environment. The authorities in several nations have been facing challenges in dumping the used e-products. Unfortunately, the e-waste dumping areas have been releasing toxic substances which has adversely impacted the livelihood of the recycling workers. To gain valuable substances from these products few companies have been focusing on extracting them through recycle process. E-waste recycling has been appreciated globally but it requires proper supervision and assistance. With the usage of sustainable measures, the goal of reusing and recycling can be achieved when e-waste management is performed by trained persons.*

## 1. INTRODUCTION

Unfortunately, the world is hit by the concern of a surge in solid waste. Among them, e-waste has been a crucial concern for the authorities worldwide. The figures have reached 53.6 million metric tons which has alarmed the world together. Sadly, 83 percent of it is not taken care which has impacted the global economy (World Health Organisation, 2023). While surrendering the e-products there must be adequate rules that promote the recycling practice among the public (Smieja et al., 2023). Since e-waste comprises useable and efficient resources it is required that it can be utilized for useful purposes. The United Nations has demarcated that there will extensive rise expected by 2050. The recycling engagement has

DOI: 10.4018/979-8-3693-4264-0.ch007

only reached 17 percent till now which predicts an irreparable loss to the earth. Emerging nations have engaged informal individuals as their workers to extract valuable material. This reflects that not only the environment is at stake but also the people involved in extracting valuable items that are dumped by the public (United Nations Environment Protection, 2022). The global per capita e-waste generation has jumped to a larger extent. Among them, e-waste classification and category have been introduced. Small as well as larger equipment and temperature-restricted equipment promote such activities. Europe is well known for generating e-waste per capita which is further followed by Oceania nations (Mohan et al., 2022; Liu et al., 2023). Currently, the dumping areas comprise small electronic and electrical appliances as well as consumer electronic products, which require proper supervision for recycling purposes (Liu et al., 2023).

However, there is a necessity to have effective e-waste management systems in the nations that are exposed to this risky business. The global e-waste monitor has drawn concern that discarded items are estimated to be 74Mt in coming years. This shows that due to a shortage of repair options and a short lifeline of the e-products, their quantity will be much larger. Although the value obtained through recoverable items can be a huge capital despite that emphasis is not made on collecting them they are dumped and sometimes burned. The opportunity of reusing and treatment is not provided in many developing nations. Due to the absence of proper implementation of rules, several nations have engaged informal workers for extracting business (Mohan et al., 2022; Forti et al., 2020). This has resulted in a hazardous impact on the health of these workers in these countries. Safeguarding the earth and individuals connected to it requires effective implementation of international guidelines which should be obeyed despite having international instruments (WHO, 2023). To have clarity about e-waste and its management there is a requirement to create awareness about greener practices and proper disposal for prevailing environment-friendly e-waste management system.

Globally many electrical and electronic devices are left aside and are broken into pieces and have been dumped. They are under the ambit of e-waste which has brought damage to the habitats when they are not treated and disposed of properly. There is a need to encourage recycling practices initiatives have been made to turn electronic appliances into reusable equipment (WHO, 2023). Due to improper disposal of e-waste, there has been a surge in health and environmental crisis. For instance, metals such as lead have been commonly used in electronic products which have badly impacted the soil. The concern is to be made on how these metals should be reused without causing any dangerous influence on the environment (Tsydenova & Bengtsson, 2011).

## 2. GLOBAL INITIATIVE

United Nations with the support of global instruments is promoting the practice of a healthy atmosphere and engaging in e-waste management. The nations while exporting the e-waste products are required to obtain the recipient's informed consent (UNEP, 2022). To safeguard the sustainability of the circular economy the mode of recycling and repairing are the best options at present in order to use the e-product. Biodiversity can be maintained when there is sustainable consumption. With an effective e-waste process, there will not be any climatic disasters. It is the responsibility of the users to limit their purchasing of e-products. For the purpose of functioning recycled products should be their priority. The value gathered from discarded items can be utilized for environment-friendly recycling activities. Companies across the globe can support the production of green e-products and provide creations that have better lifeline and performance (UNEP, 2022). With #BeatPollution UNEP has motivated the nations to coordinate with each other to tackle climate change and focus on greener practices which results in lesser environmental pollution and contamination (UNEP, 2022). It is significant to prevail responsible consumption as well as production to achieve sustainability by having effective e-waste management efforts (UNEP, 2022). Developing nations like India have recently introduced E-waste rules which have allowed the destruction of technologies in the system. In this process, it will cover air conditioning and refrigerators in order to attain sustainability (Ministry of environment, forest and climate change, 2022). This demarcates that it is promoting the e-waste recycling practice by imposing stricter technical guidelines. Also, East African Compliant Recycling Company and WEEE centre are engaged in e-waste recycling in Kenya (UNEP, 2015). With the support of local entrepreneurs and collaboration with regional and international bodies, WEEE centres are focused on the collection and dismantling of e-products (UNEP, 2015).

In 2021, Germany has been positioned as the first municipal waste recycler. In the European Union, almost 49.6 percent has been the recycling rate of municipal waste (UNEP, 2015). By 2030, annual e-waste production is estimated to reach 75 million metric tons. Asia covers half of the portion of it. China is the largest e-waste producer currently (Statista, 2023). It is expected that there will be a huge e-product waste collection of up to 64 million metric tons in 2024 (Alves, 2023).

## 2.1 Regulations and Trade: Welcoming E-Waste Recycling

Illegal e-waste trade has been a major concern of the authorities. Basel and Stockholm Convention are the greatest achievements that have promoted regulations in waste transportation. For instance, in the Basel Convention opportunity has been provided to the parties to formulate agreements to permit transboundary movement of this waste. Engagement in the waste transfer has promoted e-waste circulation through regional conventions. Also, the Aarhus Convention in Europe has initiated regulations that enhance transferring of dangerous waste. WEEE Directive and Durban Declaration in the African belt have also promoted such transfers. Although in the United States, there is no federal regulation governing e-waste any concern is addressed by States Environmental Protection Agency. This agency issues certification and regulates the e-waste recycling sector. Today many States in this nation have applied their laws for executing recycling of e-waste to prevent landfill activities. The European nations have been engaging their population in e-waste returning practices. Japan has indulged its population in recycling its electronic appliances through dealers and governmental bodies in specific areas. These home products are easily recycled under government supervision. In this nation, it is the producer's responsibility to maintain recycling and disposal of e-waste (Liu et al., 2023). Despite this, there are only a few nations that are obeying e-waste regulations. Due to the weaker implementation of these legislations, there hasn't been a positive outcome. With the absence of uniformity in the applicability of these legislations some nations are unable to achieve their goal. Less capacity and stakeholder participation have weakened the possibility of recycling at a larger level (Liu et al., 2023).

Illegal trading and e-waste smuggling still prevail in developed nations despite having restrictions on exporting. There is a need to have alterations in the laws which could prohibit such movements. The developing nations are now treated as e-waste dumping areas which has badly affected the livelihood of its people. There is urgency in the effective implementation of banning the import of specific solid waste. This will be beneficial in preventing cross-border participation of electronic waste. For instance, China has banned the import of 24 solid wastes (Liu et al., 2023).

## 2.2 E-Waste Recycling: A Greener Perspective

To achieve the 13th sustainability goal there is a need to focus on generating raw materials and lessening greenhouse gas impact (Seif et al., 2023). This will be beneficial in prevailing the 3rd sustainable goal of well-being which is valuable in achieving the goals of the Paris Agreement. A greener side can be seen in Japanese practice where recycling takes place in three steps. It includes gathering non-metals turning them into regular waste for recycling purposes and resulting in standardized treatment. This demarcates that there is huge capital invested in e-waste gathering and handling it. With the availability of an improved treatment system, e-waste recycling has been welcomed in Japan. Despite the presence of certified companies, Japan is not able to cover complete recycling as there are still recycling channels that are scattered and require assistance. China with its recent Waste Electrical Appliance Treatment Fund has encouraged enterprises to indulge in the recycling process to attain sustainability (Liu et al., 2023; Surender et al., 2023; Kumari et al., 2023). This reflects that there has been effective treatment status provided by the government and non-governmental organizations of this nation to promote e-waste recycling.

With the rise in utilization of EV batteries companies such as China Tower have replaced lead-acid batteries with EV batteries. Also, the Nissan Energy Storage program with its partnering nations is focusing on providing a lifeline to EoL batteries for second use. Powevault with Nissan and Renault which are car manufacturers to turn them into power banks for repurposing EoL EV Batteries which act as cheaper as well as greener modes. This has been beneficial in reusing these batteries (Mohan et al., 2022; Parajuly et al., 2020). With direct recycling, pyrometallurgy and hydrometallurgy recycling of the e-products has taken place globally. Tesla along with local recyclers is focused on developing their own recyclable EoL battery. With its recycling system, it engages its partners in recycling EoL batteries. Also, Umicore has established its recycling system. This has been beneficial in reducing the GHG emissions (Parajuly et al., 2020).

The European Union nations are following a government-oriented approach and the third-party organizations are coordinating with each other to have well well-organized and monitored system. There is a huge number of e-waste treatment and disposal plants in these nations that are functioning under government supervision. The municipal authorities, non-profit organizations, green producers, and protectors are working in the e-waste recycling sector under the American aspect. With governmental collaboration and partnership, the producers can conduct e-waste recycling successfully. For instance, the National Electrical Manufacturers Association has been engaging in this process. Many companies are involved in e-waste recycling in their States (Liu et al., 2023). This shows competency in these companies to execute effective recycling practices.

By investing in disposal technology these nations are focusing on e-waste safer dismantling and separation methods. Currently, in developing nations, there is also a requirement for pollutant treatment and energy recovery systems. Dismantling takes place with the help of tools as automatic dismantling can't take place. The toxic substances are removed and useless substances are crushed. After going through treatment, the substances are separated by magnetic force. Separated substances can be utilized in manufacturing new products. The non-recyclable materials are turned into fuels and the rest are utilized for energy generation. By using the bio-safe method the pollutants are treated at different stages (Liu et al., 2023). The producer needs to focus on investing in a qualified dismantling factory. With proper disassembling the useful components can be recycled and discarded can be turned into new products. There is a necessity to have controlled emissions of organic pollutants. Any form of e-product should enter into the dismantling workshop to carry out the effective recycling process. There is a need for regular effective intervention of the workers to disassemble and store the dangerous material to form it into a new product. With proper supervision transformation of these components will be beneficial for environmental and health safety (Liu et al., 2023).

By encouraging the practice of recovery technology, the transformation of valuable metals undergoes mechanical as well as physical techniques. In mechanical technique, the e-waste is divided into different attributes. They are dismantled, crushed, and separated in this technique. After this process, the metals are regenerated. Due to cost and eco-friendly this technique is utilized globally (Liu et al., 2023). With the popularity of molten salt and electrochemistry technology today there is an eco-friendlier approach toward the extraction process and has upgraded the metal and e-waste recycling activities (Liu et al., 2023). For non-metallic material, there is involvement of regeneration methodology. The concern is that in the recycling process, the main emphasis has been placed on metal recycling and some of the non-metallic materials are dumped in the landfill only. There is a requirement to have advanced technology that could recycle various types of non-metallic (Liu et al., 2023).

## 2.3 Impact: A Darker Side

Individuals and communities involved are not only dependent on metal recycling for their earning survival but have faced various difficulties in performing the recycling process. Due to the absence of proper implementation of e-waste regulations, the population involved in recycling activities has been struggling. The recycling infrastructure and training centres require proper assistance as they also involve toxic substances. To ensure public safety there is a requirement that international regulations should be eco-friendly (WHO, 2023).

With the presence of toxic pollutants, the fumes that are released after the burning of the substance have exposed the public to a hazardous atmosphere. It has been found that there have been unsound activities such as acid baths, heating, and plastic coating in the e-waste areas (WHO, 2023). Sadly 16.5 million children have been working in the waste picking practice which has resulted in affecting the child and pregnant laborer's health. Due to informal e-waste recycling engagements exposure to harmful chemicals has impacted their nervous and respiratory system (WHO, 2023). To protect them from its impact, there is a need to have effective preventive measures. Countries like the United States produce 46 pounds of e-waste per capita annually. To attain the goal of sustainability there is a requirement for effective e-waste legislation with supports financial assistance in recycling e-waste in an organized manner (Forti et al., 2020).

Sadly, around 12.9 million females and 73 million children have been engaged in informal recycling of solid waste. The Agbogbloshie dump depicts a serious concern about how e-waste dumping is impacting the environment of Africa. It is the largest dumping area of these products and many workers are in the recycling process. People are residing around this dump and are exposed to its impact. Such a population has faced health concerns and their health report depicts that the quality of their health has diminished. There have been workers who have faced various serious health issues. Among them, respiratory and neurological discomforts are common. Serious disease has been the end outcome of long working hours of these workers. For instance, the kidney and blood systems of the younger population have been adversely affected due to dangerous fumes in this area. These workers face chronic health issues such as skin, eye, and stomach issues. This means that this workplace causes inflammation and stress, and DNA is badly affected. Humans are exposed to toxic substances and their food chain is badly impacted. E-waste requires proper management in recycling to reduce occupational and health damage to the future population (Liu et al., 2023).

There have been situations where pollutants of grievous substances are released during the dismantling. The biological chains have been badly affected by the fumes of heavy substances and organic pollutants. There have been incidents of dust volatilization. Elements like nitrogen, dioxins, and sulfur are some air pollutants. The lifeline of the animals and plant kingdom has been worse affected. There has been a surge in the migration movement of the animals in these dumping areas. This shows that the natural resources surrounding these areas have caused health concerns to all the inhabitants. Unfortunately, in Vietnam, the informal child workers have been exposed 3.90 times to the fumes of heavy metals which is an alarming stage of concern. It has been observed that this population has been consuming rice having a huge number of heavy metals in it. This reflects concern that the authorities should take the initiative to reduce food pollution especially where e-waste dismantling and commercial food production take place. There have been events where it has been observed that to uncontrolled release of toxic fumes during the recycling of e-waste has impacted the humans of Guiyu in China. Learning from the mistakes of the past there is a requirement for advanced technology to promote the recycling practice in the e-waste sector. Sadly, sometimes due to complex composition, the dismantling method has resulted in the release of heavy metals in natural resources (Liu et al., 2023). Despite providing green disposal facilities there have been nations that have faced such forms of problems. For instance, there has been an e-waste storage facility provided in Thailand. Also, Norway has supported the practice of e-waste recycling facilities. In these areas, the presence of hazardous elements has been located. There is a need to take precautionary measures while the e-waste recycling process takes place. Floating dust, wastewater, etc. require proper assistance during the disassembling process (Liu et al., 2023).

## 3. CONCERN

By 2030 it is estimated that the e-waste ratio will surge by 30 percent and only 20 percent of e-waste disposal has been recorded (Tiseo et al., 2024). The customers are throwing away their older e-products, donating them and still using them. The users are required to locate a recycling centre and they can also sell these items to an organization that works for recycling purposes (Alves et al., 2023). In 2021 13.5 million metric tons of e-products were brought into the European market. Only one-third of it was recycled and was prepared for reuse. Presently, Norway has the highest per capita e-waste (Alves et al., 2023). At the same timeline, almost 11 kg per individual on average was WEEE in the European Union (Alves et al., 2023).

The recycling market of electronic is expected to reach \$65.8 billion with 12. 7 percent of CAGR and China will be \$15.3 billion in 2026. Japan as well as Canada will jump almost to 9 percent. Adopting the recycling of e-waste solutions in different sectors has promoted a 12 percent CAGR in a few nations. The Asia Pacific region will have a \$5.8 billion market and Latin America will reach 16.2 percent of CAGR during that period (Seif et al., 2023). Egypt's recycling system has a limited approach as there are limited environmental policies that are related to WEEE's disposal as well as management. There is improper e-waste mapping which has resulted in illegal dumping. Only 20 percent of e-waste is recycled in this nation which reflects the necessity to have their local treatment facility (Seif et al., 2023).

There have been circumstances where it has been observed that while executing the recycling of the e-products the authorities haven't engaged them in the fine separation process. These products are innovations of complex methodology which has left precious raw material to be turned into meaningless dust. The recycling of this product should upgrade the recycling techniques which could make utilization of this valuable raw material. Electrical as well as electronic waste should undergo advanced technology to attain their usable components. For instance, globally motivation should be given to electrochemical as well as solvent extraction activities which could be beneficial in attaining fine separation engagement. This will be helpful for the producers to utilize these rare materials gained after the recycling process in their product creation (Liu et al., 2023).

To fulfil the obligation of recycling the emerging and developing nations have been following their individual approaches. The recycling process has been facing difficulties in several nations as they are not concerned with the various stages of recycling properly. Many countries have ignored the stage of collection and disassembling the e-products adequately which has resulted in a harmful impact on the ecosystem. With effective utilization of internet technology, the authorities should acknowledge the presence of these stages and should focus on innovative measures that support global standards. For instance, a codifying system can be followed in the e-products. Online services should be provided to offer disposable of these products with a greener approach (Liu et al., 2023). This will create awareness and will motivate the sellers and purchasers about the various ways through which they can get their e-products recycled in a green and cost-friendlier manner.

Despite this recyclers are facing problems due to the lesser concentration of valuable components in e-products and these products become very difficult to recycle (Parajuly et al., 2019). There is a need to focus on "waste to wealth" technologies which can be possible if there are guidelines for executing lower-cost rescuing and recycling activities globally (Cannon et al., 2015) There is a need to reduce participation in waste crime. There should be proper allocation of various categories of e-waste that can be recycled and which can't be recycled in developing nations (Alves et al., 2023). Landfilling has resulted in global warming issues in developing nations. There has been a 14 percent release of methane from these areas. Due to the dangerous impact exporting of the e-waste and burning through incinerators has restricted approval. Excessive usage of Co and Cr in water water-splitting process has caused serious concern. The presence of Cu, Li, and Ni will be limited in the future. During the Covid-19 tenure, there has been a surge in e-waste which has affected the individual's health (Seif et al., 2023).

## 4. NEED

It is forecasted that by 2029 e-waste management market will be almost $137 billion market (Tiseo et al., 2023). There is a need to focus on the technology that supports green electronic products. To reduce e-waste, it is essential to incorporate green technologies that encourage recycling practices (Liu et al., 2023). The government globally should promote the production of eco-friendly creations. In this process, emphasis is to be made on non-dangerous and environment-friendly electronic products. With intelligent creations and energy-saving characteristics, these products will enhance higher reliability as well as longer timelines. The government should motivate the producers to indulge in greener designs which can strengthen the safe disposal of these products. The producers should focus on investment in recyclable products under this domain. This will be beneficial in reducing the total presence of e-waste in the future. To make the recycling process approachable it is essential to focus on the material utilized, the construction technique used and the design of these products to be greener. For instance, when there is disassembly and reusing of these products it is expected that the producer of these products is following the greener practices. This will be beneficial in simplifying the process of recycling. The material utilized in these products should be recyclable in nature to avoid any greater loss. Imposing a stricter ban on toxic substances in the creation of these products which facilitate recycling globally. This will result in a reduction in the presence of alkaline and acidic substances in the dumping areas (Liu et al., 2023). There is a requirement for energy conversion technology that could promote sustainable measures. The storage technology should have integration with greener practices. With the use of clean energy sources emphasis is required to be made on the reduction of carbon dioxide emissions (Seif et al., 2023).

There is a requirement to have effective regulatory measures which have distribution of e-waste. The regulatory bodies are inefficient in the prevailing healthy atmosphere. There have been circumstances where extensive dumping and recycling have been limited. There is a requirement for modern regulations that promote the practice of a three-dimensional framework (Liu et al., 2023). To have a less harmful impact the government should highlight global cooperative measures in the usage of technologies that support business obligations to safeguard the labor's interest. This will modernize the e-waste management and it will result in safeguarding individuals' health (Liu et al., 2023). The government should focus on the designing system so that recycling can be effective. By highlighting economic and geographical factors the government should formulate rules that are beneficial for the green producer. It should include preventive measures that can make the recycling activity efficient. Unfortunately, the current regulatory framework has not been competent which has led to a hazardous impact overall. These management initiatives should be applicable to the transboundary movement of e-waste which will result in an effective tool. By following a life cycle assessment there will be less dangerous impact on the habitant. The recycling engagement of these products can be improved when the government can strengthen these enterprises by investing in these activities. Through funding, there can be an effective result in the formation of e-waste recycling and will be able to address their problems. The government should provide proper guidance to the marketing and manufacturing sector which will result in innovative efforts in bringing effective e-waste recycling practices (Liu et al., 2023). For instance, Japan has implemented laws that cover 34 electronics which are covered under their recycling laws.

WEEE Declaration has encouraged the European population to recycle electrical as well as electronics (Liu et al., 2023). There is a need to harmonize national policies that enhance the recycling of e-waste to prevail sustainable waste management system (Modak et al., 2017).

There is a need to promote the usage of AI algorithms which can make the recycling process efficient. To have accuracy in the outcome machine learning can enhance the recycling involvement. Almost $400,000 in funding has been provided to various American Universities for the robotic creation of e-waste recycling. For instance, aluminum foil recycling has been the prime goal of Oculatus Consulting. Also, plastic recovery from non-recyclable municipal solid waste for composite production is the prime research focus for the University of Massachusetts Lowell to take care of (Office of Energy Efficiency & Renewable Energy, 2021). The "Recycle Cube Robot system" has been proposed by researchers in India to execute the segregation of e-waste (Shreyas et al., 2022). AI will help analyze the data that has been collected from the recycling stage and suggestions can be provided. It will promote accuracy in measuring as well as tracking waste (Fang et al., 2023).

Inappropriate e-waste recycling has led to the release of organic as well as non-organic components causing harm to the environment. The natural resources require safety from these releases. With recyclable materials, the waste management system will be more efficient. Imposing extended producer responsibility in stronger economies has focused on a recycling-based circular economy (Ghulam & Abushammala, 2023). While performing formal recycling it should be the responsibility of the competent authorities to include all the recyclable metals in this process. The regulation of export and import for recycling purposes of e-waste should have uniform standards. Globally integrated organizations are to be established worldwide to monitor the recycling of e-waste. The mining industry should jointly perform greener methods to fulfill their commercial obligations while manufacturing e-products (Ghulam & Abushammala, 2023). The Media sector should indulge in creating awareness about the significance of recyclable e-waste. Responsible media participation will encourage the public to improve their utilization of e-products. With sustainable input, the media industry should reflect on eco-friendly disposal practices. The recycling activities can be improved in the future when the media provide essential insight into the necessity to engage in sustainable measures in this process (Islam et al., 2023).

There is a need to focus on the activities of e-waste importers by ensuring proper implementation of regulations on investment. Recently it has been suggested that a deposit refund system should be applicable globally to attain the goal of zero waste and lessen environmental degradation. By setting up sustainable new facilities and transportation while performing various stages of recycling the authorities should engage in sustainable technology (Alblooshi et al., 2022). UNICOR has been recently held liable for contamination of the one of their warehouse in 2022. This reflects that globally there is a need to focus on environmental justice practices to provide fair treatment to recycling workers (Ogunseitan et al., 2023).

Effective formal e-recycling will be beneficial in providing effective service. This practice has been a great support for the community at large. By the appropriate way of the recycling process, the substances gained through landfills have been used in the production of new products. The formal worker's participation in e-waste fields results in healthier approaches. This has restricted the involvement of younger labor. Governmental and non-governmental organizations should invest in the building of worker's residences away from the workplace. There is a need to alter occupational regulations to provide a protective atmosphere at the workplace and worker's residency. By focusing on the formalization of formal e-recycling sector nations would be able to achieve their sustainability goal in the future. Financial support is required by private as well as governmental bodies to develop safer electronics (Ceballos & Dong, 2016). Formal e-recycling is actively used in developed nations. These facilities have resulted in better control over the health and environment. For instance, Colombia has established formal e-recycling facilities. Since it is a new approach few developed nations have achieved the goal of effective functioning of formal e-recycling. With the easy availability of electronic recyclers, this practice has a healthy impact on the community (Ceballos & Dong, 2016). Stronger economies have been exporting their e-waste to emerging nations. By following the zero-waste concept developing nations need to focus on the formulation of formal workers. In India Central Pollution Control Board has been highlighting the necessity to have formal e-waste recycling activities. Its emphasis is on providing proper training to the recycling workers (Dutta & Goel, 2021). The best available technologies can be beneficial for achieving resource recovery in emerging economies (Raghupathy et al., 2018). With effective legal assistance, the formal and informal e-recyclers can have organized disassembling, separation, and sorting (Akter & Hossain, 2023). Collaborating among the stakeholder's implementation of reverse logistics will be beneficial in recycling activities (Vargas et al., 2023).

The European Union Directives have highlighted the need to maintain national registers. This has motivated in prevailing effective e-waste management system. By focusing on the WEEE Directives disposal of e-product waste has been gathered by several formal recyclers under the schemes of producer's responsibility. There is a need to have an effective tracking system in developed nations. With management schemes in emerging nations, formal recyclers' participation can be improved. There has been a surge in waster pickers in developing areas which requires the legalization and incorporation of such activities. By having cooperative public-private partnerships in pickers the e-waste recycling system can become more effective (Zhang et al., 2018). Although the informal waste sector has been beneficial in e-waste recycling despite the risks involved. By using greener and cost-friendly technology in the formal and informal sector e-waste collection would be done appropriately without risking the quality of nature and the human community for recycling purposes (Dhillon & Sandhu, 2017).

## 5. CONCLUSION

To achieve the goal of a stronger economy participation of electronic as well as electrical products is necessary. There is a need for cross-sector cooperative efforts to have a holistic system to promote e-waste management in recycling practice (Liu et al., 2023). With trained labor, the process of recycling e-waste should prevail. The regulations governing e-waste must act responsibly and should engage in the recycling process in a greener way. With global cooperative behavior, nations can achieve their goal. There is a need to add e-waste and second-hand items to the Basel Convention. The authorities should formulate stricter classifications of these products to maintain their durability and safety. With controlled usage and stricter rules about the smuggling of these products, the nations can focus on environmental safety. Producers should encourage the coding, registration, and disposal system under the supervision of global organizations (Liu et al., 2023). Local awareness about proper e-waste management is required. There is a necessity to assist in technical and managerial training. With the effective involvement of the trained local informal sector, the collectors will perform their duties without placing their health and environment in danger (Cannon et al., 2015). By investing in green hydrogen production there can be generation of power. For instance, solar cells should be utilized to reduce external energy consumption to act as an external power supply. Wind and tidal energy should be used for electricity generation. The use of renewable energy can promote energy efficiency for future generations (Seif et al., 2023).

## REFERENCES

Advanced Manufacturing & Industrial Decarbonization. (2021). *Department of energy, in partnership with remade institute, awards over $16 million for 23 projects that will reduce carbon emissions across the manufacturing sector*. Office of Energy Efficiency & Renewable Energy.

Akter, N., & Hossain, M. I. (2023). Role of informal and formal actors in e-waste management. *The Business Standard*. https://www.tbsnews.net/thoughts/role-informal-and-formal-actors-e-waste-management-746346

Alblooshi, B. G. K. M., Ahmad, S. Z., Hussain, M., & Singh, S. K. (2022). Sustainability management of electronic waste: Empirical evidences from a stakeholders' perspective. *Business Strategy and the Environment*, *21*(4), 1856–1874. doi:10.1002/bse.2987

Alves, B. (2023, April 17). *Consumer behaviour with old electronics worldwide 2021*. Statista. https://www.statista.com/statistics/1337224/e-waste-consumer-behavior-worldwide/

Alves, B. (2023, December 18). *Global e-wastes statistics & facts*. Statista. https://www.statista.com/topics/3409/electronic-waste-worldwide/#topicOverview

Alves, B. (2023, November 7). *E-waste flow in the European Union 2021, by method*. Statista. https://www.statista.com/statistics/1337536/electronic-waste-flow-european-union-by-operation/

Alves, B. (2023, November 7). *E-waste per capital collection in the European Union 2021, by country*. Statista. https://www.statista.com/statistics/1337511/electronic-waste-collection-european-union-per-capita/

Cannon, T. (Ed.). (2015). *Global waste management outlook*. UNEP. https://www.unep.org/resources/report/global-waste-management-outlook

Ceballos, D., & Dong, Z. (2016). Formal e-recycling: The complexity of solving the e-waste problem worldwide. *Harvard T. H. Chan*. https://www.hsph.harvard.edu/hoffman-program/2016/08/26/formal-e-recycling-the-complexity-of-solving-the-e-waste-problem-worldwide/#:~:text=Developed%20countries%20generally%20use%20formal,such%20as%20plastic%20and%20metal

Ceballos, D. M., & Dong, Z. (2016). The formal electronic recycling industry: Challenges and opportunities in occupational and environmental health research. *Environment International*, *95*, 157–166. doi:10.1016/j.envint.2016.07.010 PMID:27568575

Dhillon, J., & Sandhu, V. (2017). Integrating the activities of informal and formal sector in e-waste management: Potential benefits and challenges. *International Journal of Latest Technology in Engineering Management & Applied Science*, *6*(7), 133–136. https://www.ijltemas.in/DigitalLibrary/Vol.6Issue7/133-136.pdf

Dutta, D., & Goel, S. (2021). Understanding the gap between formal and informal e-waste recycling facilities in India. *Waste Management (New York, N.Y.)*, *125*, 163–171. doi:10.1016/j.wasman.2021.02.045 PMID:33706255

Fang, B., Yu, J., Chen, Z., Osman, A. I., Farghali, M., Ihara, I., Hamza, E. G., Rooney, D. W., & Yap, P. (2023). Artificial intelligence for waste management in smart cities: A review. *Environmental Chemistry Letters*, *21*(4), 1959–1989. doi:10.1007/s10311-023-01604-3 PMID:37362015

Forti, V., Balde, C. P., Kuehr, R., & Bel, G. (2020). *The global e-waste monitor 2020* (1st ed.). Unitar Scycle., https://ewastemonitor.info/wp-content/uploads/2020/11/GEM_2020_def_july1_low.pdf

Ghulam, S. T., & Abushammala, H. (2023). Challenges and opportunities in the management of electronic waste and its impact on human health and environment. *Sustainability (Basel)*, *15*(3), 1837–1859. doi:10.3390/su15031837

*How disposable tech is feeding an e-waste crisis*. (2022). UNEP. https://www.unep.org/news-and-stories/story/how-disposable-tech-feeding-e-waste-crisis

IslamM.NayemA. I.BristyU. A.EshaS. A. (2023). Approaches to e-waste reduction through 4R, public awareness, and design for sustainability. *IN4OBE Global Virtual Summit 2022 Transforming Education and Empowering Learners*. [REMOVED HYPERLINK FIELD] doi:10.2139/ssrn.4458682

Kumari, P., Kumari, N., Mohan, C., Chinglenthoiba, C., & Amesho, K. T. T. (2023). Environmentally Benign Approach to Formulate Nanoclay/Starch hydrogel for Controlled Release of Zinc and Its Application in Seed Coating of Oryza Sativa Plant. *International Journal of Biological Macromolecules, 257*, 128278. doi:10.1016/j.ijbiomac.2023.128278 PMID:38029920

Liu, K., Tan, Q., Yu, J., & Wang, M. (2023). A global perspective on e-waste recycling. *Circular Economy, 2*(1), 1–16. doi:10.1016/j.cec.2023.100028

Modak, P., Pariatamby, A., Seadon, J., Tata, B. T., Borongan, G., Thawan, N. S., & Lim, M. B. (2020). *Highlights Asia waste management outlook summary for decision-makers*. UNEP. https://wedocs.unep.org/bitstream/handle/20.500.11822/27293/AsiaWMO_Sum.pdf?sequence=1&isAllowed=y

Mohan, C., Kumari, N., Jindal, R., & Gautam, R. (2022). *Application of Efficient Naturally Occurring Clay Mineral for Fuchsin Basic Dye Removal*. Springer Germany.

Mohan, C., Kumari, N., & Robinson, J. (2022). Sustainable and environmental friendly energy materials. *Materials Today: Proceedings, 69*, 494–498. doi:10.1016/j.matpr.2022.09.187

Ogunseitan, O. O. (2023). The environmental justice agenda for e-waste management. *Environment, 65*(2), 15–25. doi:10.1080/00139157.2023.2167457

Parajuly, K., Kuehr, R., Awasthi, A. K., Fitzpatrick, C., Lepawsky, J., Smith, E., Widmer, R., & Zeng, X. (2019). Future e-waste scenarios. *UNEP*. https://wedocs.unep.org/bitstream/handle/20.500.11822/30809/FutEWSc.pdf?sequence=1&isAllowed=y

Parajuly, K., Ternald, D., & Kuehr, R. (2022). *The future of electric vehicles and material resources* (1st ed.). UNEP. https://wedocs.unep.org/bitstream/handle/20.500.11822/34225/ElecVe.pdf?sequence=1&isAllowed=y

Raghupathy, L., Kruger, C., Chaturvedi, A., Arora, R., & Henzler, M. P. (2018). *E-waste recycling in India - bridging the gap between the informal and formal sector*. Greene. https://greene.gov.in/wp-content/uploads/2018/01/Krueger.pdf

Seif, R., Salem, F. Z., & Allam, N. K. (2023). E-waste recycled materials as efficient catalysts for renewable energy technologies and better environmental sustainability. *Environment, Development and Sustainability, 26*(3), 5473–5508. doi:10.1007/s10668-023-02925-7 PMID:36691418

Shreyas, M. A. S., Rajaraman, R., Harini, S., & Kiliroor, C. C. (2022). Application of artificial intelligence to enhance collection of e-waste: A potential solution for household WEEE collection and segregation in India. *Waste Management & Research, 40*(7), 1047–1053. doi:10.1177/0734242X211052846 PMID:34726090

Smieja, J. (2023, March 24). *The enormous opportunity of e-waste recycling*. Weforum. https://www.weforum.org/agenda/2023/03/the-enormous-opportunity-of-e-waste-recycling/

Surender, C., Mohan, C., & Kumar, R. (2023). Novel modification of activated charcoal sheet with N-methylpolypyrrole and silver nanoparticles for removal of hexavalent chromium in water treatment processes. *Materials Protection, 64*(4), 503–511. doi:10.5937/zasmat2304503S

Tiseo, I. (2023, August 29). *E-waste management market value worldwide 2022-2029*. Statista. https://www.statista.com/statistics/1154804/global-e-waste-management-market-value/

Tiseo, I. (2024, January 9). *Outlook on global e-waste generation 2019-2030*. Statista. https://www.statista.com/statistics/1067081/generation-electronic-waste-globally-forecast/

Tsydenova, O., & Bengtsson, M. (2011). Chemical hazards associated with treatment of waste electrical and electronic equipment. *Waste Management (New York, N.Y.)*, *31*(1), 45–58. doi:10.1016/j.wasman.2010.08.014 PMID:20869229

Vargas, D. B., Campos, L. M. S., & Luna, M. M. M. (2024). Brazil's formal e-waste recycling system: From disposal to reverse manufacturing. *Sustainability (Basel)*, *16*(1), 1–16. doi:10.3390/su16010066

Zhang, A. Y. (2018). Informal vs formal recycling; The dilemma of e-waste management. *Blavatnik School of Government*. https://www.bsg.ox.ac.uk/blog/informal-vs-formal-recycling-dilemma-e-waste-management

# Chapter 8
# The Role of Solid Waste Management in Boosting Sustainable Development Goals

**Joan Nyika**
https://orcid.org/0000-0001-8300-6990
*University of Johannesburg, South Africa*

**Megersa Olumana Dinka**
https://orcid.org/0000-0003-3032-7672
*University of Johannesburg, South Africa*

## ABSTRACT

*The management of solid waste is a growing challenge with rising trends in industrialisation and population growth, and has become a public and environmental health concern. The challenge spills over to negatively affect the realization of sustainable development goals (SDGs). This chapter focuses on how solid waste management (SWM) can be used to promote the realizations of SDGs while prioritizing on economic, social, and environmental growth. Findings show that systematic SWM has the capacity to influence most SDGs and hence the need to uptake it. This can be done by promoting circular economic models, formulating, implementing, and enforcing SWM regulations that are pro-sustainable development, improving governance of the solid waste sector through planning and adoption of better technologies to valorise waste.*

## 1. INTRODUCTION

In more recent times around the globe, urbanization and population growth trends are on an upward trajectory, which has translated to lifestyle changes that encourage the consumption of natural resources and ultimately, more solid waste generation (Gupta et al., 2023). Other authors have also reported similar trends (Nyika & Dinka, 2022a; Pujara et al., 2019). Projections show that by 2030, solid waste production globally will rise to more than 2.2 billion tons annually compared to the current rate approximated

DOI: 10.4018/979-8-3693-4264-0.ch008

at 2.01 billion tons of waste a year, most of which is untreated, unsorted and disposed (Al-Dailami et al., 2022). Such a change will have negative effects on economic, environmental, and social aspects of sustainable development (Gupta et al., 2023) and has been confirmed in several case studies.

A study based in municipalities of Chile for instance, quantified the cost of unsorted municipal waste to £297.66 per ton (Sala-Garrido et al., 2023). In Yemen, which mainly used landfill disposal for solid waste management (SWM) and particularly the vicinity of Sana landfill, negative environmental impacts in terms of natural resources (land and water) contamination, air pollution and degrading aesthetic value of land were evident (Al-Dailami et al., 2022). In South Africa (Nyika et al., 2020), Kenya (Nyika & Dinka, 2023) and India (Pujara et al., 2023), open dumping of solid waste especially in urban and sub-urban areas was a public health concern through the generation of toxic leachate. In cities of developing countries, only about 65% of generated solid waste is collected even though 20-50% of municipal budgetary allocations are slotted for SWM whereby, 80-95% of this total is spent on its collection and transportation (Abubakar et al., 2022). In low-income developing countries, solid waste collection rates go even as low as 10% contributing to grave environmental and public health risks (Abubakar et al., 2022).

With this precognition, there is a need for regulatory authorities and stakeholders of different regions of the world to combine forces in adopting strategies and technologies geared to efficient, effective, and sustainable SWM. Ng and Yang (2023) made similar suggestions and emphasized on incorporating waste valorisation in SWM plans and moving economies from linear-based to circular-based models. This chapter attempts to explore such options and links them to sustainable development and in particular, the role of the solutions to SWM in realizing specific sustainable development goals (SDGs). The aim of this chapter is therefore to examine the global solid waste production trends, the current options to SWM and their effects and finally, the link between better management and SDGs.

## 1.1 Solid Waste Production Trends

Approximately 2.01 billion tons of municipal solid waste is produced annually around the world (Al-Dailami et al., 2022). Out of this total, about 33% is managed in a conservative and unsafe manner. The mean waste generation rate per capita is 0.74 kilogram daily per person and ranges between 0.11 and 4.54 kg with higher amounts favouring developed countries (Kaza et al., 2021). According to a World Bank report, global waste generation rates are set to grow to 3.4 billion tons by 2050 owing to a population growth rate double the current rate (Kaza et al., 2021). Trends in specific regions are as shown in Figure 1 whereby, high income nations (Europe, East and Central Asia) produce and are projected to produce more solid waste compared to low-income ones (Sub-Saharan Africa, Middle East, and North Africa). Similarly, the waste generation per capita rate in the same period will also rise by 19% for high-income countries and 40% for low and middle-income countries. Currently, East Asia and Pacific region are producing 23% of the world's highest quantities of solid waste while the Middle East and North Africa are producing the lowest at 6% (Kaza et al., 2022). Waste production rates of the North Africa, Middle East, South Asia, and Sub-Saharan Africa (SSA) regions by 2050 will be the fastest at double and triple the current regions. In these regions, waste management is conservative and heavily relies on open dumping, which will have not only economic but also environmental health implications.

The composition of generated solid waste varies based on the consumption patterns of various regions. Globally, the portion of waste that is organic (food and green), glass, metal, plastic, paper/cardboard, wood, rubber/leather, and others constitute 44, 5, 4, 12, 17, 2, 2 and 14% of waste, respectively (Kaza et al., 2021). In developed countries, the portion or organic waste produced is 32% compared to dry waste (glass, metal, cardboard, paper, and plastic), which is easily recyclable at 51%. In low- and middle-income countries, organic waste dominates total waste at 53-57% although the portion, which is diverted to recycling, is less than 20%. Specific trends in waste production also show concern as discussed in subsequent sections.

## 1.2 Solid Waste Production in Urban Areas

The world's urbanization trends are on an upwards trajectory, which accelerated in the 20th century. In the first quarter of the 21st century, 50% of the population is already living in cities with predictions showing a growing rural to urban transition in the coming decades (Marcotullio & Sorensen, 2023). The trend comes with changes in lives, economies, consumption patterns and resource consumption, which ultimately influence solid waste production and management. Currently metropolises and cities are not able to cope with the ever-rising solid waste production trends and for this reason lower amount are being collected and when collected, even lower amounts are managed in an environmentally sound manner. A United Nations (UN, 2022) report noted that 82% of municipal solid waste generated was collected and out of the total, only about 55% was managed in controlled facilities (Figure 2). Collection rates in SSA and Oceania were lower compared to Asian, American, and Australian regions. A high preference to open-dumping was also established in Central and Southern Asia (Figure 2) where there was a huge gap between the wastes collected (at 72.1%) and the waste managed in controlled facilities (at 24%). The trend is of concern owing to its environmental ramifications.

*Figure 1. Waste production trends around the globe*

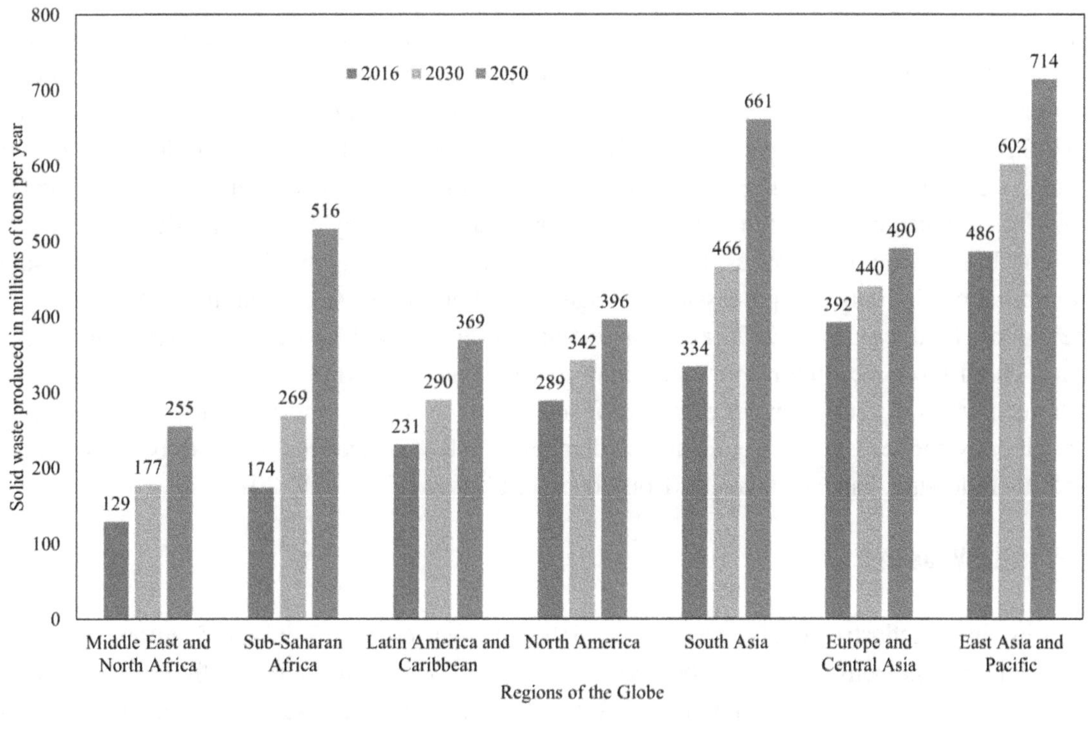

*Figure 2. Waste collection and management in urban areas of the world (UN, 2022)*

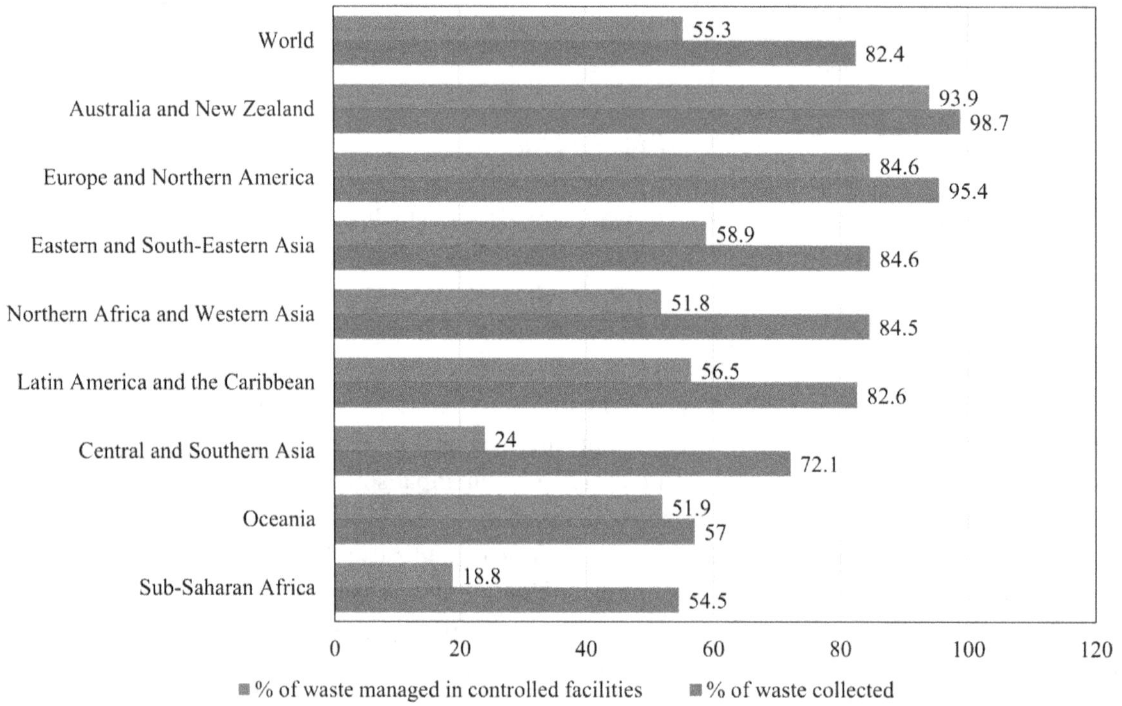


## 1.3 Food Waste Generation Patterns

Just like the general municipal waste generation trends (Figure 1) that increase based on income levels, so does food waste generation rates. Several countries based on their income and food waste generation patterns are as shown in Table 1 (Iacovidou et al., 2012). The mean food waste production was estimated at 0.43, 0.34, 0.31 and 0.26 kg per capita daily for high-, upper -middle, lower-middle and low-income countries, respectively. According to Pandey (2021), high income countries generate 56% of total food waste, 40% of which occurs at the consumption stage and is driven by consumer attitudes, values, and behaviour. In low-income countries, food losses occur post-harvesting and before reaching consumers due to limitations in transport, processing, storage, and wholesaling. According to a UN (2022), 13.3% of the world's food was lost in 2020 after harvesting. Additionally, 17% was lost at retail, food service and household levels accounting for more than 931 million metric tons of food in 2016 alone (UN, 2022). Such wastage trends counter efforts to food security especially in SSA that is the most food in-secure globally and leads to environmental challenges as 8-10% of greenhouse gases (GHGs) produced at landfills following disposal emanates from food waste (Iacovidou et al., 2012).

## 1.4 Plastic Waste

Plastics use in the industrial era is growing due to their many uses in the household, automotive, health-care, packaging, and construction sectors (Nyika & Dinka, 2022c; Kibria et al., 2023). In 2015, the per capita consumption of plastic was rated at 16, 139, 48, 136, 36, 108, and 45 kg per person in the Middle

*Table 1. Food waste fractions and generation rates of some countries based on their income levels*

| Country/ Income Class | Food waste fraction of total waste (%) | Food waste generation rate (kg/capita/day) |
|---|---|---|
| High-Income Countries | | |
| United Kingdom | 17.8 | 0.23 |
| United States | 12.7 | 0.26 |
| Spain | 36.9 | 0.61 |
| Norway | 23.8 | 0.4 |
| New Zealand | 36 | 0.48 |
| Australia | 36 | 0.68 |
| Cyprus | 36.9 | 0.69 |
| Upper-Middle income | | |
| Venezuela | 44.9 | 0.4 |
| Russian Federation | 30.1 | 0.3 |
| Malaysia | 62 | 0.5 |
| Dominican Republic | 49.9 | 0.32 |
| Brazil | 44.9 | 0.22 |
| Argentina | 44.9 | 0.35 |
| Lower-Middle Income | | |
| China | 26.2 | 0.19 |
| India | 40 | 0.2 |
| Indonesia | 62 | 0.47 |
| Sri-Lanka | 40.3 | 0.35 |
| Nicaragua | 43.8 | 0.34 |
| Low-income | | |
| Vietnam | 60 | 0.37 |
| Nepal | 40.3 | 0.2 |
| Myanmar | 54 | 0.24 |
| Cambodia | 55 | 0.29 |
| Bangladesh | 40.3 | 0.2 |

East and Africa, North America, Central Europe, Western Europe, Asia excluding Japan, Japan, and the rest of the world, respectively (Jambeck et al., 2015). China was the best producer of the plastics, yet it consumed only 25% compared to Europe and North America (Kibria et al., 2023).

Approximately 10 to 13% of inorganic municipal solid waste (MSW) comprises of plastic packaging materials, accessories, cups, containers, bags, and bottles. Out of the portion, single-used plastics contribute about 50% and end up being disposed in landfills where they are introduced in soils, air, and water to cause pollution in the environmental compartments (Kibria et al., 2023). According to the UN (2022), more than 17 million metric tons of plastics ended up in oceans in 2021 and the quantity is expected to double or triple by 2040 to become a major threat to marine life. Kibria et al. (2023) also pointed out that 60-80% of marine debris have plastic content. The quantity of plastics released to the

ment. The implementation of policies and legislation to control plastic production, consumption, and disposal, such as bans on single-use plastics, fees, and deposit refund schemes can be enforced by the regulation authorities to mitigate plastic use. Design optimization by promoting eco-design principles to reduce plastic content in products, enhance durability, facilitate recycling, and promote reuse is the major step to be exercised. An improved waste collection segregation and disposal system must be in place for efficiency. The formation of clubs and societies among youth to encourage public awareness and educate the community to adopt the environmental usage of plastics is key to a better system in place for pollution prevention.

## 3.3 Recycling of Plastics

For effective recycling of plastic, waste hierarchy, a framework is to be focused and aims to maximize value of recycling product and minimize the amount of energy waste and resources from production to end-of-life. To achieve this framework, waste should go through shortest possible recycling cycle. Plastic recycling is a systematic process involving multiple technical steps aimed at transforming post-consumer or post-industrial plastic waste into reusable materials. Deposit refund schemes, pay-as-you-go schemes, incineration, and landfill taxes have enhanced recycling. The procedure begins with the collection of plastic waste from numerous sources, followed by sorting based on material type and characteristics. After sorting, the plastic undergoes cleaning to remove impurities like labels and dirt. Subsequently, the cleaned plastic is tattered into smaller bits to increase its surface area. Depending on the desired end product, the shredded plastic is subjected to processes like melting or pyrolysis, where high temperatures break down the polymer chains. The resulting molten plastic can then be extruded or molded into specific forms. Additives may be introduced to enhance the recycled plastic's properties, and stringent quality regulatory measures are implemented to ensure the end product complies with industry standards. The following are the different recycling techniques used.

### 3.3.1 Physical or Mechanical Methods

Mechanical recycling is a collective process used to recycle plastic waste through physical techniques s such as sorting, washing, grinding, and re-granulating. Mechanical recycling is the prevailing machinery for plastic recycling, utilizing physical processes such as sorting, grinding, washing, separating, drying, and re-granulating to recuperate plastics that can be replaced for new plastics. This method is labor-demanding and results in inferior-quality plastics compared to virgin plastics. The requirement of labor in the shredding and segregation of plastics makes the process less cost-effective and less manageable. Packaging plastics like polyethylene, polypropylene, and PET usually undergo mechanical recycling. Mechanical methods are generally involved with the downcycling of plastics (Eg PET -1, PP-1). Mechanical recycling disadvantages include decreased performance than virgin ones, market average product, containing a market average of additives, less purity, and polymer chains will have degraded (Mohan et al., 2023; Lange et al., 2021).

### 3.3.2 Chemical and Thermochemical Methods

Chemical recycling technologies employ heat, chemical reactions, or both together to recycle used plastic into virgin-equivalent plastic, fuel, or other chemicals. It offers the potential to improve plastic recycling by producing raw materials of virgin quality, reducing landfill use, and creating new markets. Chemical recycling includes conversion, decomposition, and purification processes. The depolymerizations of plastics to basic monomer are an excellent method to revive the raw material thereby creating less waste (Huang et al., 2022). This technique employs a rapid heating process without requiring a catalyst. The plastic waste mixture is exposed to a combination of alternating current (AC) and direct current (DC) pulses to produce high-quality graphene (Algozeeb et al., 2020). Thermochemical recycling involves the use of heat to break down plastic waste into its basic components, such as monomers or fuel-type oils, through processes like pyrolysis. This method is particularly useful for plastics that are difficult to recycle mechanically or chemically. Thermochemical recycling, specifically pyrolysis, degrades polymeric materials by heating them in the lack of oxygen, leading to the production of monomers or fuel-type oils (Soni et al., 2021) Thermochemical recycling can handle mixed and contaminated plastic streams effectively, making it suitable for wide range of plastic waste types. It offers a sustainable solution when mechanical or chemical recycling is not economically feasible. Thermochemical recycling processes currently require high-quality waste inputs, and practical tests with mixed or contaminated waste streams are still facing challenges. Developing scalable thermochemical recycling processes is crucial for improving plastic waste recycling efficiency (Thunman et al., 2019).

### 3.3.3. Biological Methods

Biological recycling is an evolving technology that uses microorganisms, such as bacteria or fungi, to break plastic into its elementary components for reprocess. In the process of biological recycling, the environment and economy could benefit in such a way that plastic waste will be converted into more useful products of equal or better quality. The process involves using enzymes to break plastic polymers into their monomers, which can then be used to make new plastics. Most recycling research has concentrated on PET and polyurethane, partly because violating their chemical bonds is at ease (Shamisi et al., 2019) Moreover, researchers have found preliminary evidence of enzymes that can degrade polyethylene (PE). Biological recycling does not involve the identical level of sorting for plastic waste related with mechanical recycling, thereby saving time and money. It devours minimum energy than mechanical and some chemical recycling methods.

Synergistic enzymes are produced from microorganisms that exhibit excellent degradation of plastic (Cai, 2023). The multienzyme secretion capacity of microbes like Ideonella sakainesis has promising scope. PET and certain plastics could be degraded by hydrolysis using bacterial enzymes. Fungal strains like Aspergillus flavus and Pleurotus ostreatus can degrade plastics by oxidizing the molecular backbone (Ribeiro et al., 2022). Halophiles and psychrophiles show exceptional potential for plastic degradation which exists in extreme environmental conditions. Hence, an advanced approach of combining enzymatic degradation with the advanced process must be applied for effective waste reduction (Orlando et al., 2023). The biocatalytic depolymerization using enzymes as catalysts has been studied by several researchers to prove the efficiency of recycling (Zhu et al., 2022). Even though recycling is efficient, it's high-priced than generating new plastics, and companies may face elevated start-up costs in developing a recycling facility. The enzymes researchers have recognized are currently incomplete to degrading only limited types of plastic, and research is needed to report the unintentional outcomes of biological recycling (Aaerts et al., 2018). The nanomaterials can alter the metabolic cycles of microorganisms to improve the degradation of plastics. The integration of nanoparticles in the polymer matrix or the fabrication of nanocomposites overpowers the deficiencies of biodegradable polymers. Researchers showed that when

$SiO_2$ was added to an LDPE solution with Bacillus strain, the degradation was quicker compared to the conventional methods. Similarly, the use of Pseudomonas putida and supermagnetic iron nanoparticles also showed promising results (Orlando et al., 2023). A special study conducted by Fullerenes-60 aids during the degradation action of the bacteria by reducing the lag phase and the exponential phase of the growth cycles of the bacteria. Nanobarium with a diameter of 38 nm was used to reduce the life cycle of a bacterial consortium. Studies have also been conducted using protein-coated titania nanoparticles and lactobacillus plantarum species reduced the tensile strength of LDPE. Certain fungal strains like penicillium oxallicum *NS4* have been identified to degrade the morphology of HDPE and LDPE (Orlando et al., 2023). Modification of particle size and morphology can create biodegradable nanomaterials that can replace traditional non-biodegradable materials, thereby controlling the destructive effects of plastic pollution on the environment.

## 4. GREEN CHEMISTRY IN PLASTIC FABRICATION

Plastics are a widely used material in product design and manufacturing, with their application in single-use items increasing. The circular economy offers a solution by minimizing waste and maximizing resource use. Products get shared, repaired, and recycled, keeping materials circulating and reducing the need for new ones thereby helping protect our environment. Recycled plastics, despite their environmental advantages, often suffer from reduced strength compared to virgin counterparts. As consumers turn to biodegradable alternatives derived from plants or chemically treated (with oxygen/other chemicals), concerns arise about their long-term performance and potential environmental impact. The presence of biodegradable plastics in the recycling stream acts as a contaminant, jeopardizing the quality of recycled materials and potentially undermining their environmental benefits. Green chemistry principles (prevent waste, design safer chemicals, increase energy efficiency, promote renewable feedstock, reduce derivatives) offer a sustainable pathway to plastic production, dramatically lessening the burden of plastic waste.

### 4.1 Green Chemistry Principles in Plastic Fabrication

Green chemistry plays a crucial role in addressing the ecological impact of plastic fabrication. By adhering to the principles of green chemistry, which aim to diminish or eliminate the use of lethal substances in chemical products, researchers are exploring ways to make plastics more sustainable and less harmful to the environment. Researchers are working on developing green plasticizers, which are compounds synthesized from sustainable materials, biodegradable, and non-toxic, to replace traditional plasticizers like phthalates.

The development of chemical derivatives of plastics can be improvised for the preparation of biodegradable plastics. Adopting steps to develop products with minimum toxicity and maximum life will add to the cause of eco-friendly techniques. Designing safer chemicals and products aims to minimize toxicity and improve sustainability while maintaining functionality. The development of probabilistic models to predict toxic effects based on molecular properties is a crucial step to be borne in mind while designing chemical methods or physical structures for plastic materials. Consideration should also be given to the toxicity test, disintegration test, and other physicochemical parameters like tensile strength, density, and thermal resistivity. Biological interaction with both biotic and abiotic factors can also be studied during the designing process (Birbaum, 2013). Adoption of a safer ingredient list while manufacturing plastics focusing on carcinogenicity, mutagenicity, persistence, bioaccumulation, and other hazards can change the course life cycle of a plastic product in the environment. By focusing on safer chemical design, industries can create merchandises that are less likely to pose threats to human health and the environment, ultimately contributing to a more sustainable future.

The usage of less hazardous reagents and raw materials in the manufacturing process can minimize many toxic outcomes like fumes and water contaminating the surroundings. Minimal use of hazardous reagents can also avoid risks of accidents during production and degradation post-use. The application of biodegradable starch packing peanuts as an alternative to polystyrene reduces environmental impact (McCance et al., 2023).

## 4.2 Green Chemistry in Polymer Synthesis

Green chemistry promotes the use of renewable resources instead of petrochemical products, aiming to maintain competitiveness while reducing environmental impact and advocates for cleaner polymerization processes that generate minimal waste, such as polycondensation which produces water as the only by-product, optimizing operation. By utilizing renewable feedstocks like cellulose, starch, vegetable oils, or lignin, green chemistry enhances resource conservation and reduces dependence on fossil fuels in polymer synthesis. Bio-based polyesters derived from plant resources are designed to be biodegradable, aligning with green chemistry principles of sustainability and environmental preservation. The design and synthesis of chemically recyclable polymers enable their reutilization as starting monomers or new value-added chemicals, offering a closed-loop method to polymer material usage and sustainable development. Polymerization-depolymerization cycles can also be employed where chemically recyclable polymers like polyesters, poly thioesters, polycarbonates, and others have made significant achievements in achieving polymerization-depolymerization cycles for sustainable material usage. Hence, by incorporating green chemistry practices into polymer synthesis, researchers and industries can create more sustainable, environmentally friendly polymers that donate to a circular economy and reduce the ecological footprint of plastic materials.

Lignocellulose, vegetable oils, terpenes, biomass, and carbohydrates are providing sustainable feedstocks for the next generation of polymers. These novel polymers offer exciting possibilities for sustainable plastics, but their unique chemical compositions present a significant hurdle: bringing them to market requires substantial investment. While exploring new polymers is exciting, relying on sustainable existing monomers in established processes offers a faster, cheaper less groundbreaking approach to sustainable plastics. Also, registering and applying for safety certificates of these conventional polymers is not needed (Alaerts et al., 2018; Zhu et al., 2022; He et al., 2023). Methods to produce specific monomer classes like olefins, cyclic monomers, methacrylic, and bifunctional monomers are studied to synthesize them from renewable resources. One example is a synthesis of polyethylene from bio sourced ethanol (Alaerts et al., 2018; Zhu et al., 2022; He et al., 2023). Currently, the industrial production method for polylactic acid (PLA) is ring-opening polymerization (ROP) of lactone monomers and tin (II) octoate with Sn $(Oct)_2$ as the catalyst, approved by FDA. However, studies related to long-term deliverance and the physiological effects of the catalyst have not been conducted. Removing the catalyst residues in the waste plastic is another hurdle. An alternative proposed by Christina et al. is synthesis by enzymatic polymerization for the synthesis of PLA (Tournier et al., 2023). In general, avoiding toxic precursors like di isocyanates in polyurethanes synthesis, valorizing additional waste streams, mild conditions, using green solvents or solvent-free methodology, and manufacturing recyclable or biodegradable material (Kandpal et al., 2023).

## 5. PLASTIC WASTE MANAGEMENT WITH NANOTECHNOLOGY

Nanotechnology can contribute significantly to reducing plastic pollution through improvements in plastic properties and enhanced degradation mechanisms. The major advantage of nanoparticles is the ability to alter the rate of biodegradation depending on the need. Improving plastic durability, and enhancing thermal stability are some of the positive outcomes. Among all the chemical, physical, and biological methods, the use of nanomaterials has presented a promising result (Singh et al., 2008). The initial process of producing to decomposing the plastic, nanoparticles can modify the course of the plastic waste in the ecosystem. Nanotechnology can be used to manage plastic waste in several ways.

### 5.1 Constructive Approach

Studies show that over 60% of typical municipal plastic waste consists mainly of polyethylene (HDPE/LLDPE/LDPE) and polypropylene. Plastic waste is a source of carbon-based chemicals that could be converted to valuable products like graphene, carbon nanotubes, carbon nanospheres etc. by upcycling. This technique of upcycling is called constructive approach and involves creating enhanced valued nanomaterials while tackling the issue of plastic waste pollution. The process of transforming waste into valuable nanomaterials involves the following steps: 1) Collection and organizing of the plastic waste. 2) Depolymerization- formation of smaller molecules and 3) Transformation to various nanomaterials by methods like thermal decomposition, chemical vapor deposition and flash joule heating.

#### 5.1.1 Nanomaterial Synthesis

Plastic waste can be converted into valuable nanomaterials like graphene, carbon nanotubes, or nanocomposites through various techniques like pyrolysis, hydrothermal treatment, or chemical recycling. These nanomaterials can find applications in electronics, composites, and energy storage. CNTs possess high tensile strength thermal and electrical conductivity. Characteristics like high tensile strength, thermal and electrical conductivity make CNTs highly versatile, with applications ranging from building materials to advanced optical and electronic devices. Laser ablation method, arc discharge method, microwave assisted synthesis and catalytic pyrolysis method are some of the processing methods for synthesis of carbon nanotubes (Arapalli, 2004; Shi et al., 2000; Kandpal, 2023). Graphene has exceptional electrical and thermal conductivity, remarkable flexibility, incredibly low weight, and impressive resistance. Thermal decomposition with substrate, solid state method and two step pyrolysis are the methods observed in literature for graphene synthesis. Carbon nanospheres (CNS), with size of 50-100 nm, can be synthesized by the following methods: Stober synthesis method, CVD method, hydrothermal carbonization etc. Melt blending, powdered metallurgy methods are used for producing nanocomposites, which generally uses waste thermoplastics for recycling. Converting waste plastic into valuable nanomaterials, like in a circular economy, helps to minimize the environmental concerns related with plastic pollution and paves the way to establishing a circular economy.

### 5.1.2 Upcycling of Plastic Waste

By incorporating nanomaterials with specific functionalities into recycled plastics, their properties can be enhanced, encouraging their use in high-value applications and reducing reliance on virgin plastic. Some upcycling methods are biodegradation, nanocatalyst; pyrolysis, chemical vapor deposition etc are reported for PET waste. Biodegradation is considered an environmentally friendly, method to break down plastic. Biodegradation of PET by enzymes from microbes has found to break down and hydrolyze the chemical bonds at mild temperature. Nanocatalysts speed up the breakdown of plastic waste (PET) when open to light, heat, or other ecological factors which make them more manageable for prevailing recycling processes, such as thermal degradation, landfilling, and incineration. Researchers have explored catalytic pyrolysis to create both hydrogen and carbon nanotubes from waste materials. Here the design of the catalyst significantly impacts the outcome. For instance, a nickel-based catalyst (NiMnAl catalyst) synthesized by co-precipitation with plastic waste has been shown to effectively produce hydrogen and high-quality carbon nanotubes, with a yield of nearly 90%. Chemical vapor deposition using a Ni-based catalyst is another synthesis method seen in literature that gives good results. In all, high demand for value added nanomaterials with applications in sensors, batteries etc are found to attract upcycling as fines are being imposed on landfills and incinerators (Anusha et al., 2024).

## 5.2. Degradative Approach

Nanotechnology has emerged as a groundbreaking solution to the challenges posed by plastic waste management. Plastics, known for their durability and stability, present a substantial environmental hazard due to their resistance to natural degradation processes. Conventional methods of plastic degradation, such as chemical or thermal processes, are energy-intensive and generate harmful by-products, accelerating environmental issues. Plastics are difficult to degrade naturally and are often disposed of improperly, leading to long-lasting environmental pollution. Nanotechnology provides innovative methods for plastic degradation, including inducing the succession of microbial population and utilizing nanoparticles as catalysts for polymer breakdown. Nanotechnology also enables the modification of plastic polymers at the nanoscale, improving their physical properties and creating new opportunities for recycling and repurposing plastics. Recent advancements include processes that convert waste plastics into valuable outputs like graphene and hydrogen, offering new avenues for material reuse and reducing environmental impact (Algozeeb et al., 2020; Vieira et al., 2022). Nanotechnology offers a sustainable and eco-friendly alternative by leveraging nanomaterials to enhance the degradation of plastics.

### 5.2.1 Nanoparticle Additives

Nanoparticle additives in plastic waste management offer several benefits. Embedding specific nanoparticles like metal oxides or enzymes into plastic materials can promote their degradation through photocatalytic or enzymatic mechanisms. These nanoparticles can be designed to break down plastics into smaller, harmless components. Nanoparticles can influence the rate of biodegradation, allowing for tailored degradation processes based on specific needs.Nanotechnology enables the modification of waste plastics to maintain their original properties while adding new material qualities like thermal or electrical conductivity, expanding the possibilities for recycling and repurposing plastics. Flash Joule heating, also known as ultrafast Joule heating, is a cutting-edge method that rapidly heats materials to

## 6. CONCLUSION

Nanotechnology offers promising solutions for reducing plastic waste through various approaches, including developing biodegradable plastics, improving waste management techniques, and enhancing the efficiency of recycling processes. Some specific conclusions drawn from the search results include the production of biodegradable plastics, which can help shrink the environmental impacts of plastic waste, the utilization of nanocomposites for durable product production, and conversion techniques of waste into reusable products. The improvement of research techniques right from the raw material selection process to the product preparation, and the application of nano-biotechnologies can improve techniques of the manufacturing process of plastic items. The development of novel catalysts for the easier breakdown of materials can be more functionalized to enhance the degradation process.

The application of green chemistry in plastic waste reduction embraces a promise for creating more sustainable plastics items, improving recycling processes, and fostering a circular economy that minimizes environmental impact while addressing the global challenge of plastic pollution. Researchers are exploring green plasticizers synthesized from sustainable materials that are biodegradable and environmentally friendly, aiming to replace traditional plasticizers like phthalates. Switching fossil-based plastics with bio-based alternatives derived from renewable resources can reduce greenhouse gas emissions, facilitate easier recycling, and promote biodegradability. Nanotechnology can enhance waste processing and recycling efficiency, leading to better resource conservation, reduced pollution, and sustainable development by promoting the circular economy. Overall, nanotechnology presents opportunities to optimize waste reduction strategies, balancing economics and ecology while moving toward a sustainable and eco-friendly future.

## REFERENCES

Abo-Shama, U. H., El-Gendy, H., Mousa, W. S., Hamouda, R. A., Yousuf, W. E., Hetta, H. F., & Abdeen, E. E. (2020). Synergistic and antagonistic effects of metal nanoparticles in combination with antibiotics against some reference strains of pathogenic microorganisms. *Infection and Drug Resistance*, *13*, 351–362. doi:10.2147/IDR.S234425 PMID:32104007

Abu Jadayil, W., Qureshi, M. R. N. M., Ajaj, R., Aqil, E., Shawahin, G., Anver, H., & Aljeawi, S. (2022). An Empirical Investigation on Plastic Waste Issues and Plastic Disposal Strategies to Protect the Environment: A UAE Perspective. *Sustainability (Basel)*, *14*(24), 16719. doi:10.3390/su142416719

Alaerts, L., Augustinus, M., & Van Acker, K. (2018). Impact of bio-based plastics on current recycling of plastics. *Sustainability (Basel)*, *10*(5), 1487. doi:10.3390/su10051487

Algozeeb, W. A., Savas, P. E., Luong, D. X., Chen, W., Kittrell, C., Bhat, M., Shahsavari, R., & Tour, J. M. (2020). Flash graphene from plastic waste. *ACS Nano*, *14*(11), 15595–15604. doi:10.1021/acsnano.0c06328 PMID:33119255

Algozeeb, W. A., Savas, P. E., Luong, D. X., Chen, W., Kittrell, C., Bhat, M., Shahsavari, R., & Tour, J. M. (2020). Flash graphene from plastic waste. *ACS Nano*, *14*(11), 15595–15604. doi:10.1021/acsnano.0c06328 PMID:33119255

Amoabediny, G. H., Naderi, A., Malakootikhah, J., Koohi, M. K., Mortazavi, S. A., Naderi, M., & Rashedi, H. (2009). Guidelines for safe handling, use and disposal of nanoparticles. *Journal of Physics: Conference Series*, *170*, 012037. doi:10.1088/1742-6596/170/1/012037

Andrady, A. L. (2011). Microplastics in the marine environment. *Marine Pollution Bulletin*, *62*(8), 1596–1605. doi:10.1016/j.marpolbul.2011.05.030 PMID:21742351

Birnbaum, L. S. (2013). Designing safer chemicals. *Environmental Health Perspectives*, *121*(1), a9–a9. doi:10.1289/ehp.1206349 PMID:23287533

Cai, Z., Li, M., Zhu, Z., Wang, X., Huang, Y., Li, T., Gong, H., & Yan, M. (2023). Biological degradation of plastics and microplastics: A recent perspective on associated mechanisms and influencing factors. *Microorganisms*, *11*(7), 1661. doi:10.3390/microorganisms11071661 PMID:37512834

Chandran, R. R., Thomson, B. I., Natishah, A. J., Mary, J., & Nachiyar, V. (2023). Nanotechnology in Plastic Degradation. *Biosciences Biotechnology Research Asia*, *20*(1), 53–68. doi:10.13005/bbra/3068

David, M., Prakash, L., Sangeetha, J., Naik, J., Thangadurai, D., & Thimmappa, S. C. (2020). Degradation of Plastics Using Nanomaterials. In O. Kharissova, L. Martínez, & B. Kharisov (Eds.), *Handbook of Nanomaterials and Nanocomposites for Energy and Environmental Applications*. Springer. doi:10.1007/978-3-030-11155-7_74-1

Dubé, M. A., & Salehpour, S. (2014). Applying the principles of green chemistry to polymer production technology. *Macromolecular Reaction Engineering*, *8*(1), 7–28. doi:10.1002/mren.201300103

Dunn, R. A., & Welden, N. A. (2023). Management of Environmental Plastic Pollution: A Comparison of Existing Strategies and Emerging Solutions from Nature. *Water, Air, and Soil Pollution*, *234*(3), 201. doi:10.1007/s11270-023-06190-2

Evode, N., Qamar, S. A., Bilal, M., Barceló, D., & Iqbal, H. M. (2021). Plastic waste and its management strategies for environmental sustainability. *Case Studies in Chemical and Environmental Engineering*, *4*, 100142. doi:10.1016/j.cscee.2021.100142

Garofalo, E., Di Maio, L., Scarfato, P., Di Gregorio, F., & Incarnato, L. (2019). Nanotechnology-based strategy to upgrade the performances of plastic flexible film waste. *Polymers*, *11*(5), 830. doi:10.3390/polym11050830 PMID:31071960

Geyer, R., Jambeck, J. R., & Law, K. L. (2017). Production, use, and fate of all plastics ever made. *Science Advances*, *3*(7), e1700782. doi:10.1126/sciadv.1700782 PMID:28776036

Gkountela, C. I., & Vouyiouka, S. N. (2022). Enzymatic polymerization as a green approach to synthesizing bio-based polyesters. *Macromol*, *2*(1), 30–57. doi:10.3390/macromol2010003

Hannon, J. (2020). Exploring and illustrating the (inter-) disciplinarity of waste and zero waste management. *Urban Science (Basel, Switzerland)*, *4*(4), 73. doi:10.3390/urbansci4040073

Häußler, M., Eck, M., Rothauer, D., & Mecking, S. (2021). Closed-loop recycling of polyethylene-like materials. *Nature*, *590*(7846), 423–427. doi:10.1038/s41586-020-03149-9 PMID:33597754

Thunman, H., Vilches, T. B., Seemann, M., Maric, J., Vela, I. C., Pissot, S., & Nguyen, H. N. (2019). Circular use of plastics-transformation of existing petrochemical clusters into thermochemical recycling plants with 100% plastics recovery. *Sustainable Materials and Technologies*, *22*, e00124. doi:10.1016/j.susmat.2019.e00124

Tournier, V., Duquesne, S., Guillamot, F., Cramail, H., Taton, D., Marty, A., & André, I. (2023). Enzymes' power for plastics degradation. *Chemical Reviews*, *123*(9), 5612–5701. doi:10.1021/acs.chemrev.2c00644 PMID:36916764

Vaverková, M. D. (2019). Landfill impacts on the environment. *Geosciences*, *9*(10), 431. doi:10.3390/geosciences9100431

Vieira, O., Ribeiro, R. S., de Tuesta, J. L. D., Gomes, H. T., & Silva, A. M. (2022). A systematic literature review on the conversion of plastic wastes into valuable 2D graphene-based materials. *Chemical Engineering Journal*, *428*, 131399. doi:10.1016/j.cej.2021.131399

. Vital-Grappin, A. D., Ariza-Tarazona, M. C., Sandoval, I., Rojas-Guerrero, C. A., Villarreal-Chiu, J. F., & Cedillo-González, E. I. (2023). Design of Green TiO2-Based Semiconductors for Microplastic and Nanoplastic Photocatalytic Degradation.

Wang, J., Tan, Z., Peng, J., Qiu, Q., & Li, M. (2016). The behaviors of microplastics in the marine environment. *Marine Environmental Research*, *113*, 7–17. doi:10.1016/j.marenvres.2015.10.014 PMID:26559150

Wang, M., Wang, H., Gu, Y., Zhu, M., Kumar, M., Liang, J., Tie, Z., Ma, J., & Jin, Z. (2023). In Situ Generation of Flash Graphene Supported Spherical Bismuth Nanoparticles in Less than 200 ms for Highly Selective Carbon Dioxide Electroreduction. *ACS Materials Letters*, *6*(1), 100–108. doi:10.1021/acsmaterialslett.3c01183

Webb, H. K., Arnott, J., Crawford, R. J., & Ivanova, E. P. (2013). Plastic Degradation and Its Environmental Implications with Special Reference to Poly(ethylene terephthalate). *Polymers*, *5*(1), 1–18. doi:10.3390/polym5010001

Wojnowska-Baryła, I., Bernat, K., & Zaborowska, M. (2022). Plastic waste degradation in landfill conditions: The problem with microplastics, and their direct and indirect environmental effects. *International Journal of Environmental Research and Public Health*, *19*(20), 13223. doi:10.3390/ijerph192013223 PMID:36293805

Yang, H., Chen, G., & Wang, J. (2021). Microplastics in the marine environment: Sources, fates, impacts and microbial degradation. *Toxics*, *9*(2), 41. doi:10.3390/toxics9020041 PMID:33671786

Zheng, Y., Yanful, E. K., & Bassi, A. S. (2005). A Review of Plastic Waste Biodegradation. *Critical Reviews in Biotechnology*, *25*(4), 243–250. doi:10.1080/07388550500346359 PMID:16419620

Zhu, B., Wang, D., & Wei, N. (2022). Enzyme discovery and engineering for sustainable plastic recycling. *Trends in Biotechnology*, *40*(1), 22–37. doi:10.1016/j.tibtech.2021.02.008 PMID:33676748

## 2.2 Carbonaceous Sustainable Nanomaterials

Carbon, the most abundant non-metal element, exists naturally in various forms including coal, diamond, and graphite. Carbonaceous nanomaterials (CNMs) refer to carbon-based materials engineered to minimize environmental impact. These materials can be categorized based on the number of dimensions. Carbon dots (CDs) and graphene quantum dots (GQDs) fall under the classification of zero-dimensional CNMs. Carbon nanotubes (CNTs), graphene, and diamonds exemplify one, two, and three-dimensional carbonaceous nanomaterials (CNMs), respectively. These applications span biosensing, diagnosis, targeted drug delivery, bioimaging, and tissue engineering. In comparison to other nanomaterials like metal or organic nanoparticles, CNMs with substantial surface area to volume ratios offer several advantages. These include increased drug-loading capacity, enhanced bio- compatibility, simplified surface functionalization, easier immobilization of macromolecules, and physical properties such as electrical, optical, and thermal conductivity.

### 2.2.1 Nanocarbon Dots and Graphene Quantum Dots

Carbon dots (CDs) are quasi-spherical carbon nanoparticles consisting of carbon, hydrogen, nitrogen, and oxygen atoms, with sizes smaller than 10 nm. Macroscopic carbon typically exhibits low water solubility; however, the presence of hydrophilic carboxylic and amine groups on the surface of CDs facilitates their dispersion in water. GQDs are crystallized graphene disks, usually consisting of less than 10 graphene layers and with a size ranging between 2 and 20 nm. Both CDs and GQDs exhibit robust photoluminescence emission within the visible light spectrum, coupled with high photostability and resistance to photobleaching (Poon et al., 2021). Their outstanding biocompatibility, coupled with the ease of surface modification facilitated by the presence of hydrophilic groups, renders them well for various biomedical applications. Traditional methods for synthesizing CDs, such as laser ablation, electrochemical methods, pyrolysis, and exfoliation, typically necessitate expensive initial precursor materials that are non-renewable. In recent years, sustainable approaches have emerged for synthesizing CDs and GQDs by utilizing sustainable and renewable sources. This includes the utilization of organic pollutants present in the environment and diverse biomass materials like vegetables, fruits, wool, and cotton. To reduce the energy intensity of the manufacturing process, Xu et al. modified the traditional laser method by employing a femtosecond pulsed laser to synthesize CDs at room temperature. An eco-friendly electrochemical method for the production of graphene quantum dots (GQDs). This method involved utilizing a platinum wire as a cathode and a graphite rod as an anode, eliminating the need for toxic chemicals such as oxidant salts and acids. A sustainable and straightforward one-step hydrothermal carbonization method has been established for the synthesis of CDs derived from oyster mushrooms (Harper-Harris et al., 2021). These CDs exhibit selective sensing of Pb2+ ions, showcasing specific electrostatic binding to DNA molecules, antibacterial activity, and anticancer effects against MDA-MD-231 breast cancer cells. In the research conducted by Sangam et al., a catalyst-free and scalable hydrothermal method was employed to synthesize graphene quantum dots (GQDs) using sugarcane molasses, an agro-industrial waste product.

CDs, for instance, have been utilized in the selective detection of DNA of proteins and various blood serum components like cholesterol, glucose, and alcohol. Likewise, GQDs have found utility in the sensitive identification of various disease biomarkers, including enzymes, antigens, proteins,

*Table 1. Overview of carbon-based materials in biomedical uses*

| Nanomaterials with a focus on sustainability | Eco-friendly origin or Renewable Source | Biomedical Application |
|---|---|---|
| Quasi-spherical carbon nanoparticles, often referred to as Carbon dots (CDs) | Derived from organic pollutants or waste, including vegetables, fruits, wool, cotton, and oyster mushrooms. | Biosensing, particularly for DNA detection<br>Proteins and biomolecules associated with diseases<br>Cellular imaging for the monitoring of bone regeneration<br>Delivery of pharmaceutical substances |
| Crystallized graphene disks, often referred to as Graphene Quantum Dots (GDs) | | Detection of DNA, proteins, and biomolecules associated with diseases, typically known as biosensing.<br>Cellular imaging<br>Delivery of pharmaceutical substances |
| Graphene is a two-dimensional carbon material. | Derived from organic carbonaceous waste materials such as bones, bagasse from newspapers, plants, and natural sources. | Limited by the tendency to aggregate in tissue environments, affecting its application in drug delivery.<br>Delivery through the skin. |
| Graphene oxide (GOX) | | Small-scale electrodes.<br>Detection of biological substances and the creation of materials for use in devices that convert electrical signals into mechanical movement.<br>Design of detection systems for microRNA. |
| Carbon nanotubes (CNTs) | Derived from plants (leaves, seeds, roots, and stem) and waste cooking oil | Monitoring and visualizing cellular structures<br>Administering therapeutic substances<br>Inducing hyperthermia for cancer treatment<br>Manipulating at the nanoscale |

DNA, and other biomolecules. Furthermore, apart from their applications in biosensing, CDs and GQDs have been designed for bioimaging purposes owing to their exceptional biocompatibility, high photostability, and intense fluorescence. The outcomes demonstrated significant cellular uptake and effective fluorescence labeling capabilities, indicating their considerable potential for cellular bioimaging.

### 2.2.2 Graphene

Graphene, a two-dimensional carbonaceous nanomaterial, is composed of a single layer of closely packed sp2 carbon atoms organized in a hexagonal structure. Graphene displays distinctive physical characteristics in contrast to its bulk structure, featuring a substantial surface area, elevated electrical and thermal conductivity, remarkable elasticity and mechanical strength, and adjustable optical properties. Graphene oxide (GOX) is graphene subjected to chemical functionalization, incorporating oxygen-containing groups like carboxy, carbonyl, and hydroxyl (Ouyang et al., 2020). The introduction of these surface-bound oxygen groups enhances the hydrophilicity of graphene, fostering aqueous solution stability through robust repulsive interactions, including electrostatic and hydrogen bonding, in addition to $\pi-\pi$ interactions. As a result, graphene, particularly GOX, has been extensively explored for diverse biomedical applications. Traditional techniques like mechanical or chemical exfoliation of graphite have been devised for graphene synthesis. Other methods such as industrial waste, food waste, plants, agricultural waste, and natural carbonaceous materials through a process known as graphitization. A composite aerogel by combining bacterial cellulose with caffeic acid-reduced graphene oxide, aiming to design a wearable device based on bio-pressure sensing. An environmentally sustainable method was developed for the synthesis of graphene oxide (GOX) through a single-step reforming process using sugarcane bagasse waste under atmospheric conditions.

The single-atom-thick carbon layer offers an extensive surface area for attaching drug molecules and targeting ligands on both sides of graphene. Hence, nanomaterials based on graphene exhibit significant promise for drug delivery due to enhanced drug-loading efficiency. Nonetheless, graphene's tendency to aggregate in tissues poses potential challenges by inducing oxidative stress and potential toxic effects on humans. Functionalized graphene has been investigated for its potential in biosensing and tissue engineering applications.

### 2.2.3 Carbon Nanotubes

Carbon nanotubes (CNTs) consist of a graphene layer arranged in a cylindrical structure with nanoscale dimensions. Depending on the arrangement of graphene cylinders, carbon nanotubes (CNTs) can be categorized into single-walled carbon nanotubes (SWCNTs) and multi-layered carbon nanotubes (MW-CNTs). CNTs exhibit exceptional physical properties, including outstanding thermal and electrical conductivity, as well as high mechanical strength. As a result, there is a need for alternative approaches to manufacture CNTs that leverage renewable resources. Catalytic chemical vapor deposition stands out as a prevalent method for CNT production. The use of precursor materials derived from bio-sources can render the chemical vapor deposition process more energy-efficient by reducing reaction time and temperature. Duarte and colleagues employed the chemical vapor deposition (CVD) method to synthesize carbon nanotubes (CNTs) from waste cooking oil. This energy-efficient method has facilitated the swift production of carbon nanotubes (CNTs) from diverse carbon sources, catalysts, and substrates. Carbon-nanotubes (CNTs) can function as electromechanical actuators by applying potential in an electrolyte. In their work, Ru et al. employed a nanoporous film of CNTs as an electrode in ionic electroactive polymer actuators. Carbon nanotubes (CNTs) have found application in creating biosensors for the optical or electronic detection of biomolecules like DNA, glucose, and proteins. Multi-walled carbon nanotubes (MWCNTs) through the pyrolysis of a chickpea peel precursor. These MWCNTs exhibited non-toxic blue fluorescence signals when applied to human prostate carcinoma cells. The utilization of carbon nanotubes (CNTs) has been illustrated in transporting drugs with low solubility and bioavailability more effectively to the intended site through improved cellular permeation. Multi-walled carbon nanotubes (MWCNTs) by incorporating a layer of phospholipid-poly (ethylene glycol) and anti-Pgp antibodies (Zuk et al., 2022). This modification aimed to enhance the biocompatibility, blood circulation, and cancer cell targeting capabilities of the MWCNTs. Carbon nanotubes (CNTs) can serve as flexible nanotweezers, enabling concurrent adjustments and fluctuations in biomedical analytical investigations, particularly in nucleic acid-based spectroscopy.

## 3. ECO-FRIENDLY CERAMICS

Bio ceramics represent a crucial biocompatible ceramic material, possessing outstanding bioactivity, chemical stability, resistance to heat, and mechanical characteristics similar to tissues. They have been employed as a nanoporous scaffold to mend and reconstruct impaired tissues, address bone defects, and administer drug molecules. Bioceramics, including calcium phosphate and hydroxyapatite, can be obtained from readily available natural waste materials such as eggshells, bovine bone, fish bone, and seashells. These resources are abundant and cost-effective, being geographically widespread. Since these natural raw materials originate from sustainable biological systems, they lack the inherent toxicity and potential side effects associated with many synthetic materials upon exposure. With the rising global consumption of eggs and their mineral enrichment (containing calcium carbonate, calcium phosphate, and 1% magnesium), eggshell biowaste presents an opportunity for the production of bio ceramics like calcium phosphate and hydroxyapatite. Additionally, eggshells contain small quantities of biologically significant ions such as magnesium, sodium, silicon, and strontium (Colvin et al., 2003). These ions play a role in the mineralization process and contribute to the promotion of bone growth. Calcium phosphate (Ca P) stands out as a extensively studied synthetic biomaterial for bone regeneration due to its natural occurrence in human bone, along with its exceptional bioactivity, biocompatibility, and biodegradability. Numerous synthesis techniques, including sol–gel, hydrothermal, solid-state reaction, ultrasonic, and microwave methods, have been explored for producing Ca P biomaterials from eggshell biowaste. Hydroxyapatite materials have been generated through the high-temperature (1050 oC) solid-state decomposition of eggshells. Of the different synthetic methods available, microwave synthesis emerges as the most environmentally friendly approach for producing bio ceramic nanoparticles with a narrow size distribution and high throughput (Zuk et al., 2022). By adjusting microwave parameters like power and exposure time, the size, shape, and crystallinity. Bio ceramics obtained from eggshells demonstrate enhanced biological performance in comparison to synthetic counterparts.

## 4. CONCLUSION

This review outlines key design principles for creating sustainable nanomaterials using renewable resources and eco-friendly scalable manufacturing methods with minimal hazardous waste generation (zero-waste). Utilizing recycled waste as a renewable resource in the production of sustainable nanomaterials is a promising approach toward achieving zero-waste manufacturing. The review also explores sustainable design principles applied to the creation of nanocellulose, carbon, and bioceramics from renewable sources like plants, wood, fruits, biopolymers, bacteria, eggshell bio-waste, and recycled materials. Nanomaterials derived from natural resources offer various advantages over those produced from non-renewable sources, including versatile surface functionalities, biocompatibility, and biodegradability. Sustainable nanomaterials derived from these resources currently exhibit lower performance compared to those synthesized from non-renewable sources. The swift innovation in nanomaterials production from non-renewable materials has emphasized high performance without considering the potential impact on the availability of raw materials, energy resources, and the introduction of toxic chemicals to the environment and human health. Additionally, the complete life cycle of nanomaterials, from production to disposal, has been overlooked.

There are several potential avenues for future research in sustainable nanomaterials. Firstly, the development of eco-friendly strategies based on green chemistry for synthesizing sustainable nanomaterials is crucial for creating an environmentally sustainable future. Secondly, optimizing the performance of sustainable nanomaterials should consider the selection of appropriate renewable resources and manufacturing methods (evaluating energy requirements, waste production, and scalability). The choice of renewable materials should be justified based on factors such as energy efficiency, potential waste generation, material source (renewable or non-renewable), biocompatibility, biodegradability, and recyclability. Thirdly, optimizing the performance of sustainable materials in their intended environments, such as the biological setting, is essential for understanding the relationship between nanomaterial design and properties. Fourthly, conducting a comprehensive assessment of the environmental and human health impact of nanomaterials throughout their entire life cycle, from production to disposal, is crucial. Fifthly, encouraging the development of new sustainable methods for recycling and disposing of nanomaterials is important. Lastly, the establishment of new regulations and standards for the production and use of nanomaterials, prioritizing sustainability and minimizing harm to human health and the environment, is necessary. Establishing a performance matrix for sustainable nanomaterials would generate a comprehensive dataset, which could be analyzed using machine learning. In the coming years, machine-learning methods are anticipated to aid in the thoughtful creation of sustainable nanomaterials with peak performance through the analysis of extensive datasets.

The framework for designing sustainable materials serves as a link between research and commercialization. It is anticipated that leveraging natural and sustainable resources provides a prospect for producing the next generation of sustainable nanomaterials through a less energy-intensive and environmentally friendly approach, without posing risks to human health and the environment.

## REFERENCES

Abitbol, T., Rivkin, A., Cao, Y., Nevo, Y., Abraham, E., Ben-Shalom, T., Lapidot, S., & Shoseyov, O. (2016). Nanocellulose, a tiny fiber with huge applications. *Current Opinion in Biotechnology, 39*, 76–88. doi:10.1016/j.copbio.2016.01.002 PMID:26930621

Chopra, D., Gulati, K., & Ivanovski, S. (2021). Understanding and optimizing the antibacterial functions of anodized nano-engineered titanium implants. *Acta Biomaterialia, 127*, 80–101. doi:10.1016/j.actbio.2021.03.027 PMID:33744499

Colvin, V. L. (2003). The potential environmental impact of engineered nanomaterials. *Nature Biotechnology, 21*(10), 1166–1170. doi:10.1038/nbt875 PMID:14520401

Gaur, M., Misra, C., Yadav, A. B., Swaroop, S., Maolmhuaidh, F. Ó., Bechelany, M., & Barhoum, A. (2021). Biomedical Applications of Carbon Nanomaterials: Fullerenes, Quantum Dots, Nanotubes, Nanofibers, and Graphene. *Materials (Basel), 14*(20), 5978. doi:10.3390/ma14205978 PMID:34683568

Hadden, M., Martinez-Martin, D., Yong, K.-T., Ramaswamy, Y., & Singh, G. (2022). Recent Advancements in the Fabrication of Functional Nanoporous Materials and Their Biomedical Applications. *Materials (Basel), 15*(6), 2111. doi:10.3390/ma15062111 PMID:35329563

Hald Albertsen, C., Kulkarni, J. A., Witzigmann, D., Lind, M., Petersson, K., & Simonsen, J. B. (2022). The role of lipid components in lipid nanoparticles for vaccines and gene therapy. *Advanced Drug Delivery Reviews, 188*, 114416. doi:10.1016/j.addr.2022.114416 PMID:35787388

Harper-Harris, J., Kant, K., & Singh, G. (2021). Oleic Acid-Assisted Synthesis of Tunable High-Aspect-Ratio Multiply-Twinned Gold Nanorods for Bioimaging. *ACS Applied Nano Materials*, *4*(4), 3325–3330. doi:10.1021/acsanm.1c00360

Horejs, C. (2021). Lasting impact of lipid nanoparticles. *Nature Reviews. Materials*, *6*(12), 1071. doi:10.1038/s41578-021-00392-y

Jayathilaka, W. A. D. M., Qi, K., Qin, Y., Chinnappan, A., Serrano-García, W., Baskar, C., Wang, H., He, J., Cui, S., Thomas, S. W., & Ramakrishna, S. (2019). Significance of Nanomaterials in Wearables: A Review on Wearable Actuators and Sensors. *Advanced Materials*, *31*(7), 1805921. doi:10.1002/adma.201805921 PMID:30589117

Mittal, S., & Lai, J.-C. (2022). Wireless, closed-loop, smart bandage with integrated sensors and stimulators for advanced wound care and accelerated healing. *Nature Biotechnology*. PMID:36424488

Mohan, C., Kumari, N., & Robinson, J. (2022). Sustainable and environmental friendly energy materials. *Materials Today: Proceedings*, *69*, 494–498. doi:10.1016/j.matpr.2022.09.187

Mohan, C., Robinson, J., & Negi, A. (2023). Ion-Selective Electrode (ISE) Based on Polyvinyl Chloride Membrane Formed from Heterocyclic Quinazoline Compounds as Ionophore material. *Engineering Proceedings*, *48*, 10. doi:10.3390/CSAC2023-14914

Morán, J. I., Alvarez, V. A., Cyras, V. P., & Vázquez, A. (2008). Extraction of cellulose and preparation of nanocellulose from sisal fibers. *Cellulose (London, England)*, *15*(1), 149–159. doi:10.1007/s10570-007-9145-9

Ouyang, C., Zhang, S., Xue, C., Yu, X., Xu, H., Wang, Z., Lu, Y., & Wu, Z.-S. (2020). Precision-Guided Missile-Like DNA Nanostructure Containing Warhead and Guidance Control for Aptamer-Based Targeted Drug Delivery into Cancer Cells in Vitro and in Vivo. *Journal of the American Chemical Society*, *142*(3), 1265–1277. doi:10.1021/jacs.9b09782 PMID:31895985

Poon, K., Lu, Z., De Deene, Y., Ramaswamy, Y., Zreiqat, H., & Singh, G. (2021). Tuneable manganese oxide nanoparticle based theranostic agents for potential diagnosis and drug delivery. *Nanoscale Advances*, *3*(14), 4052–4061. doi:10.1039/D0NA00991A PMID:36132835

Ruggeri, M., Bianchi, E., Rossi, S., Vigani, B., Bonferoni, M. C., Caramella, C., Sandri, G., & Ferrari, F. (2020). Nanotechnology-Based Medical Devices for the Treatment of Chronic Skin Lesions, *Research to the Clinic. Pharmaceutics*, *12*(9), 815. doi:10.3390/pharmaceutics12090815 PMID:32867241

Seo, G., Lee, G., Kim, M. J., Baek, S.-H., Choi, M., Ku, K. B., Lee, C.-S., Jun, S., Park, D., Kim, H. G., Kim, S.-J., Lee, J.-O., Kim, B. T., Park, E. C., & Kim, S. I. (2020). Rapid Detection of COVID-19 Causative Virus (SARS-CoV-2) in Human Nasopharyngeal Swab Specimens Using Field-Effect Transistor-Based Biosensor. *ACS Nano*, *14*(4), 5135–5142. doi:10.1021/acsnano.0c02823 PMID:32293168

Surender, M. C., & Kumar, R. (2023). Novel modification of activated charcoal sheet with N-methyl poly pyrrole and silver nanoparticles for removal of hexavalent chromium in water treatment processes. *Materials Protection*, *64*(4), 503–511. doi:10.5937/zasmat2304503S

Tran, H.-V., Ngo, N. M., Medhi, R., Srinoi, P., Liu, T., Rittikulsittichai, S., & Lee, T. R. (2022). Multifunctional Iron Oxide Magnetic Nanoparticles for Biomedical Applications: A Review. *Materials (Basel)*, *15*(2), 503. doi:10.3390/ma15020503 PMID:35057223

Zuk, M., Podgórski, R., Ruszczyńska, A., Ciach, T., Majkowska-Pilip, A., Bilewicz, A., & Krysiński, P. (2022). Multifunctional Nanoparticles Based on Iron Oxide and Gold-198 Designed for Magnetic Hyperthermia and Radionuclide Therapy as a Potential Tool for Combined HER2-Positive Cancer Treatment. *Pharmaceutics*, *14*(8), 1680. doi:10.3390/pharmaceutics14081680 PMID:36015306

# Chapter 11
# Waste Management From Agricultural Field:
## Sustainable Practices and Framework

**Aditi Tiwari**
*Intertek India Pvt. Ltd., Gurugram, India*

**S. V. Niveditha**
https://orcid.org/0009-0007-1420-8397
*Presidency University, Bangalore, India*

**Chandra Mohan**
*K.R. Mangalalm University, Gurgaon, India*

**Anjaneyulu Bendi**
https://orcid.org/0000-0002-2538-507X
*Presidency University, Bangalore, India*

## ABSTRACT

*Prioritizing the 4Rs of reduce, reuse, recycle, and recover offers the most effective waste management approach. Several ideas to utilize agricultural wastes include biogas production, biochar production, composting, supplementary feed for animal and eco-friendly material construction. Biogas which is a renewable fuel and is produced by anaerobic fermentation of organic matter. It is primarily composed of methane, the key component of natural gas, along with carbon dioxide and trace amounts of other gases. Biochar is the process in which crop residues can be reborn through pyrolysis or other processes. This valuable soil amendment enhances fertility, water retention, and nutrient holding capacity. Composting with crop residues has a long history. Traditionally, farmers pile leftover crops in dung pits, where they decompose and become a valuable resource. The by-products produced from agricultural waste can be used as supplement to animal feed. Agricultural waste can also be used in making eco-friendly building materials such as bricks, panels, etc.*

DOI: 10.4018/979-8-3693-4264-0.ch011

## INTRODUCTION

The best management practices incorporate principle of business, goals, production, environmental quality, and sustainability available in farm resources management systems. Agriculture has been a prime resource since more than 10,000 years ago for providing humans with facilities such as foods, fuels and fibers. Agricultural resources have contributed a lot for the development of processes significant for life on earth and for human evolution. (FAO, 2017; Rivera-Ferre et al., 2013) Land and water, the lifeblood of agriculture, have fueled its remarkable growth over the last five decades. (*The International Code of Conduct for the Sustainable Use and Management of Fertilizers*, 2019)This expansion has solidified agriculture's position as the world's leading economic engine and job creator, playing a prominent role in many countries economics. ("Food and Agriculture: Driving Action across the 2030 Agenda for Sustainable Development," n.d.) This 'green revolution' in the world of agriculture have transformed world to a new level.

Agricultural wastes could be either the by-product of production and processing of agriculture products which includes vegetable, meat, fruits, poultry, dairy, or crops having some amount of useful components which can be further re-utilized for better consumption. These by-products may be in the form of solids, liquids, or slurries. Agricultural waste also known as Agro-waste may include wastes from food processing industries, waste from animal, hazardous and toxic agricultural waste, and crop residues. (Obi et al., 2016) Sustainable agriculture demands careful attention to waste management. Effective strategies are crucial for minimizing environmental impact and creating an eco-friendly-future. Agricultural waste management encompasses the practices and strategies to handle and disposal of waste produced from agricultural activities, aiming to reduce the environmental impact and promoting eco-friendly practices.

Agricultural waste disposal is a major environmental hurdle threatening the future of sustainable farming. Therefore, innovative waste management techniques that prioritize reduction, reuse, and recycling are emerging as a powerful waste management solution.

**Reduce**- The main idea behind this technique is to rationale the amount of trash produced by agricultural activities. By adapting methods and procedures that reduces waste generation, we can take a step towards sustainable agriculture development. Example of such practices includes, Precision farming, which efficiently optimize the utilization of pesticides, water, fertilizers.

**Reuse**- This idea emphasizes on the utilization of resources to their maximum efficiency. The key to sustainable agriculture waste management lies in resource reutilization. By repurposing what might

*Figure 1. Some of the techniques for waste management*

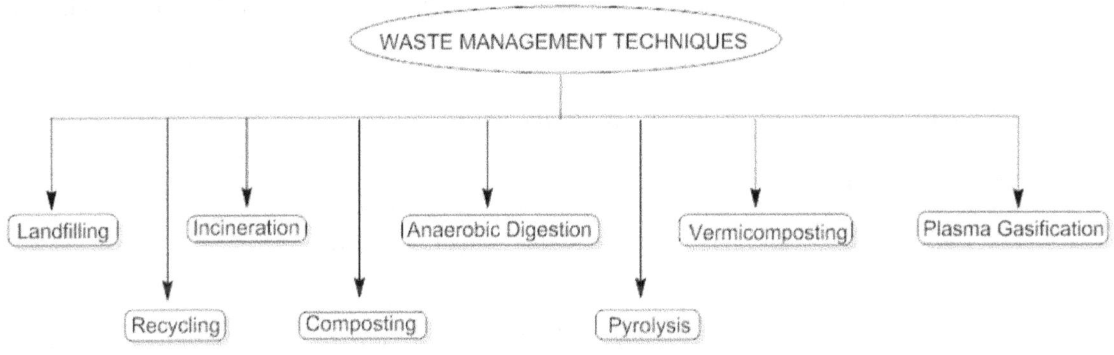

otherwise be discarded, is one of the way to significantly improve the efficiency and minimize waste. Incorporation of crop residue re-utilized as organic matter to enrich the soil structure and fertility could be the best example for this.

**Recycle**- The main strategy for this is to transform waste into valuable assets. Composting, which is a corner store of resource reuse, transforms complex organic agricultural waste such as crop residues and food scraps into nutrient-rich compost. This idea boosts soil fertility, improves water retention, and can even decrease reliance on synthetic fertilizers.

## SUSTAINABLE PRACTICES FOR AGRICULTURAL WASTE MANAGEMENT

**Zhijuan Wang** et al. prepared biochar rich in calcium from paper mill sludge for adsorbing phosphorus. The carbonization technique was adopted to prepare the biochar, where the sludge was first dried and powdered. The obtained sludge powder was then pyrolyzed under varied temperatures ranging from 350 °C to 950 °C in a limited $O_2$ and $N_2$ atmosphere. The biochar prepared at 750 °C in the $N_2$ atmosphere exhibited high phosphorus removal capacity. Using calcium biochar as the adsorbent, phosphorus removal efficiency of 82.65–99.23% was achieved through a chemisorption mechanism under varied initial pH of 3–11. The maximum adsorption capacity of biochar at 25 °C was found to be 68.5 mg $g^{-1}$. The used biochar adsorbent contained 14.5 wt% of P, which was proposed for use as fertilizer because P is an essential nutrient for plant growth **(Figure 2).** (Wang et al., 2021)

**Teresa Rodríguez-Espinosa** et al. looked into the sustainability of farming by repurposing organic waste as a source of nitrogen fertilizer in farming systems. Nine organic wastes were chosen, their nitrogen, water, and organic content were examined, and they were used in cultivation for a full six months. Among the waste, sewage sludge compost, and date palm leaf pruning exhibited a high nitrogen content of 22.6 g $kg^{-1}$ and 9.3 g $kg^{-1}$ after drying at 105°C respectively, and a low C/N ratio. However, the quantity of organic waste required was significantly greater than that of urea so the author proposed the mixing of organic waste and inorganic fertilizers to increase the crop yield **(Figure 3)**. (Rodríguez-Espinosa et al., 2023)

**G. Vinci** et al. evaluated the environmental sustainability of mushroom production in Italy by applying a Life cycle assessment (LCA) and an integrated Water-Energy-Nitrogen-Carbon-Food (WENCF) analysis. According to the LCA analysis, the mushroom production generates $2.55 \times 10^4$ kg of $CO_2$ equivalent for a functional unit of 23,000 kg of substrate. So to reduce the environmental impact recycled water, manure from horses and poultry substrate, and straw were used in the production process. According to the findings, based on the recovery the water footprint decreased by 56%, which in turn caused the energy and nitrogen footprints to decrease by 40% and 30%, respectively **(Figure 4).** (Vinci et al., 2023)

*Figure 2. Paper mill sludge-based biochar for phosphorus removal*

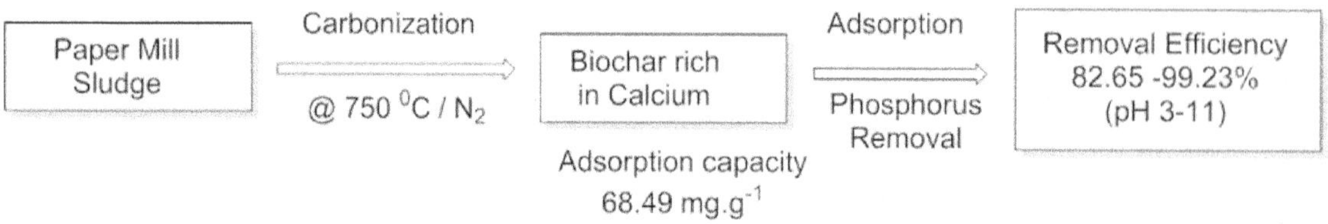

*Figure 3. Overview of sustainability idea by Espinosa et al.*

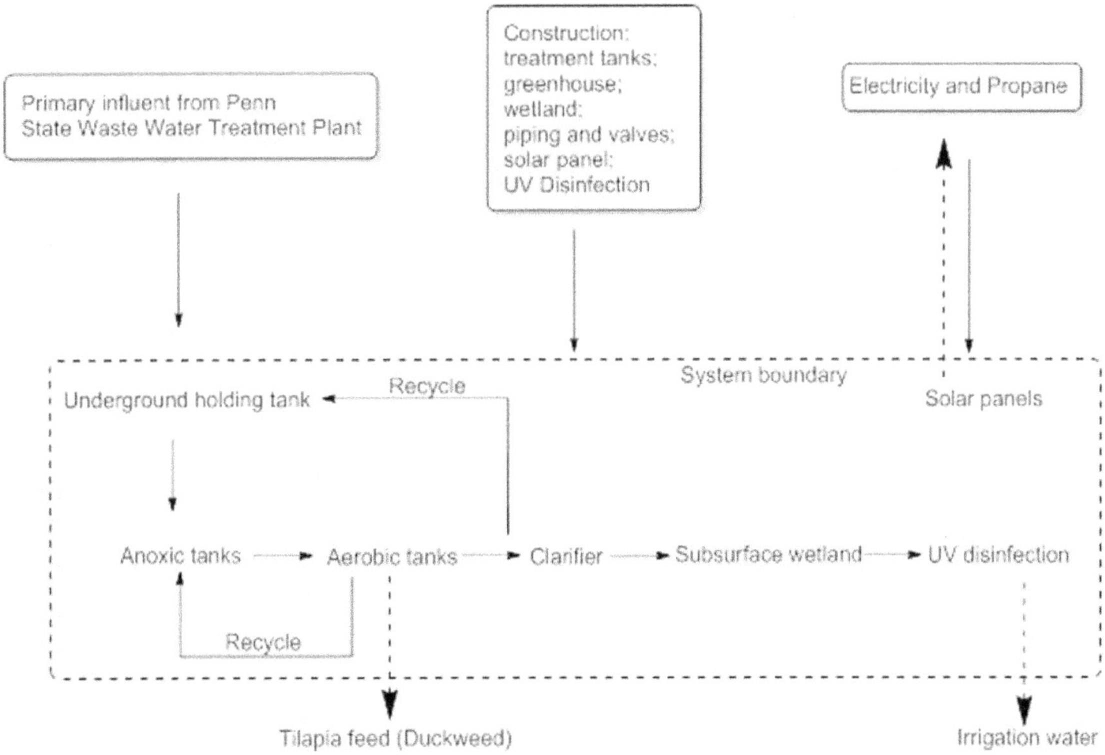


**Florence Alxandra Toth** et al. prepared an organic composite comprised of fermented chicken manure, bentonite (montmorillonite), and super-absorbent polymer (a cross-linked acrylamide and potassium polyacrylate copolymer) used in an apple orchard to enhance nutrient and water management. The 3.2 kg of composite added per tree per year showed an improvement in soil organic carbon by 4–9 g/kg and organic nitrogen by 1.8–2.8 g/kg. Apart from this it also enhanced the leaves' N, K, Ca, and Mg content and fruit weight **(Figure 5).** (Tóth et al., 2024)

**Joshua Wanyama** et al. conducted a detailed review of the application of new technologies like artificial intelligence (AI), the Internet of things (IoT), Big data analytics, Machine learning (ML), and drone technology in agriculture farming. From the review it was identified that (i) AI aids in the forecasting of water demand, and soil moisture; (ii) IoT helps in gathering data; (iii) Data analytics provides insight into irrigation management (iv) ML algorithms improve the accuracy of predictive analytics and decision support systems (v) Drone technology facilitates remote sensing and accurate data collection, that helps to monitor the soil moisture, crop health, and water efficiency. Finance, infrastructure, limited access, and knowledge are some of the constraints in the application of these technologies. However, to promote sustainable farming practices governments, research institutes, businesses, and farmer associations must fortify their alliances and partnerships with each other **(Figure 6)**. (Wanyama et al., 2024)

**S. Sumiyati** et al. conducted a comprehensive literature survey on the lifecycle assessment of recycling agricultural waste for sustainable farming. Agricultural waste recycling reduces greenhouse gas emissions, improves soil fertility, and preserves resources. Nowadays, rather than landfilling agricultural waste, it is widely used in the production of biochar, biogas, and briquette which are utilized as an energy

*Figure 4. Sustainable mushroom production*

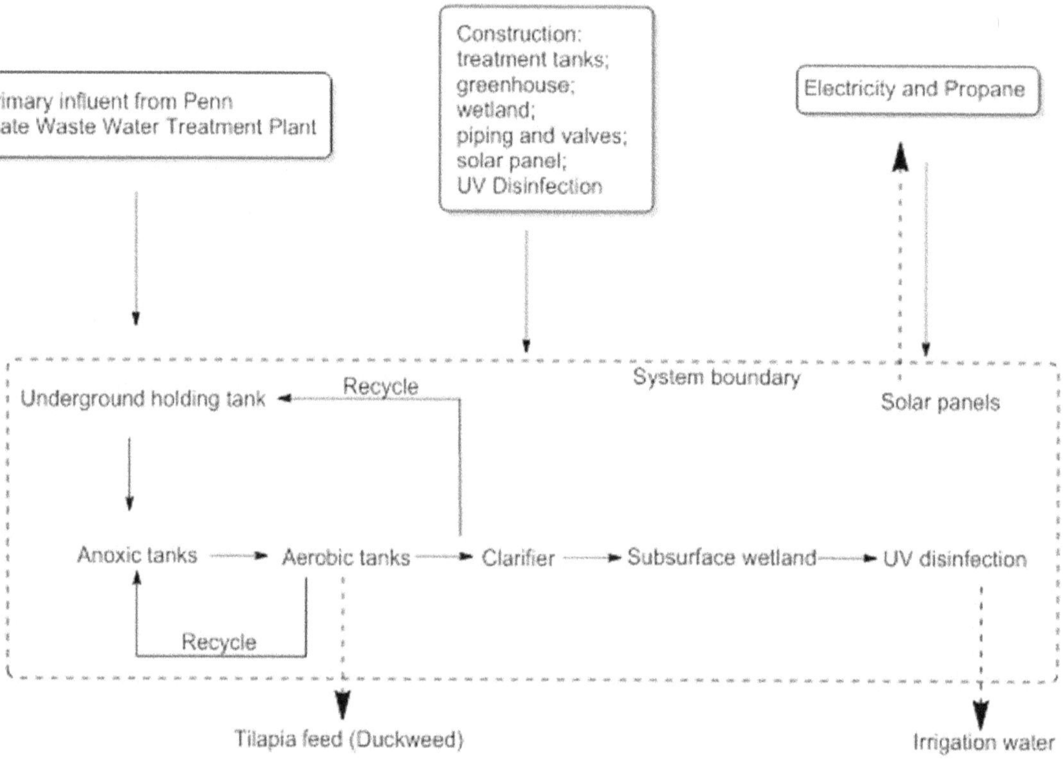


*Figure 5. Preparation of organic composite for apple orchard*

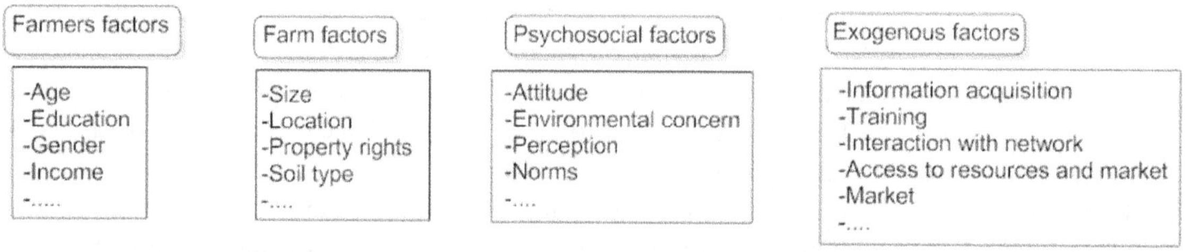


source. From the literature, it was found that composting, anaerobic digestion, and pyrolysis are various methods adopted to generate biochar which in turn promotes circular economy and sustainability **(Figure 7)**. (Sumiyati et al., 2024)

**Mustapha El Janati** et al. investigated how an oasis's soil fertility was affected by crop rotation, the use of organic fertilizer, and crop waste management. For the study, three types of cropping systems were analyzed viz. (i) date palms along with cereals, forages, and perennial crops; (ii) date palm cultivation alone; (iii) a combination of the first two systems. It was found that in (i) and (iii) cropping systems, livestock were also present which generated 19.4 and 24.2 tons of manure per farm per year. Also in (i) and (ii) cropping systems, 0.98 and 2.1 tons per hectare of dry palms were generated. The author suggested converting the dry palm waste to manure as it is rich in organic matter, which in turn helps

*Figure 6. Industrial technologies for smart irrigation*

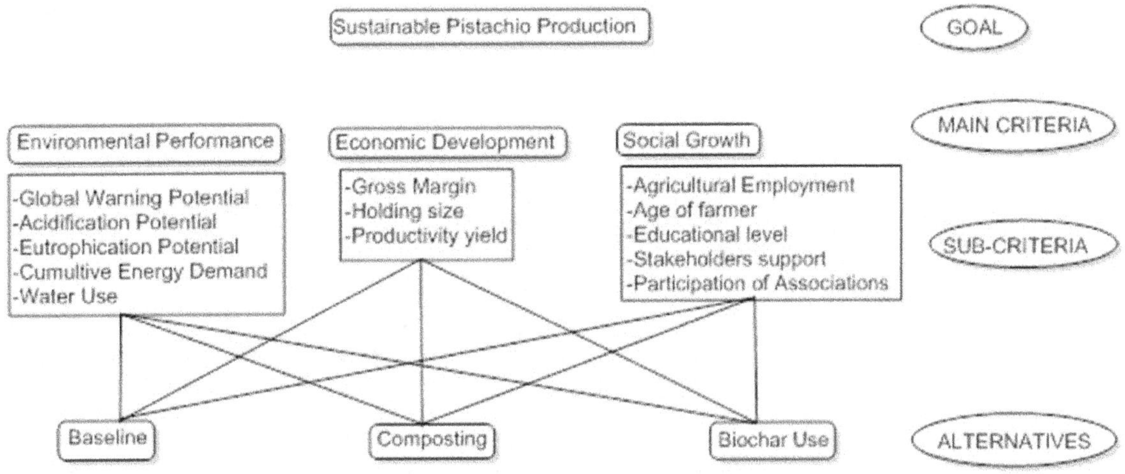


*Figure 7. Types of agricultural waste treatment technologies utilized by Sumiyati et al.*

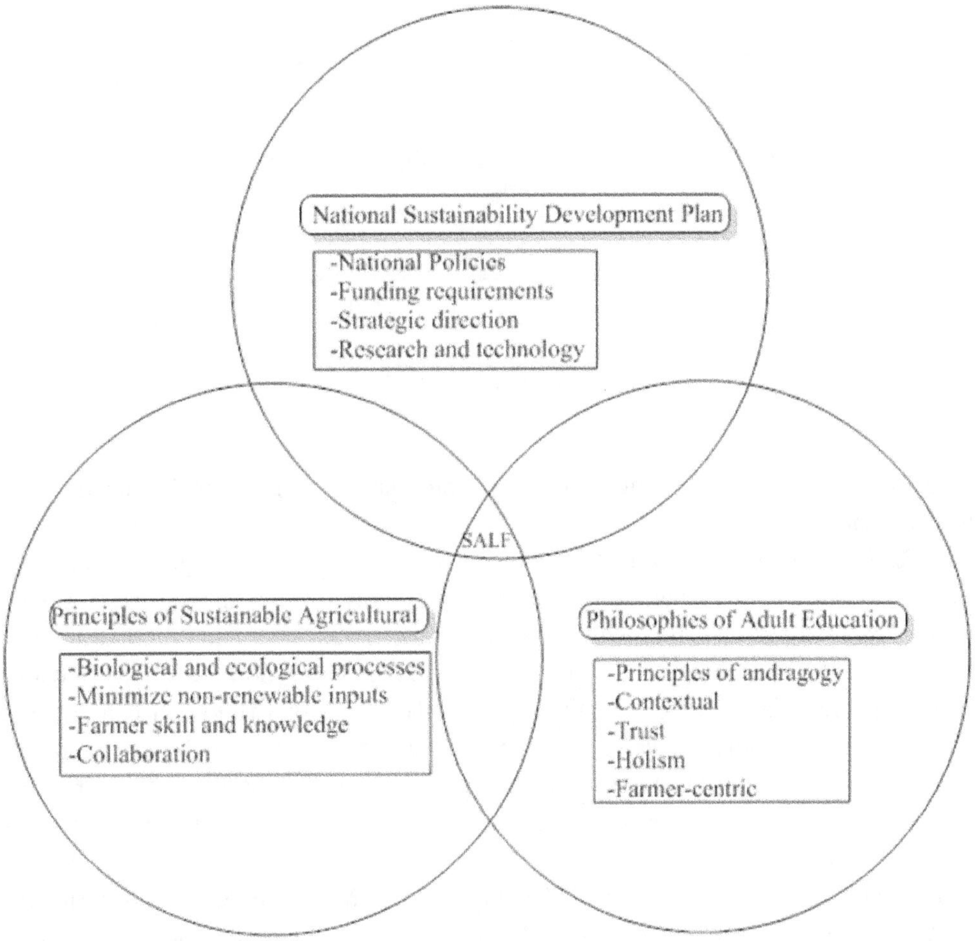

*Figure 8. Generation of manure from farming*

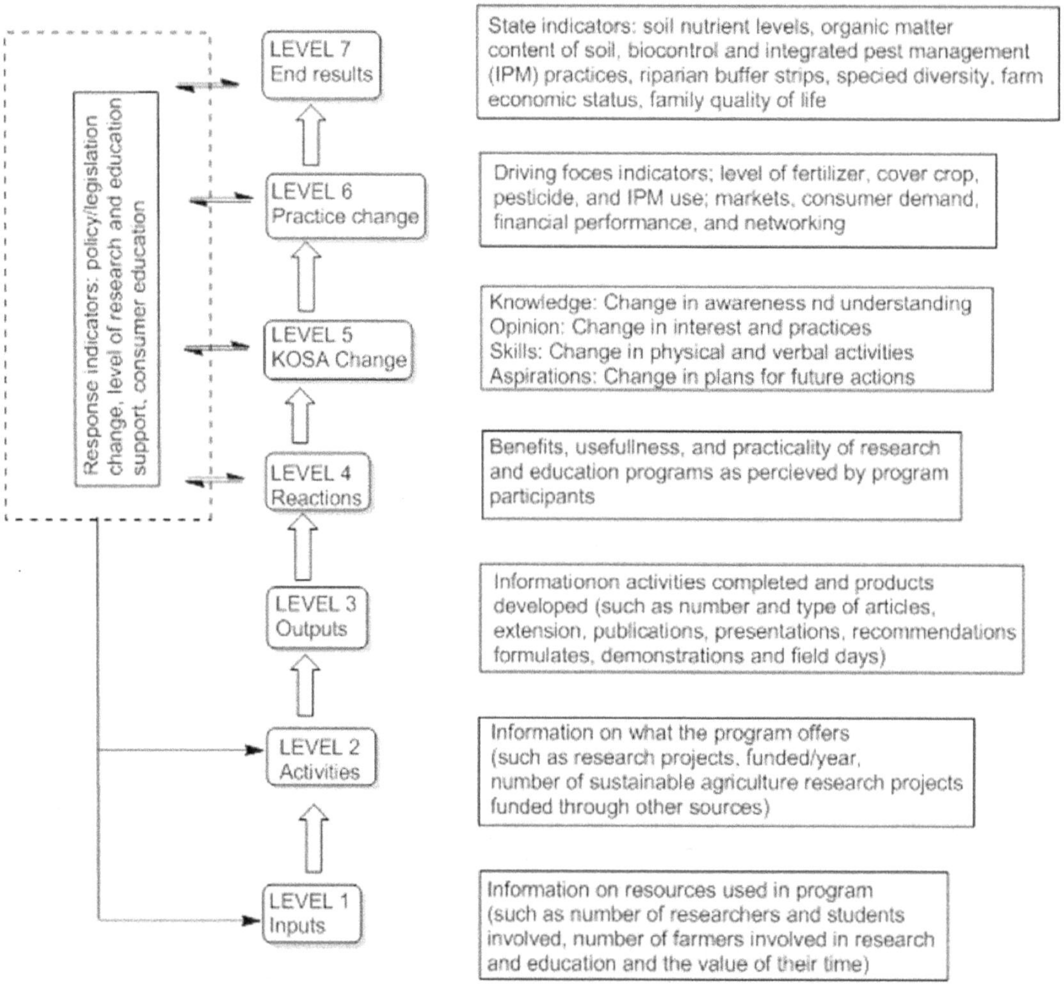


in increasing soil organic matter. Further, it promotes circular farming and makes farming sustainable **(Figure 8)** (El Janati et al., 2021).

**Nazrana Rafique Wani** et al. carried out a review of food waste management techniques. Currently, food waste is used in the generation of biofuels or biopolymers in the industries. Among various techniques, anaerobic digestion is highly performed as it generates energy and produces biogas. In this process, the complex organic food waste is broken into simpler sugar molecules, fatty and amino acids which are further fermented to generate $CO_2$, $H_2$ and $NH_3$. Microbial fluid cell is another technique adopted to extract energy from food waste. In this process, the electrochemically active microorganisms function as an electron acceptor in an electronic diode, assisting in the catalysis of the organic reaction. From the review, it was evident that proper food waste management reduces the GHG emission **(Figure 9)** (Wani et al., 2023).

**Esam Helal** et al. generated a Statistical Quality Control Model (SQCM) and corrective actions and decision model (CADM) to enhance agricultural drainage water reuse. The data was validated by collecting information from 2015 to 2019 in the Western Nile Delta region on pH, TSS, BOD, COD,

*Figure 9. Classification of different anaerobic methods for waste conversion*

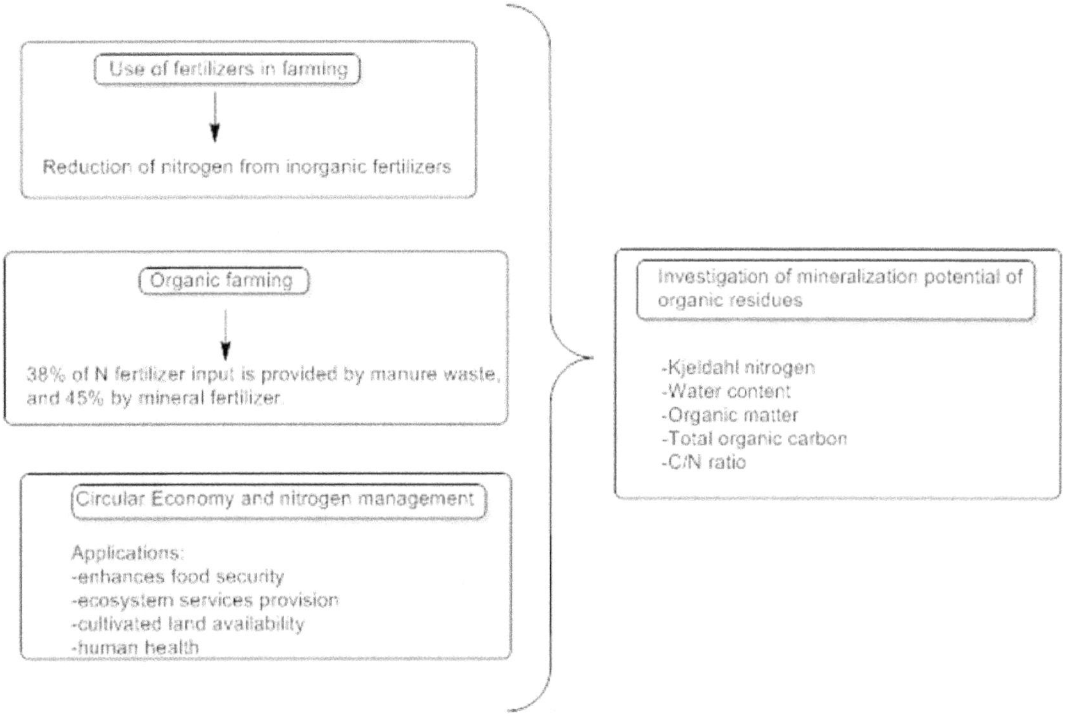


TN, Zn, etc. The findings showed that the agricultural drainage water reuse could produce safe drainage water that could be combined with fresh water after providing adequate sanitation. (Helal et al., 2021)

**Quimeng Zhang** et al. examined the sustainability of the greening process in sandy land for time scale of 20 years (2000 – 2020) using 3 factors viz. land use pattern, soil moisture content, and other factors influencing greening stability. For this analysis, Remote sensors and geo detectors were used to collect the land cover data, soil moisture, and vegetation quantity. The results revealed that the greening process and ecological state were steadily improving in a systematic way. It was found that precipitation played a vital role in vegetation **(Figure 10)** (Zhang et al., 2024).

**Patrizia Lombardi** et al. performed a critical review on agricultural waste management using Multi-criteria decision analysis (MCDA). In circular agricultural practice, the two key terms were water reuse and manure treatment. Among various techniques anaerobic digestion, composting, and thermal treatment of agricultural waste were widely used. It was discovered that biogas, fertilizers, and energy were the results of these processes. (Lombardi & Todella, 2023)

**B. Roman** et al. employed Eco-machines, an environmentally friendly wastewater treatment system that uses less energy and chemicals to treat wastewater. Eco-machine uses multiple tanks with microorganisms, plants, and macroinvertebrates to treat wastewater. In this process, apart from treating wastewater, duckweed is also grown. The treated water is used for the irrigation process and duckweed is used for livestock feed. The impact of microbial activity was not considered in this process. However, it was found to have 4 – 5 times less environmental impact than conventional wastewater treatment **(Figure 11)** (Roman & Brennan, 2021).

*Figure 10. The roadmap for sustainable approach by Qiumeng Zhang et al.*

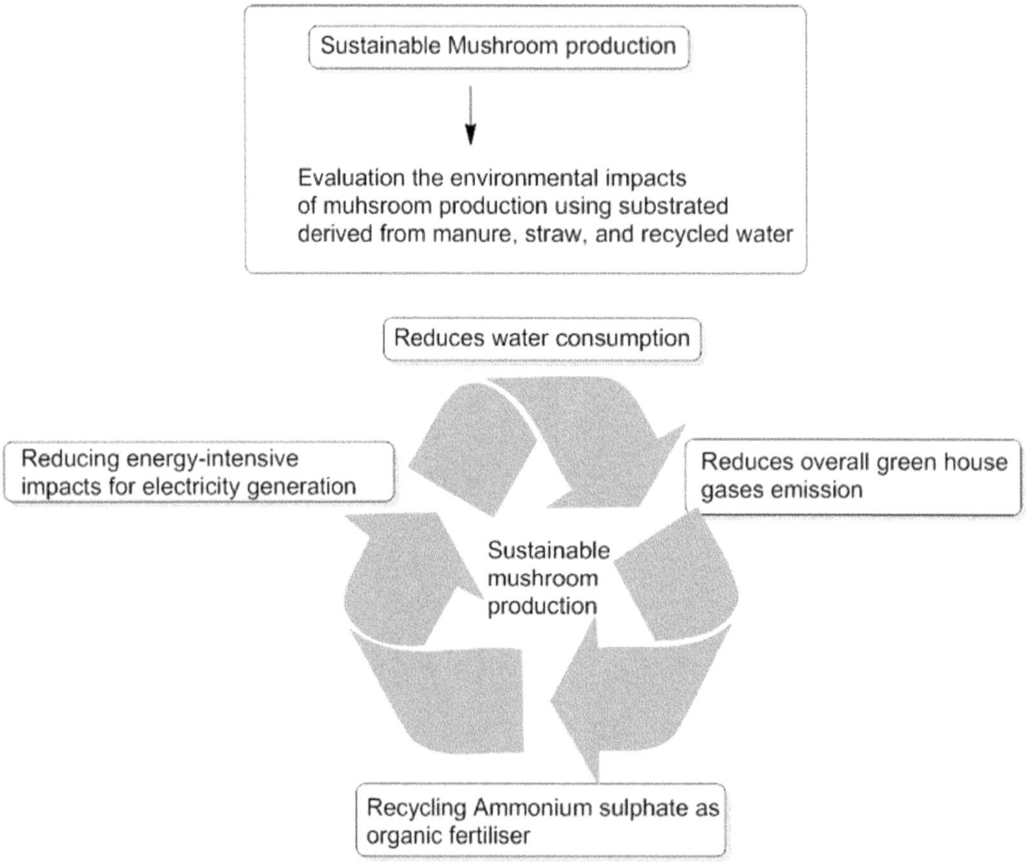


*Figure 11. The sustainable approach for waste water utilization*

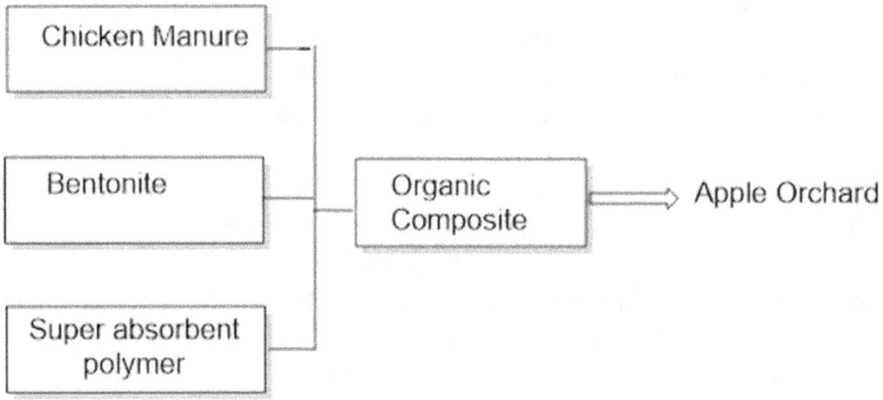


**Bella Tsachidou** et al. applied powdered biogas residue to the farming land to enhance fertility, lessen the use of chemical fertilizers, and mitigate climate change through carbon sequestration. A detailed investigation was performed to understand the stability, decomposition, and fate of biogas residue as biofertilizers. Apart from this, the carbon release from the biofertilizer and also along with the addition of chemical fertilizer is studied. The incubation period for the study was 270 days. It was found that biogas residue withstood 62% of initial TS and 42% of OM indicating a slow release of nutrients that improved soil fertility. The structure composition, such as the formation of lignin, was used to identify C sequestration. From the study, it was found that after incubation, the lignin fraction increased by 45% indicating that the biogas residue has good carbon sequestration capacity. The author proposed that applying biofertilizers improves sustainable farming practices and helps to slow down global warming. (Tsachidou et al., 2021)

**Chhatarpal Singh** et al. conducted a field experiment for studying the impact of rice husk biochar on the physico-chemical properties of soil and how they affect the species of methanotrops and their functional pmoA genes in the soil. The results showed that the pmoA genes when present in abundant amount marks a vast variation in the physico-chemical properties of paddy soil. The rice-husk biochar predominantly enhances the physical and chemical properties of soil by enhancing the methanotrops spcies and thereby increasing pmoA genes in the paddy soils. The bacterias of methanotrops majorly included *Methylovulum* and *Methylobcacter* from the *Methylococcaceae* family **(Figure 12)** (Singh et al., 2021).

**Tiefigue Pierrette** et al. reported some of the ideas for farmers to adopt and implement in agricultural practices which will prove to be sustainable agricultural practices. The study shows that farmers agreement to the sustainable agricultural practices are mainly affected by many factors which are depicted in **Figure 12.** This article is centred on providing approaches for sustainable agricultural practices which can be easily adopted by farmers which includes economical, psychological, agricultural expansion,

*Figure 12. Overview of rice husk biochar for sustainable use*

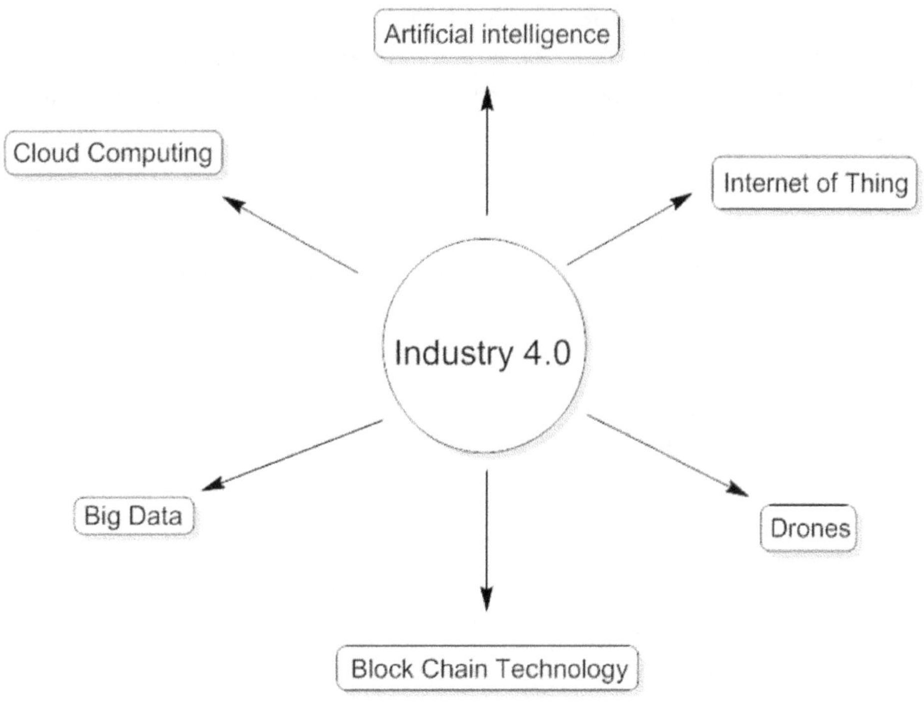

anthropology and many more. This article also highlights the concerns and problems of farmers in adopting such practices and gap in the literature. (Coulibaly et al., 2021)

**Georgia Bartzas** and **Kostas Komnitsas** performed a multi-dimensional analysis for evaluating sustainability of agricultural production which can be done at regional level. Evaluating the sustainability of agricultural production is a complex task which needs measuring and analysing various economic, social, and environmental factors at all levels. To address this challenge, this article was published which proposes a comprehensive method combining Life Cycle Analysis (LCA), Environmental Risk Assessment (ERA), on-site farm surveys, and regional assessments using the Analytical Hierarchy Process (AHA) aiming to identify the most suitable agricultural practices at the regional level **(Figure 14)**. (Bartzas & Komnitsas, 2020)

**Chris Radcliffe** reported a sustainable approach for agricultural practices and indigenous farmers. Developing countries hold a hidden and powerful weapon in the fight against hunger, that is time-tested practices for indigenous farmers. These farmers have adopted to sustainable methods since generations, but traditional agricultural extension programs haven't been effectively tapping into this knowledge. This article proposes a new sustainable approach involving extension officers as mere instructors to facilitators. This shift also includes acknowledging and appreciating indigenous knowledge and culture, while integrating the principle of adult learning (andragogy) into the extension process. The suggested approach known as the Sustainable Agriculture Learning Framework, focuses to support extension projects empowering indigenous farmers with sustainable practices, ultimately leading to improves food security **(Figure 15)**.

**Murari Suvedi** et al. proposed a conceptual framework for sustainable agriculture. They proposed many models named, CIPP Evaluation Model (Aziz et al., 2018), Development Agency Monitoring and Evaluation Model, Economic Evaluation Model (Thirtle, 1995), Logical Framework Model (Thirtle, 1995), Research Project Management Model (Thirtle, 1995), Program Evaluation Model (Bennett, n.d.), and the Driving Force -State-Response Model (Moldan et al., 1997). Out of all these models, Program Evaluation Model, and the Driving Force -State-Response Model were found to be most suited for assessing research and education programs for agricultural practices. The authors worked on combining these two models to produce a hybrid conceptual idea for the sustainable agriculture **(Figure 16)**. (Suvedi et al., 2003)

*Figure 13. Factors affecting the adoption of sustainable agricultural activities*

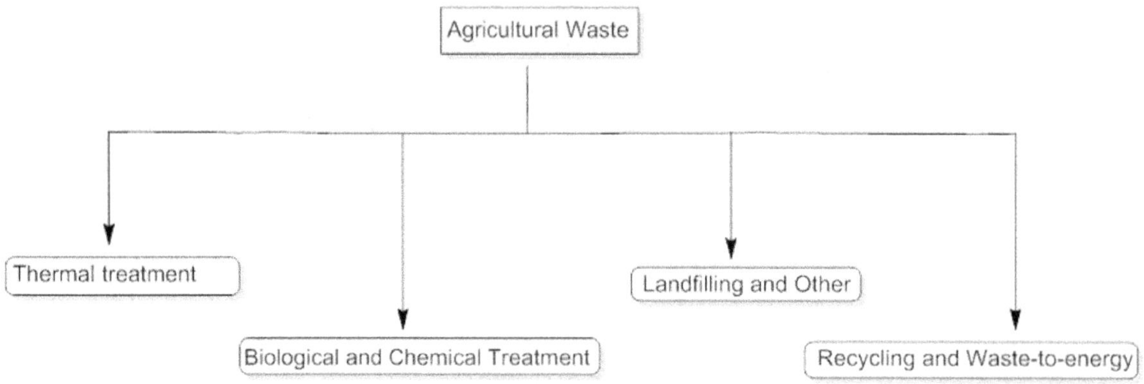

*Figure 14. Overview of two-staged analytical hierarchy process*

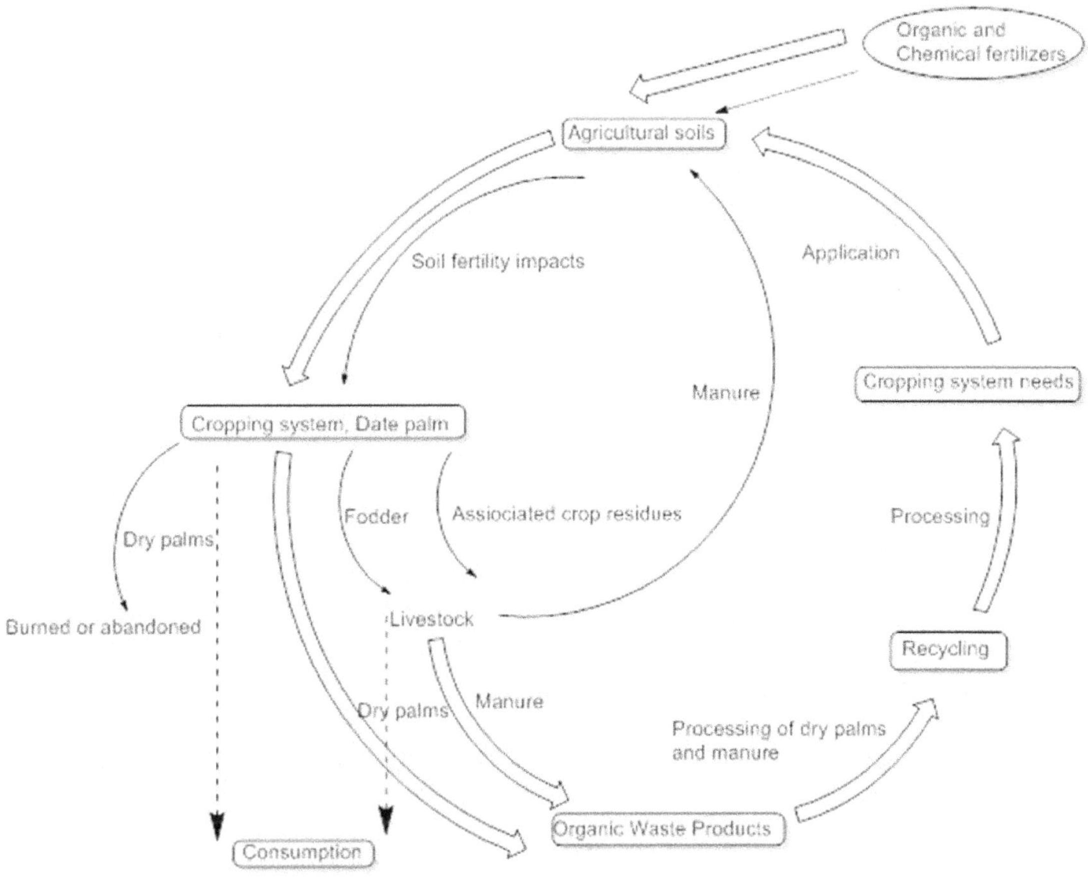


## CONCLUSION AND FUTURE PERSPECTIVES

Waste accumulation has increased due to quick urbanization, industrialization, and population expansion. We must improve waste collection, transportation, recycling, and disposal practices to reduce environmental impact. Numerous methods are used in waste management, such as recycling, composting, pyrolysis, anaerobic digestion, landfilling, incineration, and plasma gasification. This chapter describes the various management techniques used in the agricultural field. Resource conservation, environmental protection, and sustainable development depend on efficient agricultural waste management. We can reduce waste production, increase resource efficiency, and build a more resilient and sustainable agriculture sector by implementing integrated waste management techniques and encouraging creative solutions. Practical cooperation among farmers, policymakers, researchers, and industry partners is crucial in promoting a circular economy in agriculture and effecting positive change.

*Figure 15. Objective of sustainable agriculture learning framework (SALF)*

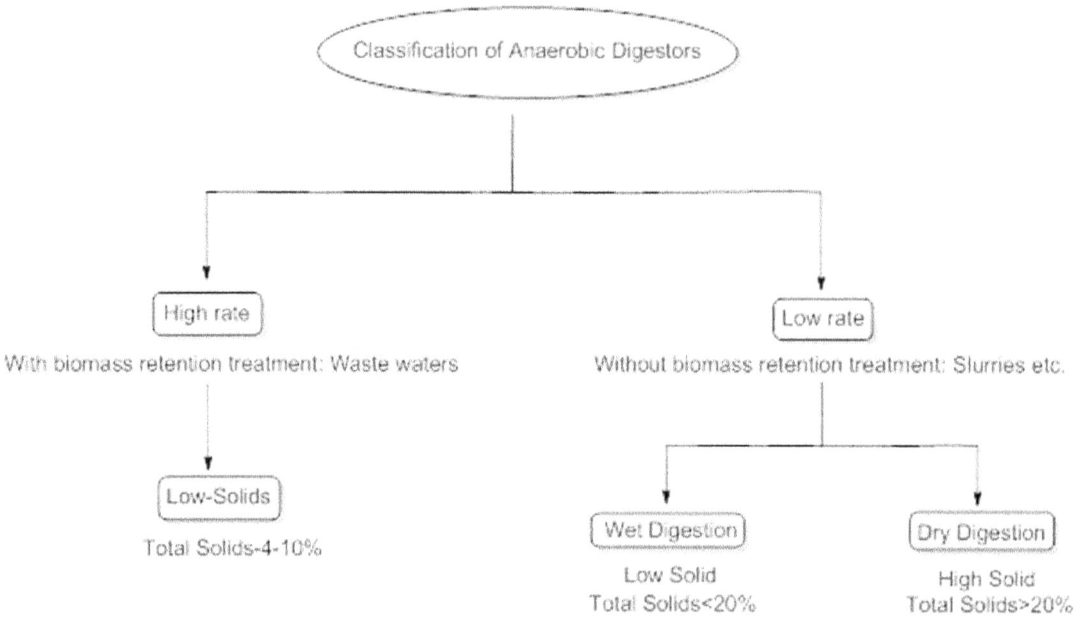


*Figure 16. Reported approach for sustainable agriculture*

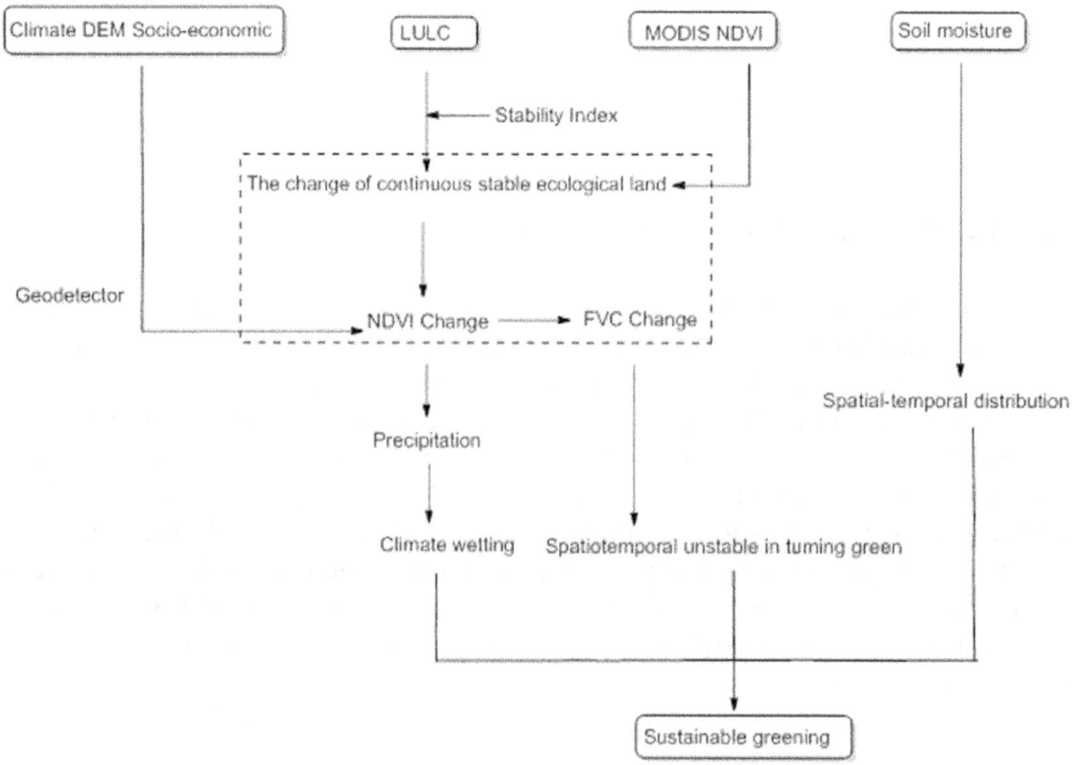

## CONSENT FOR PUBLICATION

The authors have decided to publish the work as one of the book chapter in the book entitled "Exploring Waste Management in Sustainable Development Context".

## CONFLICT OF INTEREST

The authors declared that there is no conflict of interest, financial or else.

## ACKNOWLEDGEMENT

Aditi Tiwari conveys her cordial gratefulness to the management of Intertek India Pvt. Ltd., Gurugram. Haryana, India. The authors B. Anjaneyulu and S V Niveditha truly thank the management of Presidency University, Bangalore, Karnataka India for providing the conveniences to write and submit this chapter for publication.

## REFERENCES

Aziz, S., Mahmood, M., & Rehman, Z. (2018). Implementation of CIPP Model for Quality Evaluation at School Level: A Case Study. *Journal of Education and Educational Development*, *5*(1), 189. doi:10.22555/joeed.v5i1.1553

Bartzas, G., & Komnitsas, K. (2020). An integrated multi-criteria analysis for assessing sustainability of agricultural production at regional level. *Information Processing in Agriculture*, *7*(2), 223–232. doi:10.1016/j.inpa.2019.09.005

Bennett, C. F. (n.d.). *Analyzing impacts of extension programs.*

Coulibaly, T. P., Du, J., & Diakité, D. (2021). Sustainable agricultural practices adoption. *Agriculture (Pol'nohospodárstvo)*, *67*(4), 166–176. doi:10.2478/agri-2021-0015

El Janati, M., Akkal-Corfini, N., Bouaziz, A., Oukarroum, A., Robin, P., Sabri, A., Chikhaoui, M., & Thomas, Z. (2021). Benefits of Circular Agriculture for Cropping Systems and Soil Fertility in Oases. *Sustainability (Basel)*, *13*(9), 4713. doi:10.3390/su13094713

FAO (Ed.). (2017). *Towards zero hunger 1945-2030.* Food and Agriculture Organization of the United Nations.

Food and agriculture: Driving action across the 2030 Agenda for Sustainable Development. (n.d.). *FOOD AND AGRICULTURE.*

Helal, E., Shaban, M., Abd El-Karim, E., & Ellayen, A. (2021). A comprehensive management model to maximize the benefits of agriculture drainage water reuse. *Water Environment Research*, *93*(9), 1722–1733. doi:10.1002/wer.1559 PMID:33725400

Lombardi, P., & Todella, E. (2023). Multi-Criteria Decision Analysis to Evaluate Sustainability and Circularity in Agricultural Waste Management. *Sustainability (Basel), 15*(20), 14878. doi:10.3390/su152014878

Moldan, B., Billharz, S., & Matravers, R. (Eds.). (1997). *Sustainability indicators: A report on the project on indicators of sustainable development.* Wiley.

Obi, F., Ugwuishiwu, B., & Nwakaire, J. (2016). AGRICULTURAL WASTE CONCEPT, GENERATION, UTILIZATION AND MANAGEMENT. *Nigerian Journal of Technology, 35*(4), 957. doi:10.4314/njt.v35i4.34

Rivera-Ferre, M., Ortega-Cerdà, M., & Baumgärtner, J. (2013). Rethinking Study and Management of Agricultural Systems for Policy Design. *Sustainability (Basel), 5*(9), 3858–3875. doi:10.3390/su5093858

Rodríguez-Espinosa, T., Papamichael, I., Voukkali, I., Gimeno, A. P., Candel, M. B. A., Navarro-Pedreño, J., Zorpas, A. A., & Lucas, I. G. (2023). Nitrogen management in farming systems under the use of agricultural wastes and circular economy. *The Science of the Total Environment, 876*, 162666. doi:10.1016/j.scitotenv.2023.162666 PMID:36894085

Roman, B., & Brennan, R. A. (2021). Coupling ecological wastewater treatment with the production of livestock feed and irrigation water provides net benefits to human health and the environment: A life cycle assessment. *Journal of Environmental Management, 288*, 112361. doi:10.1016/j.jenvman.2021.112361 PMID:33756385

Singh, C., Tiwari, S., Rai, P. K., & Singh, J. S. (2021). Sustainable management of paddy crop residues: Effects on methanotrophs diversity and value for soil health restoration. *Land Degradation & Development, 32*(14), 4121–4131. doi:10.1002/ldr.3982

Sumiyati, S., Samadikun, B. P., Widiyanti, A., Budihardjo, M. A., Al Qadar, S., & Puspita, A. S. (2024). Life cycle assessment of agricultural waste recycling for sustainable environmental impact. *Global Journal of Environmental Science and Management.* doi:10.22034/gjesm.2024.02.30

Suvedi, M., Den Biggelaar, C., & Morford, S. (2003). Conceptual Framework for Evaluating Sustainable Agriculture. *Journal of Crop Production, 9*(1–2), 433–454. doi:10.1300/J144v09n01_05

Thirtle, C. (1995). Monitoring and evaluating agricultural research: A sourcebook. D. Horton, P. Ballantyne, W. Peterson, B. Uribe, D. Gapasin and K. Sheridan. CAB International in association with the International Service for National Agricultural Research, Wallingford, Oxon OX10 8DE, 1993, 219 pp. *Public Administration and Development, 15*(2), 185–185. doi:10.1002/pad.4230150210

Tóth, F. A., Magyar, T., Tamás, J., & Nagy, P. T. (2024). Improving the Nutrient Management of an Apple Orchard by Using Organic-Based Composites Derived from Agricultural Waste. *Horticulturae, 10*(2), 172. doi:10.3390/horticulturae10020172

Tsachidou, B., Hissler, C., Noo, A., Lemaigre, S., Daigneux, B., Gennen, J., Pacaud, S., George, I. F., & Delfosse, P. (2021). Biogas residues in the battle for terrestrial carbon sequestration: A comparative decomposition study in the grassland soils of the Greater Region. *Journal of Environmental Management, 286*, 112272. doi:10.1016/j.jenvman.2021.112272 PMID:33677337

Vinci, G., Prencipe, S. A., Pucinischi, L., Perrotta, F., & Ruggeri, M. (2023). Sustainability assessment of waste and wastewater recovery for edible mushroom production through an integrated nexus. A case study in Lazio. *The Science of the Total Environment*, *903*, 166044. doi:10.1016/j.scitotenv.2023.166044 PMID:37572921

Wang, Z., Miao, R., Ning, P., He, L., & Guan, Q. (2021). From wastes to functions: A paper mill sludge-based calcium-containing porous biochar adsorbent for phosphorus removal. *Journal of Colloid and Interface Science*, *593*, 434–446. doi:10.1016/j.jcis.2021.02.118 PMID:33765625

Wani, N. R., Rather, R. A., Farooq, A., Padder, S. A., Baba, T. R., Sharma, S., Mubarak, N. M., Khan, A. H., Singh, P., & Ara, S. (2023). New insights in food security and environmental sustainability through waste food management. *Environmental Science and Pollution Research International*, *31*(12), 17835–17857. doi:10.1007/s11356-023-26462-y PMID:36988800

Wanyama, J., Bwambale, E., Kiraga, S., Katimbo, A., Nakawuka, P., Kabenge, I., & Oluk, I. (2024). A systematic review of fourth industrial revolution technologies in smart irrigation: Constraints, opportunities, and future prospects for sub-Saharan Africa. *Smart Agricultural Technology*, *7*, 100412. doi:10.1016/j.atech.2024.100412

Zhang, Q., Jia, B., Li, T., & Li, W. (2024). Research on the sustainability of “greening” process in the Mu Us Sandy Land based on the spatiotemporal stability of ecological land. *PLoS One*, *19*(2), e0292469. doi:10.1371/journal.pone.0292469 PMID:38319930

# Chapter 12
# Agricultural Waste Management Through Crop Residue Management:
## Challenges, Solutions, and Technological Advancements

**Chethan Patil N. D.**
https://orcid.org/0009-0001-2977-6422
*Lovely Professional University, India*

**Suryakanta Kashyap**
*National Dairy Research Institute, India*

**Sapna Jarial**
https://orcid.org/0000-0003-2660-9036
*Lovely Professional University, India*

## ABSTRACT

*Crop residue management (CRB) is a global environmental, economic, and health concern. CRB, a common practice in developing countries, is a long-term agricultural waste management strategy. It harms air quality, soil health, and human well-being. It persists due to labor shortages, economic constraints, and limited access to alternative technologies despite its widespread negative consequences. CRB management includes surface retention, mulching, and farm mechanization are examples of on-field management methods; off-field methods include baling, livestock feed supplementation, composting, mushroom cultivation, biochar production, biogas generation, and industrial applications. Modern machinery, such as the Happy Seeder, enables zero-tillage farming and crop residue management. Long-term solutions are required to address this issue. An integrated approach is required to agricultural waste management, combining technological innovation with policy support and community engagement, to reduce the adverse effects of crop residue burning while promoting sustainable agricultural practices.*

DOI: 10.4018/979-8-3693-4264-0.ch012

## 1. INTRODUCTION

Agricultural wastes are the leftover materials produced during the cultivation and processing of agricultural items such as fruits, vegetables, meat, dairy products, and crops. These residues are wastes of agricultural production and processing, containing chemicals that may be helpful to people but are not economically valuable enough to collect transport, and process for use. The nature of agricultural waste varies according to the agricultural activity and can be liquids, slurries, or solids. Agro-waste, also known as animal waste (such as manure and animal carcasses), food processing waste (with only 20 percent of maize being canned and the remaining 80 percent becoming waste), crop waste (such as corn stalks, sugarcane bagasse, and discarded fruits, vegetables, and pruning), and hazardous and toxic agricultural waste (including pesticides, insecticides, and herbicides). Although precise statistics are difficult, agricultural waste is widely acknowledged to account for a sizable proportion of overall waste generated in developed countries. As agricultural production grows, so does the amount of livestock waste, agricultural waste, and agro-industry byproducts. As developing countries continue to intensify their farming practices, there is a significant increase in global agricultural waste. It is estimated that roughly 998 million tonnes of agricultural waste are produced annually worldwide (Agamuthu et al., 2009).

Rising food consumption in developing countries has led to tremendous growth in global food production. Rice-wheat farming systems cover 13 million hectares in South Asia (India, Pakistan, Nepal, Bangladesh, and Bhutan). Rice-wheat (RW) farming systems produce around 20 per cent of overall cereal production and 40 per cent of wheat production in India (Chaudhary et al., 2024). The Indo-Gangetic Plains (IGPs) include more than 85 per cent of South Asia's RW system. India's IGPs account for around 20 per cent of the country's total geographical area (329 Mha) and approximately 27 per cent of the country's net cultivated land, producing almost 50 per cent of total food consumption (Das et al., 2024).

The continuous rice-wheat cropping system, intensive tillage for both crops, crop residue (CR) removal, and use of excessive water and chemicals have resulted in environmental and ecological imbalance. The CR is made up of plant materials that remain after crops have been harvested and threshed. These residues were once thought to be waste, but with more knowledge and research, it is becoming clear that they are not waste, but rather an important natural resource. CR recycling could be used to convert surplus farm waste into useful materials. These items can help with nutrient replenishment, soil fertility enhancement, and ecological balance, all of which improve crop yield. Many farmers use wheat straw as animal feed, but rice straw remains a problem due to its high silica concentration (Padmakumar et al., 2024).

According to Ministry of New and Renewable Energy (MNRE), India generates an average of 500 million tonnes (Mt) of crop residue each year. According to the same study, the majority of crop residue is used for feed, fuel, and other domestic and industrial applications. Table 1 summarises the general

*Table 1. Crop residue generated from various crop*

| Source | Composition |
|---|---|
| Rice | Husk, bran |
| Wheat | Bran, straw |
| Maize | Stover, husk, skins |
| Millet | Stover |
| Sugarcane | Bagasse, molasses, sugarcane tops |

categories of crop residues produced by the major cereal crops. However, there is still a 140 MT surplus, of which 92 MT are burned annually. It is also worth noting that the amount of agricultural waste burned in India is significantly larger in volume than the total amount of agricultural waste produced in other countries in the area. Paddy straw in-situ burning is a common management method in north India, while rice straw is used for composting and other purposes in the rest of the country. One of the most serious side effects of residue burning is air pollution. In 2000, India's $CH_4$, CO, $N_2O$, and NO emissions from paddy and wheat straw burning were 110, 2306, 2, and 84 Gg, respectively. The $P_2O_5$, N, $K_2O$, and S content in paddy straw are 2.3, 5.5, 25, and 1.2 kg/tonne, respectively. Besides, straw also contains 50–70 percent of the micronutrients absorbed by rice, as well as 400 kg of carbon (Ghodke et al., 2024). In addition to nutrient loss, residue burning in the field affects a variety of soil parameters, including pH, soil temperature, soil moisture, soil organic matter content, and available phosphorus. Farmers use rice straw as a quick and inexpensive way to clear their fields for wheat planting. Burning removes 90 percent of the nitrogen, 60 percent of the sulphur, 20-25 percent of the P and K, and all organic matter (all C) from the straw (Bhuvaneshwari et al., 2019). Removing all the straw from the field results in a negative K balance because the straw retains 80-85 percent of the K absorbed by the rice and wheat crops. Every year in Punjab, approximately 16 Mt of rice stubble is expected to be burned in a matter of weeks. The resulting air pollution poses serious health risks to both humans and animals (Shyamsundar et al., 2019). India is the world's second-largest producer of rice and wheat, both of which generate significant waste. Agricultural waste can be useful in a variety of agricultural and industrial applications. However, the costs of collection, processing, and transportation can be significantly higher than the revenue generated by the waste's beneficial use. Crop residues are an important component of agricultural waste that, due to their organic nature, can be used to benefit society. Large amounts of crop residue combined with unsustainable management practices have serious environmental consequences that go far beyond India. In this chapter, we will discuss the major challenges associated with crop residue burning, as well as various management strategies, alternate applications, and mechanized options for reducing CR burning.

## 2. GENERAL PRACTICE IN INDIA

India is a farming country with a variety of farming practices that are based on agro-climatic zones. Vast farming practices include rice and wheat cropping patterns in Haryana, Punjab, Rajasthan, Uttar Pradesh, Bihar & Madhya Pradesh. After the harvesting season, these areas are also known for burning straw and stubble. Even though the government began to ban the practice in the 1990s, farmers in Punjab and Haryana, in particular, burn an estimated 35 million tonnes of crop residue from their paddy fields each year in late September and October. This strategy is a low-cost way to get rid of straw while shortening the period between harvesting and growing the second (winter) crop.

## 3. EFFECTS OF STUBBLE BURNING

### 3.1. Environmental Effects

Burning crop residues has several negative environmental consequences. Crop residue burning has the most negative consequences because it emits greenhouse gases (GHGs), which contribute to global climate change. It raises levels of Particulate Matter (PM) and other air pollutants, causing health issues, a loss of agricultural land diversity, and a decline in soil fertility. Burning crop stubble in an open field has an impact on soil fertility, reducing the total nutrients in the soil.

## 3.2. Air Pollution

Crop residue burning produces GHG emissions, CO, $NH_3$, $NO_x$, $SO_x$, non-methane organic compounds (NMHCs), volatile organic compounds (VOCs), semi-volatile organic compounds (SVOCs), and PM, all of which result in the loss of organic carbon, nitrogen, and other nutrients that would otherwise be preserved in soil. Every year, about 22 Mt of $CO_2$, 0.92 Mt of CO, and 0.03 Mt of $SO_2$ are produced in Punjab from approximately 15 Mt of rice residues. According to one study, GHG emissions account for 91.6 percent of total air emissions produced by the combustion of 98.4 Mt of agricultural residue, with CO, $NO_x$, NMHCs, and SVOCs accounting for the remaining 8.4 percent. When stubs are burned, aerosols are released; open burning of rice straw in India, Thailand, and the Philippines emits high levels of $SO_2$, $NO_x$, $CO_2$, CO, and $CH_4$.

Adverse effect of burning is not limited to human only. Burning crop waste also endangers the lives of milk-producing animals. Animals can die from air pollution because high $CO_2$ and CO levels in the blood affect normal haemoglobin, resulting in deaths. More than 60,000 livestock in rice-growing areas are at risk of air pollution from rice stubble burning. According to household surveys, paddy husk burning causes various other problems, including higher health costs in polluted areas. Open field burning impacts the longevity of animals, birds, and insects. Burning occasionally reduces visibility and heightens the severity of traffic accidents. The answer to whether households are aware of the adverse effects of rubbish burning is 'yes' in roughly 90 per cent of cases; however, almost no household would take precautions to combat pollution-related diseases.

## 3.3. Effect on Soil Fertility

According to the Department of Agriculture, Punjab, soil has low nitrogen, low to medium phosphorus, and moderate to high potassium levels. Furthermore, soil organic carbon levels have fallen to an alarmingly low level. To get the production of about 7 t/ha of rice and 4 t/ha of wheat, more than 300 kg of nitrogen, 30 kg of phosphorus, and 300 kg of potassium per hectare are exhausted from soil. According to the Department of Soil Sciences, Punjab Agricultural University, burning of residue leads to loss of soil organic carbon. Furthermore, the $CO_2$ and soil nitrogen balances change quickly, converting nitrogen to nitrate. It may cause the soil to lose 0.824 million tonnes of nitrogen, phosphorus, and potassium (NPK) each year. Burning agricultural residues raises soil temperatures and depletes microorganism and flora populations. Additionally, repeated burns can reduce the bacterial population by more than 50percent. Long-term burning also leads to loss of soil from depth of 0-15cm, total nitrogen, biomass, mineralized nitrogen, and organics. The residue burning will raise dirt temperatures to 35.8-42.2 °C at a depth of 10 mm, with semi-permanent effects extending up to 15 cm of the soil depth. Furthermore, frequent burning reduces the nitrogen and carbon content of the soil, kills beneficial microflora and fauna, and depletes organic matter. Crop residue burning can disrupt the soil's carbon-nitrogen equilibrium (Lohan et al., 2018; Verma et al., 2024; Kumar and Joshi, 2018). According to the National Policy for Crop Residues (NPMCR), open incineration of one tonne of stubble would result in the loss of all organic carbon, 5.5 kg of nitrogen, 2.3 kg of phosphorous, 25 kg of potassium, and 1.2 kg of sulphur in the soil. Crop residue that incorporates into the soil enriches it with C, N, P, and K. Burning rice and wheat residues results in a loss of approximately 80percent of nitrogen, 25percent of phosphorus, 21percent of potassium, and 4-60percent of soil sulphur, though it does destroy unwanted bugs and diseases carried by the soil (Singh et al., 2024). Crop residue burning contributes to the depletion of essential crop nutrients. Crop residues retain approximately 25percent of nitrogen and phosphorus, making 50percent of sulphur and 75percent of cereal potassium intake viable nutrient sources (Surender et al., 2023; Mor et al., 2022).

## 4. REASON BEHIND BURNING OF STUBBLE

North Indian farmers burn approximately half of their paddy straw in their fields. One of the most important reasons is that farmers want to save labour costs and time for succeeding wheat crops. There is a short time between paddy harvesting and wheat sowing. Farmers prefer burning because waiting for stubble to decompose in the soil takes longer. Because of its high silica content, rice straw is considered useless as fodder for non-basmati rice crops (Mor et al., 2022). Farmers believe that burning will kill all harmful insects and weeds, so using insecticides will reduce crop expenses. Some farmers also realised that burning stubble in a field reduced tractor fuel costs. The ash produced by stubble burning is a potassium source while reducing soil acidity. Another reason for stubble burning is that lower-income farmers cannot afford expensive implements such as *rotavators, happy seeders, and super seeders*, which help manage stubbles in situ. These tools also require tractors with more horsepower to operate. Previously, massive labour in Punjab and Haryana assisted farmers in harvesting manually, leaving no stubble on the field. However, to save time, farmers are now using combine harvesters, which leaves stubble on the ground, and due to a labour shortage, burning is preferred as a quick solution.

*Table 2. Reason behind stubble burning in various states in India*

| Constraints | States of India |
|---|---|
| Short time between harvesting kharif rice and sowing rabi crops. | Punjab, Haryana, Western Uttar Pradesh |
| Low labour supply and high labour expenses | Punjab, Haryana, Western Uttar Pradesh, West Bengal, Tamil Nadu, Andhra Pradesh |
| Difficulty with land levelling after residue incorporation | All states |
| Small-scale land holdings to implement sustainable technologies | West Bengal, Tamil Nadu, Odisha |
| Rice straw management practises that are comparatively expensive in contrast to burning | All states |
| Insufficient provision of custom hiring services in a timely manner for straw management equipment (e.g., Happy-sower, Baeler rotavator, Zero-thrust seed drill chopper). | West Bengal, Tamil Nadu, Andhra Pradesh. |
| Farmers' disregard for the additional labour that is required to compost straw. | Punjab, Haryana, Western Uttar Pradesh |
| Farmer ignorance regarding in-situ straw management from a technical standpoint | All states |
| Rice straw has limited economic value for farmers under the current circumstances. | All states |
| Demand reduction for rice straw as a cattle feed | All states |
| Extreme decline in the demand for rice straw utilised in thatching | All states |
| Ex situ straw management lacks economically viable alternatives. | Punjab, Haryana, Western Uttar Pradesh |
| Logistics and storage constraints prevent the expeditious handling of the enormous quantity of straw. | All states |

## 5. CROP RESIDUE MANAGEMENT

Crop residue can be managed in two ways (i) on the field and (ii) off-field management methods.

### 5.1 On-Field Management

The key elements of CR are lignin, cellulose, hemicellulose, as well as micro and macronutrients. CR degradation is affected by their lignin, polyphenol, cellulose content, crop dependent C/N ratio, and atmospheric and soil conditions (texture, humidity). When cereal leftovers with a high C/N ratio (60:1 to 100:1) are placed into the soil, they degrade slowly, immobilizing soil nitrogen. This can be useful in zero tillage (ZT) systems, as it provides mulch that protects the soil from erosion and evaporation, but it also means that the following crop will have fewer nutrients available. The C/N ratio of the organic matter being degraded by soil microorganisms is affected by whether N is mineralized or immobilised. Inadequate N in the substrate permits the species to draw on mineral N in the soil, resulting in N immobilisation. Net immobilisation occurs when the C/N ratio of the residue is more than 20:1. The residue C/N ratio will drop as decomposition progresses due to decreasing C (respiration as $CO_2$) and increasing N (N immobilised from soil solution), and a new equilibrium will be attained, followed by N mineralization.

Farmers are increasingly employing in-situ crop residue application since it is natural. The net supply of nitrogen from crop residue to subsequent crops is determined by the time it takes to decompose before the next crop is planted, the quality of the residue, and the soil environment. Crops planted soon after grain residue is incorporated into such soils may become N-deficient, necessitating external N delivery to suit the needs of microorganisms and crops. Rice straw can be treated in-situ to minimise N deficit due to N immobilisation by allowing 10- 20 days between its integration and the seeding of the wheat crop. The net immobilisation process is transitory in the early years of zero tillage (ZT) adoption, but N immobilisation in conservation agriculture (CA) systems reduce the risk of leaching and denitrification losses of soil mineral N in the long run.

#### 5.1.1 Surface Retention and Mulching

The advantages of keeping crop residue on the soil surface include: i) reduced weed emergence, ii) cheaper weedicide costs, iii) improved physical, chemical, and biological properties of soils, iv) plant nutrient recycling, and v) reduced fertilizer use in subsequent crops. A practice where straw remnants from a previous crop are left on the soil surface without integration is known as direct drilling in surface mulched residues. Previously, seeding rice residues with ZT seed drills is required the removal of loose straw following mixed harvesting and partial or complete burning of the residues removed. With the recent ‘ Happy Seeder ‘ system, RW farmers in the Indo-Gangetic plains (IGP) can now drill wheat directly into the rice residue. The collection of leftovers on the surface helps protect the rich surface soil from wind and water erosion. Surface preservation of any or all residues may be the best option in some instances. Residues break down slowly on the surface, accumulating organic carbon and total nitrogen in the top 5-15 cm of the soil, delaying erosion and leaving leftovers on the surface improved soil $NO_3$- content by 46 percent, N absorption by 29 percent, and yield by 37 percent when compared to burning. On the other hand, retention provides a safe habitat for both dangerous and beneficial species and a carbon source for heterotrophic $N_2$-fixation, enhanced microbial activity, soil C and N, and lower rice N fertilizer requirements.

### 5.1.2 Farm Mechanization and Crop Residue Management

Despite being aware of the negative consequences of crop burning, farmers continue to use it due to a lack of economically viable and suitable machinery and options for disposing of leftovers. However, a possible option is using sophisticated technology to avoid burning, such as straw management systems, fitted combine harvesters, and happy seeders for direct drilling. Manufacturers have improved crop residue management technologies in recent years to enable finer chopping, broader and more equal spreading, and more excellent seed soil contact for improved crop emergence and yield. Farm machinery based on resource conservation technologies (RCTs) holds more promise for managing paddy leftovers to improve soil health and productivity, reduce pollution, and achieve sustainable agriculture. Advanced technologies of zero-till seed-cum-fertilizer drill/seed planters (happy seeder, spatial zero seed cum fertilizer drill) are available in the country for direct sowing of the consecutive crop in loose and anchored straw load up to 10 t ha-1. These technologies are beneficial for managing agricultural leftovers, weed management, and soil moisture and nutrient conservation. In NW India, the happy seeder method represents a breakthrough in paddy-wheat crop rotation. A Straw Management System (SMS) is mounted to the back side of the combine harvester just beneath the straw walkers and below the chaffer sieves for uniform distribution of paddy straw after paddy harvesting by a combine harvester. The straw from the straw walkers on the combine harvester is fed into the machine from one side and expelled through the housing's exit. The spinning discs disseminate the loose residues dropping from the harvester straw walker behind the harvester. The chopped material is blown off in a tangential direction and redirected by a deflector to evenly distribute the residues across the breadth of the combine harvester.

## 5.2. Off-Field Managements

### 5.2.1. Baling and Removing the Straw

Agricultural straw can be used for a variety of purposes, including livestock feed, fuel, construction materials, livestock bedding, mushroom composting, bedding for cucumber, melons, and other vegetables, and mulching for cucumber, melons, and other vegetable orchards and crops. Straw reapers are used to harvest rice straw from the field in some parts of North-Western India, and wheat straw collection is gaining popularity as a cheaper alternative to rice for animal feed. Straw baler machines are a promising commercial technology for removing and collecting straw after combine harvesting, as well as using the wastes for off-farm operations. However, depending on the height of the plant chopped by combines, these balers recover only 25-30percent of the possible straw yield after combining (Kumar, 2016). Balers make rectangular or round bales from loose straw. From a combined harvested field, the machine can recover 200–250 bales weighing 15–30 kg (depending on moisture and field condition (Meena et al., 2022)

### 5.2.2. Crop Residues as Livestock Feed

CR is traditionally used as animal feed in India (or supplemented with chemicals). Unfortunately, crop leftovers are unappealing and indigestible, so they cannot be fed to cattle alone. Rice residues are poor cattle feed due to their high silica content (4- 7 percent). Compared to other crop straws, it has more silica (12-16 percent) and less lignin (3-5 percent). (Goswami et al., 2020). Rice straw can be made more nutritious in many ways. Crop residues have been treated physically, chemically, and biologically to weaken lingo-cellulose connections and improve nutrition. Despite the cost and effort, 75 percent of wheat straw is chopped into small pieces for animal fodder using a special cutting machine. Rice straw stems have less silica than leaves but are more digestible, so cut the rice crop close to the ground to feed livestock. For animal nutrition, wastes must be processed and enriched with urea, molasses, and green fodder.

### 5.2.3. Crop Residues as Compost / Mechanized Composting

Crop wastes are utilized as animal bedding materials and then placed in dung pits to make compost. Each kilogram of straw absorbs roughly 2-3 kg of urine in the animal shed, enriching it with nitrogen (Birla et al., 2020). Crop straw compost can be reinforced with P utilizing a locally available low-grade rock phosphate source, resulting in a value-added compost with 1.5 percent nitrogen, 2.3 percent phosphorus, and 2.5 percent potassium. On the other hand, mechanized composting can significantly improve the bio-physical composting processes. The compost product can be used as a growing medium for vegetables and other crops or applied as a soil amendment on rice fields. It enhances the nutrient (*i.e.*, nitrogen and carbon contents) and organic matter content of the soil.

### 5.2.4. Production of Mushroom Crop

Mushroom cultivation is a profitable agribusiness that produces food from rice and wheat straw while also encouraging environmentally responsible waste management. *Volvariella volvacea*, or paddy straw mushroom, is considered one of the easiest mushrooms to cultivate due to its short 14-days incubation period. Paddy straw is an important component of mushroom cultivation in Punjab, whereas most farmers use wheat straw as a raw material. The estimated cost of these processes is Rs. 510 per quintal for paddy straw (raw material) and Rs. 810 per quintal for wheat straw. Rice straw can produce 5-10percent mushroom products (50-100 kg mushroom every tonne of dry rice straw) (Naresh et al., 2021).

### 5.2.5. Biochar Production and Utilization

Biochar is a carbon-rich soil amendment that increases soil fertility, carbon storage, and water filtration. It is produced through the thermal decomposition of organic materials or biomass at temperatures ranging from 500 to 700°C in the presence of a small amount of oxygen. Hydrothermal carbonization (HTC) is a novel improved carbonization method. HTC of lignocellulosic biomass is a process that completely breaks down the plant cell wall, allowing for the rapid conversion of biomass into a carbon-rich, lignin-like product (hydrochar). Hydrochar has a much higher heating value than basic material. Biochar can be made from rice straw and has a lot of potential (Danesh et al., 2023).

Using biochar as a soil amendment has a lower carbon footprint than using it for fuel. Furthermore, carbon sequestration using biochar reduces the risk of climate change caused by GHG emissions in the atmosphere. Despite its enormous potential, biochar production requires energy for carbonization and transportation of rice straw and biochar products. Additional research is needed to demonstrate that biochar production from rice straw is viable in terms of energy balance and economic benefits.

### 5.2.6. Use of Rice Straw for Biogas Production

Paddy residue biomass is an efficient source of energy. Anaerobic digestion, gasification, and pyrolysis processes, providing an immediate reduction in $CO_2$ levels in the environment. About 300 $m^3$ of biogas may be produced by anaerobic digestion of one tonne of rice residue. The process produces a high-quality gas with a 55–60 percent methane content, and the leftover slurry can be used as manure. One tonne of paddy biomass may provide 300 kWh of electrical energy through gasification (Kumar et al., 2015). It ensures a non-destructive approach to extract high-quality fuel gas from agricultural waste and harvest manure to be recycled in soil.

## 5.3. Industrial Uses of Rice Straw

The vast opportunities lie in the industrial use of rice straw, which is largely unexplored around the world and in India. The main industrial applications for rice straw include:

## 5.4. Other Methods

### 5.4.1 Crop Residues as Surface Mulch in Other Crops

Rice straw mulch can benefit a variety of crops. This practise has been shown to improve crop yields at comparable irrigation regimes and save irrigation water and fertiliser nitrogen in several crops, including forage maize, sugarcane, sunflower, soybean, chickpea, potato, and chillies, by reducing the evaporation (E) component of Evapotranspiration, acting as a barrier to vapour flow, and moderating soil temperature.

*Figure 1. Various industrial use of rice straw*

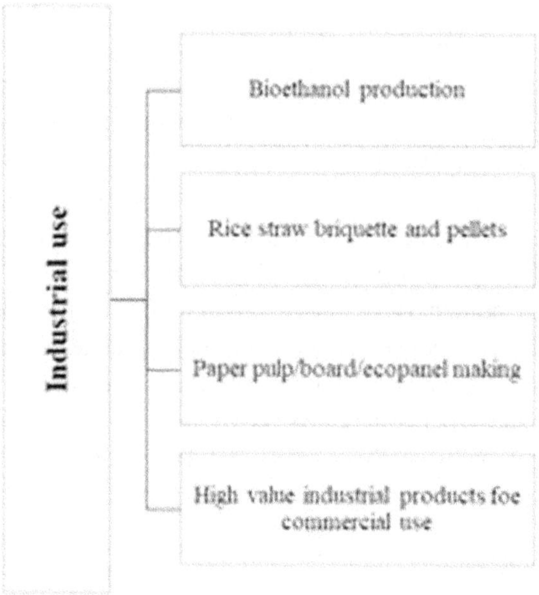

The response is stronger in high-temperature, low-rainfall years, particularly in coarse-textured soils. More soil water in the profile, particularly in the root zone, resulted in improved stand establishment and early seedling vigour in mulched plots. Straw mulching reduced fertiliser N use for comparable crop yields of 25 kg N $ha^{-1}$ in Japanese mint, 50 kg in fodder maize, and 30 kg N $ha^{-1}$ in chilli. Soil N mineralization increases as soil temperature rises and water content rises in mulched soil compared to unmulched soil. However, this approach has failed to gain attraction among farmers due to a lack of labour and the high cost of collecting and applying straw mulch.

### 5.4.2 Diversification of Crops

To address groundwater concerns and low yields, state and federal governments in the Green Revolution region have begun to shift away from paddy agriculture. A centrally supported scheme launched in 2013-14 in a few areas aims to diversify at least 5 percent of paddy acreage to more locally suitable crops (e.g. maize, millets, and oilseeds). Despite the scheme's numerous provisions (cluster demonstrations, knowledge training, farm machinery subsidies, and so on), most farmers have found it unprofitable to shift from paddy to other Kharif crops. The strong market support for paddy, as well as the yield advantages over alternative crop options for Kharif cultivation, have hampered the efficient translation of crop diversification policy goals into field results.

### 5.4.3 Improved Short-Term Paddy Varieties

To reduce water usage in paddy agriculture, research groups developed short-duration paddy cultivars in the Western Indo-Gangetic Plains. Traditional varieties (those that mature in 160 days) continue to produce higher yields, though 135–145 days varieties are becoming more popular. Some believe that adopting such types will extend rice harvest and wheat sowing, allowing farmers to clear fields and reduce residue burning.

## 6. TECHNOLOGICAL INTERVENTIONS AND BEST PRACTICES POSSIBLE TO DEAL WITH THE ISSUE OF STUBBLE BURNING

Crop residue can be managed through both in situ and ex situ agricultural methods. Ex-situ techniques include removing waste from the field and converting it into compost, as well as baling rice residue for power plants. However, there are drawbacks to ex-situ crop residue management that are not always economically or environmentally feasible. Composting is not a financially viable option for farmers due to a lack of labour and high labour costs. Baling is also not a viable alternative because the baler costs more than ten lakhs and has a working period of 10 to 15 days, leaving it unused for the rest of the year, and depreciation costs are not recouped. Furthermore, removing residues from the field without recycling them is detrimental to soil health.

Waste management at the point of production is one example of in-situ procedures. There are technologies available, such as the Rotavator and Mulcher, but they are ineffective and may result in higher production costs and delayed wheat planting. The use of a super straw management system (SMS) and a Turbo Happy Seeder reduces residue management costs while also preparing the field for the next crop. Crop residue management in situ with technology benefits farmers while also providing a realistic solution to avoid residue burning. It saves time by doing three things at once: shredding the harvested crop, spreading stubble across the swath, and sowing wheat seeds. Scientific studies show that it saves approximately 10 lakh gallons of water on the first day of seeding a crop and increases a farmer's profit by Rs 20,000 to 25,000 per hectare each year (Bhattacharyya et al., 2021, Singh and Arya, 2021). As a result, farmers gradually reduce their nitrogen fertiliser use. Finally, it reduces greenhouse gas emissions from agricultural lands.

## 7. CONCLUSION

The management of agricultural waste, particularly crop residue, is a critical component of sustainable agricultural practises. The extensive discussion presented here emphasises the issue's multifaceted nature, as well as the numerous challenges and opportunities associated with agricultural waste management. To conclude, agricultural waste management is a complex but manageable challenge that requires collaboration from all stakeholders. By implementing innovative strategies and adopting sustainable practises, we can effectively address the problem of stubble burning and pave the way for a more sustainable future for agriculture and the environment.

## REFERENCES

Agamuthu, P. (2009, November). Challenges and opportunities in agro-waste management: An Asian perspective. In *Inaugural meeting of first regional 3R forum in Asia, 11*(12). University of Malasiya.

Bhattacharyya, P., Bisen, J., Bhaduri, D., Priyadarsini, S., Munda, S., Chakraborti, M., & Nimbrayan, P. (2021). Turn the wheel from waste to wealth: Economic and environmental gain of sustainable rice straw management practices over field burning in reference to India. *The Science of the Total Environment, 775*, 145896. doi:10.1016/j.scitotenv.2021.145896

Bhuvaneshwari, S., Hettiarachchi, H., & Meegoda, J. N. (2019). Crop residue burning in India: Policy challenges and potential solutions. *International Journal of Environmental Research and Public Health, 16*(5), 832. doi:10.3390/ijerph16050832 PMID:30866483

Birla, D., Singh, D., Yadav, B., Kumar, A., & Rai, S. (2020). Modern Strategies for Crop Residue Management. *Agriculture and Food: E-Newsletter, 2*(11), 172–175.

Chaudhary, C., Yadav, D. B., Yadav, A., Chaudhary, A., & Hooda, V. S. (2024). Influence of crop residue and nitrogen management on nutrient uptake, yield, and economics of rice-wheat cropping system. *Journal of Plant Nutrition, 47*(3), 376–391. doi:10.1080/01904167.2023.2277399

Danesh, P., Niaparast, P., Ghorbannezhad, P., & Ali, I. (2023). Biochar production: Recent developments, applications, and challenges. *Fuel, 337*, 126889. doi:10.1016/j.fuel.2022.126889

Das, S., Biswas, S., Ramakrishnan, B., Das, T. K., Purakayastha, T. J., Gawade, B. H., & Sinha, K. (2024). Biological soil health with conventional and qPCR-based indicators under conservation agriculture based rice-wheat cropping system in Indo-Gangetic Plain. *Applied Soil Ecology, 193*, 105128. doi:10.1016/j.apsoil.2023.105128

. Ghodke, P. K., Anthony, C., & Sharma, A. K. (2024). Bioprocessing of Agricultural and Forest Waste. *Sustainable Production Innovations: Bioremediation and Other Biotechnologies*, 367-393.

Goswami, S. B., Mondal, R., & Mandi, S. K. (2020). Crop residue management options in rice–rice system: A review. *Archives of Agronomy and Soil Science*, *66*(9), 1218–1234. doi:10.1080/03650340.2019.1661994

. Kumar, P., & Joshi, L. (2013). Pollution caused by agricultural waste burning and possible alternate uses of crop stubble: a case study of Punjab. *Knowledge systems of societies for adaptation and mitigation of impacts of climate change*, 367-385.

Kumar, P., Kumar, S., Joshi, L., Kumar, P., Kumar, S., & Joshi, L. (2015). Policies for restricting the agriculture residue burning in Punjab. Socioeconomic and Environmental Implications of Agricultural Residue Burning: A Case Study of Punjab, India, 117-131.

Kumar, R. (2016). Crop residue generation and management in Punjab state. *Indian Journal of Economics and Development*, *12*(1a), 477–483. doi:10.5958/2322-0430.2016.00109.8

Lohan, S. K., Jat, H. S., Yadav, A. K., Sidhu, H. S., Jat, M. L., Choudhary, M., Peter, J. K., & Sharma, P. C. (2018). Burning issues of paddy residue management in north-west states of India. *Renewable & Sustainable Energy Reviews*, *81*, 693–706. doi:10.1016/j.rser.2017.08.057

. Meena, H. N., Singh, S. K., Meena, M. S., Narayan, R., & Sen, B. (2022). *Crop residue: waste or wealth?*

Mor, S., Santosh, B., & Narsi, R. B. (2022). Crop residue burning: Impacts, management techniques and policies. *Annals of Biology*, *35*(2), 245–251.

Naresh, R. K., Dhaliwal, S. S., Chandra, M. S., Malhotra, S. K., Harish, J., Singh, P. K., Kumar, V., Baliyan, A., & Gawdiya, S. (2021). Alternative uses, in and off-field managements to reduce adverse impact of crop residue burning on environment: A review. *International Journal of Environment and Climate Change*, *11*(1), 100–118. doi:10.9734/ijecc/2021/v11i130347

Padmakumar, V., Ravi, D., Rupam, B., Kanwar, S., Ramakrishna, R., Prasad, K. V. S. V., & Jones, C. S. (2024). Genetic variation for grain yield and straw quality traits in 55 diverse rice varieties in Assam. *Animal Nutrition and Feed Technology*, *24*(1), 93–103. doi:10.5958/0974-181X.2024.00007.6

Shyamsundar, P., Springer, N. P., Tallis, H., Polasky, S., Jat, M. L., Sidhu, H. S., & Somanathan, R. (2019). Fields on fire: Alternatives to crop residue burning in India. *Science*, *365*(6453), 536–538. doi:10.1126/science.aaw4085 PMID:31395767

Singh, G., & Arya, S. K. (2021). A review on management of rice straw by use of cleaner technologies: Abundant opportunities and expectations for Indian farming. *Journal of Cleaner Production*, *291*, 125278. doi:10.1016/j.jclepro.2020.125278

Singh, S., Ranguwal, S., & Guleria, A. (2023). Determinants of crop residue burning practices in Punjab. *Agricultural Research Journal*, *60*(2), 289–294. doi:10.5958/2395-146X.2023.00043.1

Verma, V., Joshi, A., Kalra, S., Singh, A., Singh, A., & Mohan, C. (2024). Novel Nanocatalysts for Sustainable Hydrogen Production from Renewable Resources. *E3S Web of Conferences*, *511*. SCOPUS. doi:10.1051/e3sconf/202451101006

# Chapter 13
# Articulating Deep Learning and Internet of Things (IoT) in the Hospitality Industry as a Catalyst for Sustainable Consumption and Food Wastage:
## Unveiling Waste Management Lensing SDG 1 (No Poverty) and SDG 2 (Zero Hunger)

**Bhupinder Singh**
https://orcid.org/0009-0006-4779-2553
*Sharda University, India*

**Vishal Jain**
https://orcid.org/0000-0003-1126-7424
*Sharda University, India*

**Christian Kaunert**
https://orcid.org/0000-0002-4493-2235
*Dublin City University, Ireland*

## ABSTRACT

*A new age of innovation in the hospitality sector is being ushered in by the confluence of deep learning (DL) and the internet of things (IoT), which offers a comprehensive solution to sustainability issues. Due in large part to its complex supply networks and high resource consumption, the hospitality sector contributes significantly to food waste and environmental damage. In order to provide a more sustainable and socially responsible approach to waste management, this chapter explores how the combination of DL and IoT systems might play a crucial role in it. DL algorithms are used to optimize a number of areas of hospitality operations because to their capacity to evaluate massive datasets. These models*

DOI: 10.4018/979-8-3693-4264-0.ch013

*are used in the food service industry for customized menu improvement, inventory control and demand forecasting. These technologies can help to achieve SDGs 1 and 2 by creating ecologically and socially responsible hospitality industries through the promotion of sustainable consumption patterns and effective waste management.*

## 1. INTRODUCTION

Despite advances in technology, there is a troubling rise in food waste as a result of intricate cuisines and culinary practices (Barrett et al., 2023). Even while cutting down on food waste is an important sustainable development aim, the numbers that are available today show that there is a lot of food wasted (Labadi et al., 2021). Food waste has been the subject of much research, especially in the hotel industry where it accounts for 12% of all waste (Araujo, 2023). Many causes of food waste have been investigated; however, the studies have a tendency to be disjointed and lack a thorough methodology (Shams et al., 2023). The previous research has mostly concentrated on industrialized nations, ignoring the more significant problem in developing economies (NaThalang, 2019). Research about food waste in Taiwanese hotels is scarce (Fitchuk, 2021). Adopting efficient waste management techniques may lower expenses, enhance brand recognition, and show a dedication to environmental sustainability for both hoteliers and culinary managers (Sartor, 2021).

When DL and IoT are combined, smart decision-making is made possible and it helps hospitality industries to adapt their operations in real-time to meet demand and avoid waste (Voorhoeve, 2021). In line with SDGs 1 and 2, this chapter explores the societal effect of these technologies (Singh, 2023). The cost reductions in the hospitality sector are facilitated by DL and IoT by improving operational efficiency and cutting waste (Hunter et al., 2020). In line with SDG 1's objective of ending poverty, these savings might be put toward community development, social projects and programs that reduce poverty (Bisaria, 2022). Within the framework of Sustainable Development Goal 2, attaining zero hunger is directly impacted by the decrease in food waste using DL and Internet of Things applications (Singh, 2023). In order to alleviate food insecurity and encourage sustainable consumption habits, this chapter highlights how smart technologies make it possible for resources to be distributed more fairly (Atser et al., 2019).

### 1.1 Background of Study

The advent of complex cuisines and technical advancements in culinary arts has made food acquisition easier in recent years (Sharma & Singh, 2022). There has been a worrying increase in waste to go along with this greater convenience in gratifying gastronomic cravings (Tice, 2024). This waste encompasses less obvious forms in addition to obvious kitchen waste (Toivonen, 2020). The European Union has included the reduction of food waste as a critical sustainable development target in its 2030 agenda, acknowledging the significance of tackling this issue (Singh, 2022). Collaboration between institutions, the public and governments is necessary to accomplish this goal (Miconi, 2022). As part of proactive measures, the Pacific Asia Travel Association in Asia has started regional campaigns to decrease food waste with the purpose of increasing public awareness of the larger objectives of environmental sustainability in the travel and tourist industry (Vind, 2023).

## 1.2 Hospitality Industry and Its Impact on Global Economy- Overview

The hotel industry has contributed significantly to the development of jobs, cross-cultural exchanges and foreign exchange, all of which have helped the world economy (Singh, 2022). In addition to improving a nation's economic status internationally, this industry is essential to the preservation of that nation's unique artistic, cultural, and handicraft traditions (Spiess-Knafl, 2022). The hospitality sector is a major player in the national economy and is well-known both domestically and internationally (Kilibarda et al., 2019). It satisfies tourists' curiosity in various cultures and artistic mediums in addition to meeting their needs for adventure, discovery and relaxation (Munir, 2022). Because food waste and consumption have a significant negative impact on the environment, it is imperative that the hospitality sector ensure food sustainability (Goh & Jie, 2019). Due to their vast operations and daily preparation of several meals, hotels, restaurants and food service enterprises greatly contribute to water use, food waste and greenhouse gas emissions (Singh, 2019). It becomes essential to adopt sustainable eating habits in order to lessen these impacts (Goh et al., 2022). The industry can encourage good change by supporting plant-based substitutes, cutting waste, and giving local and organic procurement first priority (Aditya & Kurniawati, 2023). There are multifarious methods for creating a more sustainable food system and examines the importance of food sustainability in the hospitality industry with inculcating IoT and Deep Learning (Kasavan et al., 2022).

## 1.3 Sustainable Development Goals (SDGS) 1 and 2- Significance

A key goal of the 2030 Agenda for Sustainable Development is to abolish severe poverty for all people by that year (Katuwal & Afzal, 2022). In the decades before, remarkable strides had been achieved in lowering severe poverty, which is defined as surviving on less than $2.15 per person per day purchasing power parity. Poverty reduction was losing progress (Filimonau et al., 2022). In the UN Environment and Development Commission's "Our Common Future" document, the Brundtland Commission first defined sustainable development (SD) as "the development that meets the needs of the present without compromising the ability of future generations to meet their own needs" (Ozdemir & Guicer, 2018). So, addressing humankind's desire for a better existence while honoring the constraints placed on us by nature is the aim of sustainable development goals. Sustainable development includes social development, economic growth and environmental protection as interdependent and mutually supporting elements (Dhir et al., 2020). Each of these elements has contributed significantly to initiatives pertaining to funding, global development and innovation in recent years (Kasavan et al., 2019). In terms of social development, the emphasis has shifted to incorporate high-quality education in addition to eliminating poverty and advancing population well-being (Luu, 2020).

## 1.4 Objectives of the Chapter

This chapter has the following objectives to-

- Achieve the Sustainable Development Goals (SDG) 1 (No Poverty) and SDG 2 (Zero Hunger).
- Examines the transformational influence of Deep Learning and IoT technologies on sustainable consumption and waste reduction in the hospitality industry.
- Exploring the application of deep learning and IoT in sustainability and waste management.

*Figure 1. The objectives of the chapter*

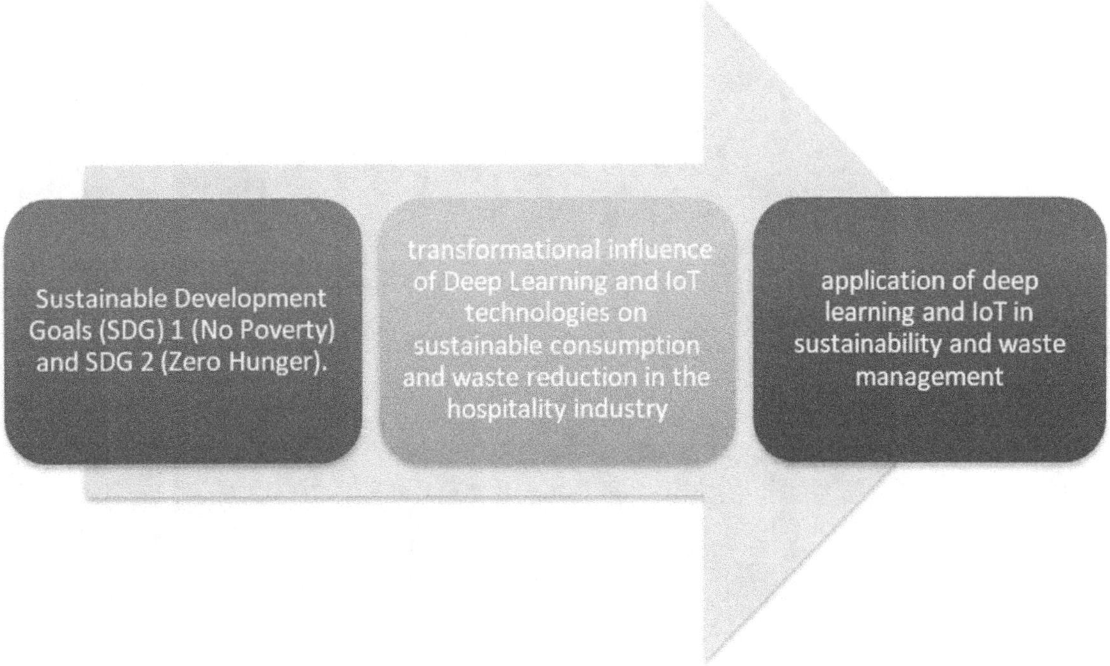


### 1.5 Structure Of The Chapter

This chapter comprehensively explores the diverse aspects of Deep Learning and Internet of Things (IoT) in Hospitality Industry Catalyst Sustainable Consumption and Food Wastages: Unveiling Waste Management Lensing SDG 1 (No Poverty) and SDG 2 (Zero Hunger). Section 2 elaborates the Addressing Food Wastage and Poverty: In-Context of Hospitality Industry. Section 3 expresses the Role of Deep Learning and IoT in Achieving Sustainability Goals (1 and 2). Section 4 lays down the Sustainable Consumption: Concept of Sustainable Consumption in Hospitality Industry. Section 5 highlights the Deep Learning and Internet of Things (IoT) in Hospitality. Section 6 discusses the Food Wastage and Poverty. Finally, Section 7 Conclude the Chapter with Future Scope.

## 2. ADDRESSING FOOD WASTAGE AND POVERTY: IN-CONTEXT OF HOSPITALITY INDUSTRY

Food wastage poses a notable challenge within the hospitality sector, where hotels, restaurants, and catering services produce a considerable volume of discarded food annually (Chalak et al., 2018). As sustainability awareness rises and the focus on waste reduction intensifies, there exist various strategies that businesses can adopt to diminish food waste and establish a more sustainable operational model (Ruiz Molina et al., 2022). There are many impactful approaches to address food waste in the hospitality industry as-

*Figure 2. The structure of the chapter*

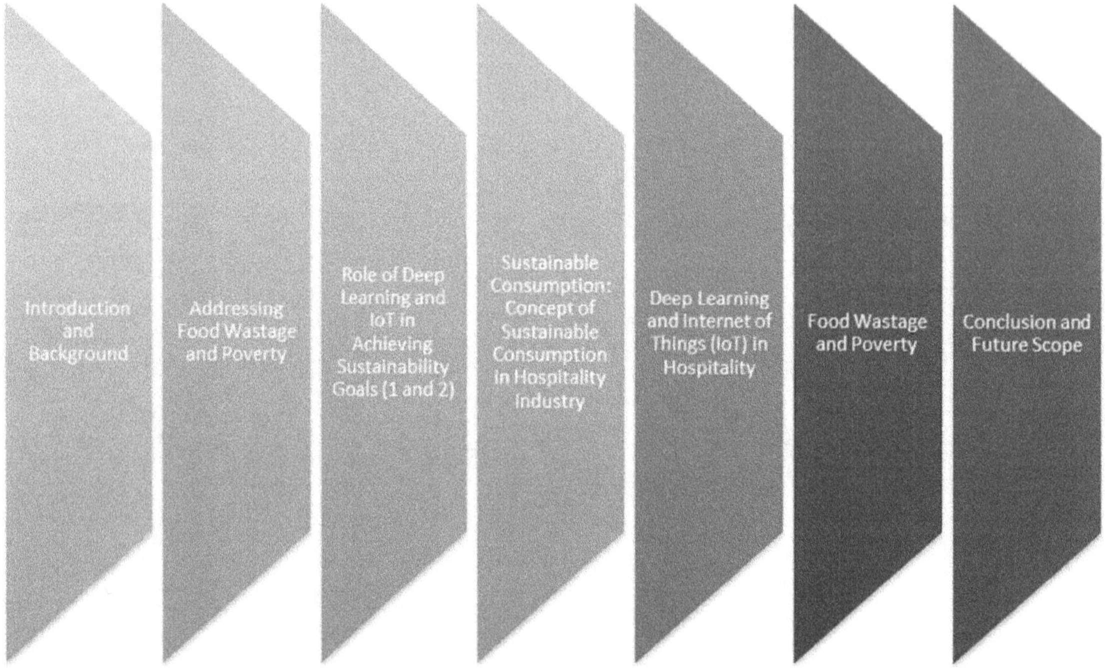


1. *Thoughtful Menu Design:* In menu engineering, choices concerning offers are made by assessing the profitability and popularity of menu items. Businesses can change their menus by eliminating some products or lowering portion sizes by identifying meals with high waste or low demand. This strategy lowers food waste while simultaneously increasing profitability and cost savings (Lee & Huang, 2023).
2. *Precise Management of Inventory:* In the hospitality sector, inadequate inventory management stands out as a major contributor to food waste (Lagioia et al., 2023). Businesses can mitigate this issue by closely overseeing inventory levels and adopting a system that monitors ingredient usage. This approach provides enhanced insight into stock levels, enabling better-informed purchasing decisions. It guards against over-purchasing and ensures the timely utilization of perishable items before their expiration. The use of technologies such as- inventory management software which facilitates a streamlined process and enabling businesses to identify and addresses areas where waste reduction is possible (de Visser-Amundson, 2022).
3. *Collaborating with Neighborhood Nonprofits or Food Banks:* Forming partnerships with nearby food banks or nonprofit organizations is a practical way to reduce food waste and have a beneficial social impact. Businesses may donate extra food to those in need instead of throwing it away. Collaborating with these groups ensures that surplus food is put to good use, reducing waste and helping to combat hunger in the neighborhood (Islam et al., 2023).
4. *Training and Awareness among Employees:* Implementation effectiveness depends critically on staff members receiving the necessary training and education about the importance of reducing food waste. Increasing awareness of the problem and providing training on proper portioning, storage

techniques and creative ways to repurpose leftovers encourage workers of organizations to take an active role in waste reduction initiatives (Sharma, 2020).

5. *Introducing Systems for Tracking and Monitoring Food Waste:* Monitoring and tracking food waste is a crucial part of spotting trends, assessing progress, and putting focused strategies into action to reduce waste. By implementing a food waste monitoring system, companies may gather information about the kinds and amounts of food that are thrown away. This gives them the ability to make data-driven decisions and carry out successful waste reduction efforts (Filimonau et al., 2019).

## 3. ROLE OF DEEP LEARNING AND IOT IN ACHIEVING SUSTAINABILITY GOALS (1 AND 2)

Addressing the issues of global environmental sustainability and food security is the goal of the second Sustainable Development Goal (SDG 2), Zero Hunger (Meier et al., 2021). It undertook an integrated literature analysis, structured around three major academic perspectives as- political economics and policy science, nutrition and public health and ecology and agricultural sciences. Artificial intelligence (AI) is gaining traction and is expected to have long- and short-term effects on a number of areas, including global productivity, inclusion and equality, environmental results and more. AI has the ability to have both

*Figure 3. Shows the impactful approaches to address food wastage in hospitality industry*

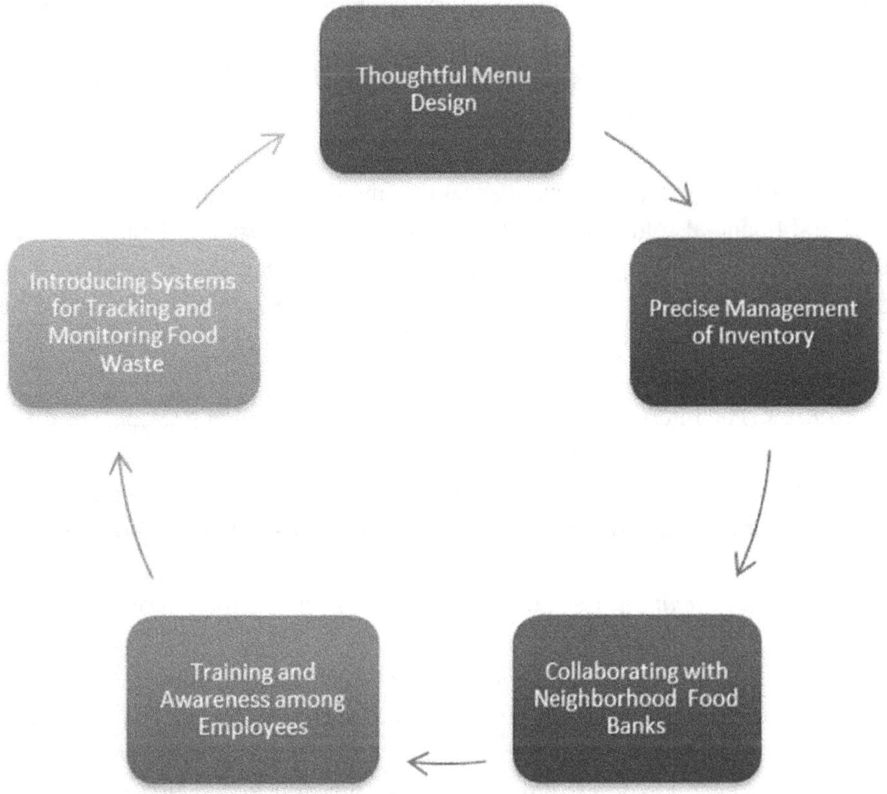

beneficial and harmful consequences on sustainable development (Scholz et al., 2020). AI was defined as any software technology that possesses the following abilities: automatic knowledge extraction, pattern recognition from data, interactive communication like- social robots or chat bots, decision-making like medical diagnosis systems, perception- audio, visual, textual and tactile, prediction like- weather forecasts and logical reasoning like theory development from premises (Tomaszewska et al., 2021). Machine learning is one of the many subfields covered by this wide description. It is multifaceted, with contributing variables including social marginalization, unemployment, and the increased vulnerability of some populations to diseases, disasters, and other obstacles that reduce their output. Although there are many reasons for it, the root cause is essentially the interdependence of human well-being. Growing inequality causes social discord and hinders economic progress, which heightens political and social tensions (Okumus et al., 2020). It may potentially result in instability and disputes under some circumstances. During the past 20 years, organic waste especially food waste has become a major challenge that requires careful attention. This kind of organic waste has had a significant influence on social, economic, and environmental issues worldwide (Camilleri et al., 2021). Food waste is the third-largest worldwide emitter of greenhouse gases, accounting for around 3.3 gigatons of emissions annually. The food industry's economic worth of wasted agricultural goods, determined at producers pricing of fisheries and seafood excluded. The amount of food that is wasted or lost is equal to the area of land needed to produce 25% of the water used for agriculture yearly (Omar et al., 2023).

The United Nations set out to eradicate poverty in all its manifestations by 2030 with the goal of "No Poverty," the first Sustainable Development Goal (SDG). Beyond only financial hardship, poverty is a complex problem that includes restricted access to essential services, chances for education, and work. With its capacity to link objects and enable data transfer across systems, the Internet of Things (IoT) offers revolutionary potential in tackling this complex problem (Filimonau et al., 2020). The integration of IoT technology into poverty alleviation measures has the potential to significantly influence the nation that has had economic difficulties in recent times (Amicarelli & Bux, 2021). The diverse factors for inculcating DL and IoT addressing SDG 1 and 2 as-

1. *Improving Access to Finance:* The lack of access to financial services is a major barrier to escaping poverty. Digital financial solutions that are enabled by the Internet of Things such digital wallets and mobile banking have the potential to provide underprivileged communities with chances to save, invest, and obtain credit. The use of digital solutions might have a revolutionary effect, as certain isolated and rural regions may encounter difficulties in obtaining traditional financial services (Principato et al., 2018).
2. *Job Opportunities:* The Internet of Things may lead to the creation of new jobs. Skilled laborers in a variety of fields, including system administration, data analysis, device installation, and maintenance, will be required as the nation makes investments in IoT infrastructure. Starting training programs can equip those from low-income backgrounds with the skills they need to succeed (Filimonau & Uddin, 2021).
3. *Improving Access to Education:* Lack of access to education is a major contributing factor to poverty. IoT can help with this problem by making remote learning easier, especially in rural areas. People may escape the cycle of poverty and improve their career chances by accessing high-quality education using digital resources, smart devices, and online platforms (Martin-Rios et al., 2022).
4. *Enhancing Methods of Agriculture:* Like in many other regions of the world, Greece has a large number of disadvantaged villages that rely mostly on agriculture. IoT-driven precision agriculture

may help farmers maximize yields, reduce losses and boost revenue by using sensors to track crop growth, weather, and soil health. This becomes especially important in areas of Greece that are vulnerable to drought or other weather-related issues (Huang et al., 2021).

5. *Health Surveillance:* Health problems can make poverty worse by making it harder to work or by resulting in high medical costs. IoT-enabled wearables can monitor health indicators, providing early alerts and reducing the need for frequent doctor visits (Habib Ur Rehman, 2020).
6. *Rural and Island Communities:* These communities which sometimes face more difficult economic circumstances than metropolitan ones, stand to gain significantly from the deployment of IoT. Whether it's streamlining agriculture or offering digital banking and education, IoT can bring small areas into the country's larger economic structure (Despoudi et al., 2021).
7. *Working Together in the Public and Private Sectors:* The government can collaborate with private businesses to lead Internet of Things projects that target poverty reduction. These kinds of partnerships may guarantee that the most disadvantaged groups benefit directly from technology breakthroughs (Synani et al., 2021).

The United Nations' second Sustainable Development Goal (SDG) "Zero Hunger" aims to eradicate hunger, attain food security, enhance nutrition and encourage sustainable agriculture by 2030. Hunger and malnutrition are not solely concerns for developing nations; they can also pose challenges in developed countries like Greece, particularly in the aftermath of economic crises (Legrand et al., 2022). The Internet of Things (IoT), with its extensive network of interconnected devices, presents inventive solutions to tackle these challenges and ensure food security on a global scale as-

1. *Accurate Farming:* Precision agriculture is a very promising use of IoT in the context of food security. Farmers can make educated decisions about irrigation, fertilization, and pest management by using sensors that track soil moisture, nutrient levels, and weather. This reduces the impact on the environment and conserves resources while also increasing agricultural yields (Sharma et al., 2023).
2. *Supply Chain Simplifying:* Reducing food waste is a major challenge for the whole food supply chain. IoT devices can track produce in real time as it is being transported, guaranteeing ideal storage conditions and timely delivery. This guarantees that a larger amount of food reaches customers and helps prevent spoiling (Said et al., 2023).
3. *Data analysis and remote monitoring:* Large agricultural holdings may be monitored using drones and remote sensors, which provide real-time data on crop health, insect infestations and environmental conditions. With analyzing this data, farmers can ensure that crops are healthy and yields are steady by putting timely interventions into place (Hassoun et al., 2023).
4. *Fisheries and Aquaculture:* It has a significant fishing sector because of its long coastline. Applications for the Internet of Things (IoT) may track fish populations, water quality and eating habits in aquaculture, which helps maintain marine ecosystems and promote sustainable fish farming (Jauhar et al., 2023).
5. *Increasing Customer Knowledge and Predicting Needs:* IoT is a tool for informing customers about the provenance, freshness, and nutritional value of the food they eat. Retailers may also improve the accuracy of demand estimates, avoiding overstocking and lowering food waste by analyzing customer purchase habits (Rane et al., 2023).

## 4. SUSTAINABLE CONSUMPTION: CONCEPT OF SUSTAINABLE CONSUMPTION IN HOSPITALITY INDUSTRY

In the hotel business, sustainable consumption refers to the responsible and ecologically conscious use of resources in the production and consumption of goods and services (Oriekhoe et al., 2024). It means making decisions that minimize negative effects while promoting long-term sustainability by considering the economic, environmental, and social effects of hospitality operations. This includes advocating for diversity and inclusion, adopting zero-carbon and nature-based solutions, managing energy, water and trash efficiently and making sure those workers are safe (Verma et al., 2023). When making decisions about procurement and consumption, sustainable consumption also entails taking into account the characteristics of traditional food items, such as sensory quality, menu diversity, freshness, local production and seasonality (De Bernardi et al., 2023). Food waste is a global problem that requires a comprehensive approach to effectively address its many issues. Understanding the consequences food waste has on the environment, the economy, and society can help us create a more fair and sustainable future (Vargas-Sanchez, 2023).

Reducing food waste and creating a more resilient, efficient and equitable food system may be achieved by putting creative ideas into action that touch all stages of the food supply chain and encourage cooperation between many stakeholders. It will take immediate steps to prevent further waste and ensure that everyone has access to wholesome food (Abatan et al., 2024). In the hotel industry, sustainable consumption refers to running operations in a way that takes social, environmental, and economic sustainability into account (Delina et al., 2023). When making judgments about the purchase and consumption of traditional food products, this method entails considering the market, production and technological characteristics of traditional food products. The industry is called upon to use natural solutions and regeneration-oriented techniques that are in line with environmental guidelines and ecosystem restoration concepts (Stewart & Gorman, 2024).

## 5. DEEP LEARNING AND INTERNET OF THINGS (IOT) IN HOSPITALITY INDUSTRY: SAVE FOOD NO WASTE

In order to maximize services and minimize waste, the Quick Service Restaurant (QSR) business, which occupies a sizable and expanding portion of the worldwide food market, requires efficient resource management (Mishra & Singh, 2023). In order to transition away from human-dependent methods and toward efficient and error-resistant processes, it is imperative to use the potential of the Internet of Things (IoT) (Nwokolo et al., 2023). Edge computing becomes an essential component in an Internet of Things architecture, offering advantages including low latency, less redundancy, efficient resource utilization, improved security and real-time decision making (Mishra & Singh, 2023). Specifically, the focus is on detecting anomalies and outliers and estimating production service levels, which are accomplished by using lightweight clustering and classification approaches. Experiments carried out in a real restaurant setting show how well the IoT-based automation system predicts production levels and sounds an alarm when abnormal waste circumstances are present (Rahimi-Midani, 2023).

Although a substantial amount of waste happens in households, the majority of it is concealed in the larger food supply chain (Workie et al., 2023). More than sixty percent of food waste is produced in commercial environments, which include industry, food service, retail and farming (Buhalis et al.,

2023). Because of the sheer size of these operations, waste creation happens much faster, with most of it happening before a food item ever reaches the produce department of a grocery store or the menu of a restaurant (Ali et al., 2024). In order to preserve their freshness, perishable food products are purposefully kept dormant throughout the supply chain. Afterwards, depending on when they are expected to arrive at retail, they are triggered using particular gasses at various times (Hong et al., 2024). This procedure takes place in the backroom storage, warehousing and transportation phases. Maintaining a sensitive ambient environment is essential to product safety and shelf-life extension. To maintain quality, perishable foods need to be kept at exact gas and humidity levels in addition to certain temperatures (Ramadass et al., 2023). An efficient, waste-free operation is influenced by a number of factors, including product placement, rotation schedule, refrigeration size and type and ventilation. Maintaining food safety and freshness is a difficult endeavor that requires advanced technology and sophisticated procedures to maximize results and reduce waste (Glebova et al., 2023).

## 6. FOOD WASTAGE AND POVERTY

The contradictory juxtaposition of habitual food waste, especially in wealthy nations, and food poverty, which affects a large percentage of the world's population, has attracted a lot of scholarly and popular attention (Kannan et al., 2023). Food waste prevention and reduction are the responsibilities of all parties involved in the food supply chain, including farmers, food producers, caterers, retailers and consumers (Machado & Hatakeyama, 2023). Many see redistribution of food excess as a partial answer to reduce food waste and alleviate food poverty; others, on the other hand, condemn philanthropic activities as inadequate remedies that might relieve governments of their responsibility to protect citizen's right to food (Ordonez de Pablos et al., 2023).

IoT Sensing-as-a-Service offers an improved framework that goes beyond simple tracking of perishable food items, going beyond the deployment of physical sensing and monitoring equipment (Legrand et al., 2023). IoT Sensing-as-a-Service empowers enterprises to leverage improved inventory performance data, resulting in efficient waste reduction and increased loss prevention efforts, by integrating AI-enabled digital decisioning and digitalized job management (Gupta et al., 2023). Better insight into the state of food quality is made possible by the interconnectivity of refrigeration units, display cases, transportation pallets and other equipment across the supply chain. It is possible to remotely monitor temperature, humidity and gas settings in real time to make sure that protocols are being followed by installing Internet of Things (IoT) sensors at several touchpoints within food storage equipment (Taneja et al., 2023). The sensors automatically track the functionality of the equipment, anticipating and identifying repair needs. When the unprocessed physical data gathered from every IoT device in the supply chain is combined with a prescriptive analytics platform, a significant revolution takes place (Oriekhoe et al., 2024). With creating ongoing feedback loops, this technology alerts staff members to any unusual temperature fluctuations that might compromise the integrity of the product. This crucial data is converted into prescriptive processes and descriptive insights, so in the case of a deviation, the platform provides simple remedial measures (Abatan et al., 2024).

The food value chain is a complex web of relationships between different companies involved in food production, delivery, processing and sale (Stewart & Gorman, 2024). Along this chain, however, significant amounts of food are lost or wasted, which has negative effects on the environment, society and the economy. The food value chain may become more sustainable and efficient by utilizing the creative solutions that the Internet of Things (IoT) offers to these problems (Hong et al., 2024). IoT devices and sensors may be integrated at various points in the value chain to gather, process and use real-time data for operations optimization, loss minimization and productivity enhancement. IoT technology in agriculture enable farmers to regulate irrigation precisely, allocate resources optimally and take early action to reduce crop losses by providing them with real-time data on soil moisture, temperature, and crop health (Ali et al., 2024). With continually monitoring the health and behavior of livestock, IoT-based livestock tracking systems enable better animal management practices and herd health by enabling for the early detection of any problems and the implementation of preventative measures to stop accidents or disease outbreaks (Tapo et al., 2024). Farmers may make more educated decisions about planting, harvesting and pest management by utilizing IoT integration of weather predictions and satellite imaging (Gao et al., 2024).

IoT devices track important variables including temperature, humidity, and product movement in food processing plants (Koskela-Huotari et al., 2024). Systems for real-time monitoring and control assist prevent spoiling, maximize production efficiency and spot any quality problems (El Zein et al., 2024). IoT-enabled quality control procedures guarantee adherence to food safety guidelines and standards, which helps to minimize food loss. IoT-based smart packaging solutions are essential for lowering food loss because they provide customers and supply chain participants with improved product tracking, expiry date monitoring and quality indicators (Menzies et al., 2024). Smart packaging gives consumers the power to make educated decisions and enables the best possible inventory management along the value chain by providing up-to-date information about product freshness, storage conditions and handling guidelines (Zhang et al., 2024).

## 7. CONCLUSION AND FUTURE SCOPE

Deep Learning and the Internet of Things (IoT), holds the potential to significantly expedite its progress toward achieving the Sustainable Development Goals (SDGs). Beyond merely revitalizing ancient ideals of democracy and justice, IoT can address contemporary challenges like no poverty and zero hunger, offering a transformative perspective. The immense potential of IoT extends to reshaping various sectors, including agriculture, healthcare, governance and tourism. With seamlessly integrating IoT into its national strategies, Greece has the opportunity to craft a compelling narrative centered around innovation, sustainability, and inclusive growth, reflecting its enduring spirit in global collaborations for a sustainable future. It is possible to minimize food loss in the hospitality business by implementing a variety of measures, including accurate inventory management, menu engineering, employee training, partnerships with neighborhood groups and the use of food waste tracking systems. Businesses may reduce their influence on the environment, save money, and support a more sustainable operational model by adopting these practices.

## REFERENCES


Abatan, A., Adeyinka, M. A., Sodiya, E. O., Jacks, B. S., Ugwuanyi, E. D., Daraojimba, O. H., & Lottu, O. A. (2024). *The role of IT in sustainable environmental management: A global perspective review.*

Aditya, A., & Kurniawati, K. (2023). Food Waste Management Challenges and Strategies in The Hotel Industry in Jakarta. *Journal of Social Research, 2*(9), 2983–2989. doi:10.55324/josr.v2i9.1295

Ali, R., Irfan, S., & Hamid, S. (2024). Emerging themes in food tourism: A systematic literature review and research agenda. *British Food Journal, 126*(1), 372–408. doi:10.1108/BFJ-11-2022-0939

Amicarelli, V., & Bux, C. (2021). Food waste measurement toward a fair, healthy and environmental-friendly food system: A critical review. *British Food Journal, 123*(8), 2907–2935. doi:10.1108/BFJ-07-2020-0658

Araujo, C. M. L. D. (2023). *The effects of retail sustainable strategies on the sustainability of the food value chain* [Doctoral dissertation].

. Atser, G., Tahirou, A., Dixon, A., Ekeleme, F., Alloh, J., & Akande, A. (2019). *Transforming Oyo State to an Agribusiness Hub:(Our vision, Our Plan) 2019-2023.* Oyo State.

Barrett, M., Marino, M., Brkic, F., & Pratesi, C. A. (2023). *How to Create a Sustainable Food Industry: A Practical Guide to Perfect Food.* Taylor & Francis. doi:10.4324/9781003449744

Bisaria, P. (2022). *Environmental issues and corporate social responsibility.*

Buhalis, D., O'Connor, P., & Leung, R. (2023). Smart hospitality: From smart cities and smart tourism towards agile business ecosystems in networked destinations. *International Journal of Contemporary Hospitality Management, 35*(1), 369–393. doi:10.1108/IJCHM-04-2022-0497

Camilleri, M. A. (2021). Sustainable production and consumption of food. Mise-en-place circular economy policies and waste management practices in tourism cities. *Sustainability (Basel), 13*(17), 9986. doi:10.3390/su13179986

Chalak, A., Abou-Daher, C., & Abiad, M. G. (2018). Generation of food waste in the hospitality and food retail and wholesale sectors: Lessons from developed economies. *Food Security, 10*(5), 1279–1290. doi:10.1007/s12571-018-0841-0

De Bernardi, P., Bertello, A., & Forliano, C. (2023). Circularity of food systems: A review and research agenda. *British Food Journal, 125*(3), 1094–1129. doi:10.1108/BFJ-05-2021-0576

de Visser-Amundson, A. (2022). A multi-stakeholder partnership to fight food waste in the hospitality industry: A contribution to the United Nations Sustainable Development Goals 12 and 17. *Journal of Sustainable Tourism, 30*(10), 2448–2475. doi:10.1080/09669582.2020.1849232

. Delina, L. L., Perez, O. A., Afable, S. B., & Steuer, B. (2023). Do global public health crises change people's behaviors towards sustainability? Evidence of the COVID-19 and sustainability nexus from Hong Kong. *Cleaner and responsible consumption, 10*, 100132.

Despoudi, S., Bucatariu, C., Otles, S., & Kartal, C. (2021). Food waste management, valorization, and sustainability in the food industry. In *Food waste recovery* (pp. 3–19). Academic Press. doi:10.1016/B978-0-12-820563-1.00008-1

Dhir, A., Talwar, S., Kaur, P., & Malibari, A. (2020). Food waste in hospitality and food services: A systematic literature review and framework development approach. *Journal of Cleaner Production, 270*, 122861. doi:10.1016/j.jclepro.2020.122861

El Zein, B., Elrashidi, A., Dahlan, M., Al Jarwan, A., & Jabbour, G. (2024). *Nano and Society 5.0: Advancing the Human-Centric Revolution.*

Filimonau, V., Coşkun, A., Derqui, B., & Matute, J. (2022). Restaurant management and food waste reduction: Factors affecting attitudes and intentions in restaurants of Spain. *International Journal of Contemporary Hospitality Management, 34*(3), 1177–1203. doi:10.1108/IJCHM-07-2021-0899

Filimonau, V., Krivcova, M., & Pettit, F. (2019). An exploratory study of managerial approaches to food waste mitigation in coffee shops. *International Journal of Hospitality Management, 76*, 48–57. doi:10.1016/j.ijhm.2018.04.010

Filimonau, V., Matute, J., Kubal-Czerwińska, M., Krzesiwo, K., & Mika, M. (2020). The determinants of consumer engagement in restaurant food waste mitigation in Poland: An exploratory study. *Journal of Cleaner Production, 247*, 119105. doi:10.1016/j.jclepro.2019.119105

Filimonau, V., & Uddin, R. (2021). Food waste management in chain-affiliated and independent consumers' places: A preliminary and exploratory study. *Journal of Cleaner Production, 319*, 128721. doi:10.1016/j.jclepro.2021.128721

Fitchuk, C. (2021). *Creating a Sustainable Business Strategy for a SME.*

Gao, J. Q., Li, D., Qiao, G. H., Jia, Q. R., Li, S. R., & Gao, H. L. (2024). Circular economy strategies in supply chains, enhancing resource efficiency and sustainable development goals. *Environmental Science and Pollution Research International, 31*(6), 8751–8767. doi:10.1007/s11356-023-31551-z PMID:38180660

Glebova, E., Gerke, A., & Book, R. (2023). The transformational role of technology in sports events. In *Sports Management in an Uncertain Environment* (pp. 169–187). Springer Nature Singapore. doi:10.1007/978-981-19-7010-8_8

Goh, E., & Jie, F. (2019). To waste or not to waste: Exploring motivational factors of Generation Z hospitality employees towards food wastage in the hospitality industry. *International Journal of Hospitality Management, 80*, 126–135. doi:10.1016/j.ijhm.2019.02.005

Goh, E., Okumus, B., Jie, F., Djajadikerta, H. G., & Lemy, D. M. (2022). Managing food wastage in hotels: Discrepancies between injunctive and descriptive norms amongst hotel food and beverage managers. *British Food Journal, 124*(12), 4666–4685. doi:10.1108/BFJ-05-2021-0513

Gupta, M., Choudhary, B., & Sharma, D. (2023). Emersion and Immersion of Technology in the Development of Smart Cities: A Bibliometric Analysis. In *Enabling Technologies for Effective Planning and Management in Sustainable Smart Cities* (pp. 303–318). Springer International Publishing. doi:10.1007/978-3-031-22922-0_12

Hassoun, A., Aït-Kaddour, A., Abu-Mahfouz, A. M., Rathod, N. B., Bader, F., Barba, F. J., Biancolillo, A., Cropotova, J., Galanakis, C. M., Jambrak, A. R., Lorenzo, J. M., Måge, I., Ozogul, F., & Regenstein, J. (2023). The fourth industrial revolution in the food industry—Part I: Industry 4.0 technologies. *Critical Reviews in Food Science and Nutrition, 63*(23), 6547–6563. doi:10.1080/10408398.2022.2034735 PMID:35114860

Hong, J., Jaegler, A., & Gergaud, O. (2024). Mobile applications to reduce food waste in supply chains: A systematic literature review. *British Food Journal, 126*(2), 509–530. doi:10.1108/BFJ-09-2022-0742

Huang, Y., Ma, E., & Wang, D. (2021). Message framing strategies, food waste prevention, and diners' repatronage intentions: The mediating role of corporate social responsibility. *Journal of Sustainable Tourism, 29*(10), 1694–1715. doi:10.1080/09669582.2020.1867151

Hunter, D., Monville-Oro, E., Burgos, B., Roel, C. N., Calub, B. M., Gonsalves, J., & Lauridsen, N. (Eds.). (2020). *Agrobiodiversity, school gardens and healthy diets: Promoting biodiversity, food and sustainable nutrition*. Routledge. doi:10.4324/9780429053788

Islam, J. U., Nazir, O., & Rahman, Z. (2023). Sustainably engaging employees in food wastage reduction: A conscious capitalism perspective. *Journal of Cleaner Production, 389*, 136091. doi:10.1016/j.jclepro.2023.136091

Jauhar, S., Pratap, S., Lakshay, Paul, S., & Gunasekaran, A. (2023). Internet of things based innovative solutions and emerging research clusters in circular economy. *Operations Management Research : Advancing Practice Through Research, 16*(4), 1968–1988. doi:10.1007/s12063-023-00421-9

Kannan, H., Rodriguez, R. V., Paprika, Z. Z., & Ade-Ibijola, A. (Eds.). (2023). *Exploring Ethical Dimensions of Environmental Sustainability and Use of AI*. IGI Global. doi:10.4018/979-8-3693-0892-9

Kasavan, S., Mohamed, A. F., & Halim, S. A. (2019). Drivers of food waste generation: Case study of island-based hotels in Langkawi, Malaysia. *Waste Management (New York, N.Y.), 91*, 72–79. doi:10.1016/j.wasman.2019.04.055 PMID:31203944

Kasavan, S., Siron, R., Yusoff, S., & Fakri, M. F. R. (2022). Drivers of food waste generation and best practice towards sustainable food waste management in the hotel sector: A systematic review. *Environmental Science and Pollution Research International, 29*(32), 48152–48167. doi:10.1007/s11356-022-19984-4 PMID:35579831

. Katuwal, S., & Afzal, M. (2022). *A framework for managing food wastage: Strategies to promote recycling and sustainability in the hotel industry*.

Kilibarda, N., Djokovic, F., & Suzic, R. (2019). Food waste management–reducing and managing food waste in hospitality. *Scientific journal. Metals Technology, 60*(2), 134–142.

Koskela-Huotari, K., Svärd, K., Williams, H., Trischler, J., & Wikström, F. (2024). Drivers and hinderers of (un) sustainable service: A systems view. *Journal of Service Research*, *27*(1), 106–123. doi:10.1177/10946705231176071

Labadi, S., Giliberto, F., Rosetti, I., Shetabi, L., & Yildirim, E. (2021). Heritage and the sustainable development goals: Policy guidance for heritage and development actors. *International Journal of Heritage Studies*.

Lagioia, G., Amicarelli, V., Strippoli, R., Bux, C., & Gallucci, T. (2023). Sustainable and circular practices in the hotel industry in Southern Italy: Opportunities, barriers and trends in food waste management. *British Food Journal*.

Lee, H. H., & Huang, P. Y. (2023). Food Waste and Environmental Sustainability of the Hotel Industry in Taiwan. *Sustainability (Basel)*, *15*(21), 15459. doi:10.3390/su152115459

Legrand, W., Chen, J. S., & Laeis, G. C. (2022). *Sustainability in the hospitality industry: Principles of sustainable operations*. Taylor & Francis.

Legrand, W., Kuokkanen, H., & Day, J. (Eds.). (2023). *Critical Questions in Sustainability and Hospitality*. Taylor & Francis. doi:10.4324/9781003218425

Luu, T. T. (2020). Reducing food waste behavior among hospitality employees through communication: Dual mediation paths. *International Journal of Contemporary Hospitality Management*, *32*(5), 1881–1904. doi:10.1108/IJCHM-09-2019-0779

Machado, M., & Hatakeyama, K. (2023, July). Digital Transformation as an Enabler of Sustainability in Supply Chain and Logistics: Evidence from the Field. In *2023 Portland International Conference on Management of Engineering and Technology (PICMET)* (pp. 1-6). IEEE. 10.23919/PICMET59654.2023.10216881

Martin-Rios, C., Demen Meier, C., & Pasamar, S. (2022). Sustainable waste management solutions for the foodservice industry: A Delphi study. *Waste Management & Research*, *40*(9), 1412–1423. doi:10.1177/0734242X221079306 PMID:35176923

Meier, T., von Borstel, T., Welte, B., Hogan, B., Finn, S. M., Bonaventura, M., Friedrich, S., Weber, K., & Dräger de Teran, T. (2021). Food waste in healthcare, business and hospitality catering: Composition, environmental impacts and reduction potential on company and national levels. *Sustainability (Basel)*, *13*(6), 3288. doi:10.3390/su13063288

Menzies, J., Sabert, B., Hassan, R., & Mensah, P. K. (2024). Artificial intelligence for international business: Its use, challenges, and suggestions for future research and practice. *Thunderbird International Business Review*, *66*(2), 185–200. doi:10.1002/tie.22370

Miconi, M. (2022). *The approach to sustainability policies in family firms: how this phenomenon has impacted management remuneration*.

Mishra, P., & Singh, G. (2023). *Sustainable Smart Cities: Enabling Technologies, Energy Trends and Potential Applications*. Springer Nature. doi:10.1007/978-3-031-33354-5

Mishra, P., & Singh, G. (2023). Introduction: Importance of Sustainable Smart City. In *Sustainable Smart Cities: Enabling Technologies, Energy Trends and Potential Applications* (pp. 1–37). Springer International Publishing.

Munir, K. (2022). Sustainable food waste management strategies by applying practice theory in hospitality and food services-a systematic literature review. *Journal of Cleaner Production*, *331*, 129991. doi:10.1016/j.jclepro.2021.129991

Nwokolo, S. C., Eyime, E. E., Obiwulu, A. U., & Ogbulezie, J. C. (2023). Africa's Path to Sustainability: Harnessing Technology, Policy, and Collaboration. *Trends in Renewable Energy*, *10*(1), 98–131. doi:10.17737/tre.2024.10.1.00166

Okumus, B., Taheri, B., Giritlioglu, I., & Gannon, M. J. (2020). Tackling food waste in all-inclusive resort hotels. *International Journal of Hospitality Management*, *88*, 102543. doi:10.1016/j.ijhm.2020.102543

OmarI.HassanH.JayaramanR.SalahK.OmarM. (2023). Using Blockchain Technology to Achieve Sustainability in the Hospitality Industry by Reducing Food Waste. *Available at* SSRN 4530352. doi:10.2139/ssrn.4530352

Ordóñez de Pablos, P., Almunawar, M. N., & Anshari, M. (Eds.). (2023). *Perspectives on the Transition Toward Green and Climate Neutral Economies in Asia*. IGI Global. doi:10.4018/978-1-6684-8613-9

Oriekhoe, O. I., Ashiwaju, B. I., Ihemereze, K. C., Ikwue, U., & Udeh, C. A. (2024). Review Of Technological Advancements In Food Supply Chain Management: A Comparative Study Between The Us And Africa. *International Journal of Management & Entrepreneurship Research*, *6*(1), 132–149. doi:10.51594/ijmer.v6i1.713

Oriekhoe, O. I., Ashiwaju, B. I., Ihemereze, K. C., Ikwue, U., & Udeh, C. A. (2024). Review Of Technological Advancements In Food Supply Chain Management: A Comparative Study Between The Us And Africa. *International Journal of Management & Entrepreneurship Research*, *6*(1), 132–149. doi:10.51594/ijmer.v6i1.713

Ozdemir, G., & Güçer, E. (2018). Food waste management within sustainability perspective: A study on five star chain hotels. *Journal of Tourism & Gastronomy Studies*, *6*(1), 280–299. doi:10.21325/jotags.2018.187

Principato, L., Pratesi, C. A., & Secondi, L. (2018). Towards zero waste: An exploratory study on restaurant managers. *International Journal of Hospitality Management*, *74*, 130–137. doi:10.1016/j.ijhm.2018.02.022

Rahimi-Midani, A. (2023). *Deep Technology for Sustainable Fisheries and Aquaculture*.

Ramadass, R., Sathyanarayana, N., & Fathima, Y. (2023). Improving citizen lifestyle in green smart city. In *Green Blockchain Technology for Sustainable Smart Cities* (pp. 29–63). Elsevier. doi:10.1016/B978-0-323-95407-5.00007-4

RaneN.ChoudharyS.RaneJ. (2023). Metaverse as a cutting-edge platform for attaining Sustainable Development Goals (SDGs). SSRN 4644035. doi:10.2139/ssrn.4644035

Ruiz Molina, M. E., Belda-Miquel, S., Hytti, A., & Gil-Saura, I. (2022). Addressing sustainable food management in hotels: Proposing a framework and examining hotel groups. *British Food Journal, 124*(2), 462–492. doi:10.1108/BFJ-12-2020-1171

Said, Z., Sharma, P., Nhuong, Q. T. B., Bora, B. J., Lichtfouse, E., Khalid, H. M., & Hoang, A. T. (2023). Intelligent approaches for sustainable management and valorisation of food waste. *Bioresource Technology, 377*, 128952. doi:10.1016/j.biortech.2023.128952 PMID:36965587

Sartor, D. (2021). *An Exploratory Study of Woolworths Consumer's Post-Covid Food Waste Behaviour.*

Scholz, P., Linderová, I., & Konečná, K. (2020). Green Management Tools as a Way to Sustainable Behaviour in the Hotel Industry: Case Study from Czechia. *Sustainability (Basel), 12*(23), 10027. doi:10.3390/su122310027

Shams, S. R., Vrontis, D., Weber, Y., Tsoukatos, E., & Iaia, L. (Eds.). (2023). *Technology, Business and Sustainable Development: Advances for People, Planet and Profit.* Taylor & Francis. doi:10.4324/9781003293187

Sharma, A., & Singh, B. (2022). Measuring Impact of E-commerce on Small Scale Business: A Systematic Review. *Journal of Corporate Governance and International Business Law, 5*(1).

Sharma, T. (2020). What a waste: confronting consumer food waste behavior in hospitality settings. In *Advances in hospitality and leisure* (pp. 167–176). Emerald Publishing Limited.

Sharma, Y., Sijariya, R., & Gupta, P. (2023). How Deep Learning Can Help in Regulating the Subscription Economy to Ensure Sustainable Consumption and Production Patterns (12th Goal of SDGs). In *Deep Learning Technologies for the Sustainable Development Goals: Issues and Solutions in the Post-COVID Era* (pp. 1–20). Springer Nature Singapore. doi:10.1007/978-981-19-5723-9_1

Singh, B. (2019). Profiling Public Healthcare: A Comparative Analysis Based on the Multidimensional Healthcare Management and Legal Approach. *Indian Journal of Health and Medical Law, 2*(2), 1–5.

Singh, B. (2022). Relevance of Agriculture-Nutrition Linkage for Human Healthcare: A Conceptual Legal Framework of Implication and Pathways. *Justice and Law Bulletin, 1*(1), 44–49.

Singh, B. (2022). COVID-19 Pandemic and Public Healthcare: Endless Downward Spiral or Solution via Rapid Legal and Health Services Implementation with Patient Monitoring Program. *Justice and Law Bulletin, 1*(1), 1–7.

Singh, B. (2023). Blockchain Technology in Renovating Healthcare: Legal and Future Perspectives. In Revolutionizing Healthcare Through Artificial Intelligence and Internet of Things Applications (pp. 177-186). IGI Global.

Singh, B. (2023). Federated Learning for Envision Future Trajectory Smart Transport System for Climate Preservation and Smart Green Planet: Insights into Global Governance and SDG-9 (Industry, Innovation and Infrastructure). *National Journal of Environmental Law, 6*(2), 6–17.

Spiess-Knafl, W. (2022). *Artificial Intelligence and Blockchain for Social Impact: Social Business Models and Impact Finance*. Taylor & Francis. doi:10.4324/9781003218913

Stewart, J., & Gorman, C. (2024). Authentic Assessment and Embedding Sustainable Literacy and Technology Skills in Tourism Education. *Irish Journal of Academic Practice*, *11*(2), 10.

Stewart, J., & Gorman, C. (2024). Authentic Assessment and Embedding Sustainable Literacy and Technology Skills in Tourism Education. *Irish Journal of Academic Practice*, *11*(2), 10.

Synani, K., Abeliotis, K., & Lasaridi, K. (2021). *Food waste in the catering and hospitality sector: generation factors and innovative practices.*

Taneja, A., Sharma, R., Khetrapal, S., Sharma, A., Nagraik, R., Venkidasamy, B., Ghate, M. N., Azizov, S., Sharma, S., & Kumar, D. (2023). Value Addition Employing Waste Bio-Materials in Environmental Remedies and Food Sector. *Metabolites*, *13*(5), 624. doi:10.3390/metabo13050624 PMID:37233665

Tapo, A. A., Traore, A., Danioko, S., & Tembine, H. (2024). Machine Intelligence in Africa: a survey. *arXiv preprint arXiv:2402.02218.*

Tice, P. H. (2024). *The Race to Zero: How ESG Investing will Crater the Global Financial System.* Encounter Books.

. Toivonen, A. (2020). *Sustainable space tourism: an introduction* (*Vol. 3*). Channel view publications.

Tomaszewska, M., Bilska, B., Tul-Krzyszczuk, A., & Kołożyn-Krajewska, D. (2021). Estimation of the scale of food waste in hotel food services—A case study. *Sustainability (Basel)*, *13*(1), 421. doi:10.3390/su13010421

Vargas-Sanchez, A. (2023). Toward a circular tourism industry: The importance of a start-up ecosystem. *Worldwide Hospitality and Tourism Themes*, *15*(6), 625–632. doi:10.1108/WHATT-09-2023-0111

Verma, A., Grobbelaar, W., & Shukla, V. K. (2023). Smart Innovation in the Hospitality and Tourism Industry. In *Computational Intelligence in Urban Infrastructure* (pp. 94–116). CRC Press. doi:10.1201/9781003218715-9

Vind, D. L. (2023). *Danish Design Heritage and Global Sustainability*. Taylor & Francis. doi:10.4324/9781003261018

. Voorhoeve, J. (2021). Fighting Poverty and Violence: How Basic Human Needs Could Be Met.

Workie, E., Kumar, V., Bhatnagar, A., He, Y., Dai, Y., Tong, Y. W., & Fu, C. (2023). Advancing the bioconversion process of food waste into methane: A systematic review. *Waste Management (New York, N.Y.)*, *156*, 187–197. doi:10.1016/j.wasman.2022.11.030 PMID:36493662

Zhang, X., Sheng, Y., & Liu, Z. (2024). Using expertise as an intermediary: Unleashing the power of blockchain technology to drive future sustainable management using hidden champions. *Heliyon*, *10*(1), e23807. doi:10.1016/j.heliyon.2023.e23807 PMID:38226273

# Chapter 14
# Revolutionizing Agricultural Waste Management With Nanotechnology for a Sustainable Future

**Asha Anish Madhavan**
*Amity University, Dubai, UAE*

**Godwin Joseph Nallathambi**
*Amity University, Dubai, UAE*

**Ankita Maria Paul**
*Amity University, Dubai, UAE*

## ABSTRACT

*Agricultural operations, essential for human survival, annually produce between 10 to 14 billion tons of waste, endangering environmental stability. This waste encompasses remnants from crops, agro-industrial activities, livestock, and aquaculture. Common disposal methods like incineration and landfills pose health and environmental risks. Effective waste management is imperative for sustainable agriculture, transforming waste into valuable resources like compost, biogas, and biochar. Nanotechnology contributes to waste management by creating nanostructured materials from agricultural waste and generating bioactive compounds for diverse applications. Nano-bioremediation enhances soil vitality and eliminates pollutants, while nano-biosensors enable precise waste monitoring. Smart delivery systems for agrochemicals reduce environmental impact, and nano-enabled plant disease management enhances crop health. Green nanomaterial synthesis from agricultural biomass promotes resource efficiency. These advancements in nanotechnology are fostering sustainable agricultural waste management practices, reducing the environmental footprint of agriculture, and promoting a circular economy.*

DOI: 10.4018/979-8-3693-4264-0.ch014

## 1. INTRODUCTION

Rapid global population growth and changing food consumption behaviors contribute to the generation of approximately 1.3 billion tons of food loss and waste annually, equating to one-third of the food produced for human consumption (Capanoglu et al., 2022). These trends necessitate corresponding adjustments in agricultural output, potentially exerting additional pressure on natural resources. Recognizing its critical role in addressing food security, poverty reduction, and economic stability, governments, donors, and international organizations are now placing renewed emphasis on this sector. Globally, the agricultural sector generates a substantial 140 billion metric tons of biomass annually.

This significant quantity of biomass presents an opportunity for conversion into valuable energy and essential raw materials (Dethier & Effenberger, 2012). The estimated potential energy yield from converting 50 billion tons of agricultural biomass waste could significantly reduce dependence on fossil fuels, mitigate greenhouse gas emissions, and provide access to renewable energy for approximately 1.6 billion individuals in developing countries currently lacking electricity (Sofowora, 2023). The quantity and composition of agricultural waste are influenced by several factors, including the type of agricultural product, processing methods employed, and the intended utilization of the final product. This waste poses significant environmental hazards, with implications across various domains. Poor waste management practices can lead to water pollution, soil degradation, and air pollution, including through activities such as open burning, exacerbating respiratory and cardiovascular diseases among affected populations. Industrial agriculture practices, including monoculture farming and intensive animal husbandry, further exacerbate environmental degradation and public health concerns. These practices deplete soil nutrients, increase reliance on harmful chemical pesticides, and result in the pollution of air and waterways (Bharagava, 2020).

## 2. TYPES OF AGRICULTURAL WASTE

Agricultural production has experienced a more than threefold increase over the past five decades due to the expansion of arable land, advancements in agricultural technology stemming from the Green Revolution, and the rapid growth of the global population (Duque-Acevedo, et al., 2020). The intensification of industrial processes and rapid population growth, alongside increased demand for consumer goods, has led to a surge in waste generation (Boechat, et al 2017). Proper waste classification is essential for effective recycling policies, enabling rational allocation of resources and appropriate handling based on origin and hazardousness as represented in Fig. 1. By understanding the different types of agricultural waste and the potential risks and opportunities associated with each, we can develop more sustainable practices for managing this waste stream.

### 2.1 Animal Waste

Animal waste like dairy manure and poultry litter are valuable resources for biomass conversion, particularly in producing bio-energy and bio-fertilizers. Regulatory frameworks that focus on waste management also promote the use of animal manure (Koul et al., 2022). The dairy industry, with its mechanized processes and significant amounts of cow manure, is well-suited for utilizing biomass. Poultry litter, rich in methane, can be efficiently converted into bio-fuel and bio-energy through anaerobic digestion,

*Figure 1. Different types of agriculture waste*

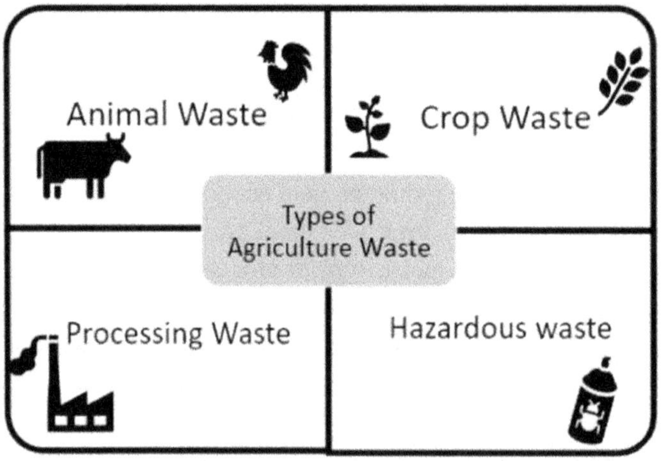


yielding higher methane yields compared to other livestock waste. Moreover, its high organic nitrogen content makes the resulting digestate a premium organic fertilizer (Koul et al., 2022).

## 2.2 Crop Waste

Crop residues, encompassing leaves, corn stovers, seed pods, cocoa pods, sugarcane bagasse, nut shells, palm waste and cotton gin trash constitute a substantial portion of agricultural waste globally. Annual crop residue production is estimated at 2802 million metric tons (Koul et al., 2022). However, only a small proportion of wheat, corn, and rice residues are currently earmarked for bioethanol production or as animal fodder, with the bulk being disposed of or subjected to open-field burning. This practice poses significant health and environmental risks, impeding agricultural sustainability. Rice straw generation alone contributes 731 million metric tons annually worldwide, with Asia contributing 667.6 million metric tons annually (Koul et al., 2022). Projections indicate that by 2030, India's rice straw production will soar to 221.8 million metric tons annually. Wheat straw generation is estimated at approximately 1–3 tons per acre annually, with global production hovering around 354.34 million metric tons annually (Koul et al., 2022).

## 2.3 Processing Waste

By-products and residues are generated during various stages of industrial processing (Sadh et al., 2018). This category encompasses various materials produced across different manufacturing sectors, including food processing, chemical production, pharmaceuticals, textiles, and more (Sadh et al., 2018). Horticulture-based food industries contribute significantly to waste generation, with fruit peel, seeds, and pomace being generated during the preparation and processing of various food products such as cider, fruit juice, jams, jellies, sauces, kinds of ketchup, and pickles. For instance, apple pomace alone accounts for approximately 25%–30% of total biomass from global apple production, which amounts to around 767.75 million metric tons annually (Sadh et al., 2018). This extensive waste generation is observed globally, with significant quantities being produced in countries like China (32.0 million metric tons), the US (15.0 million metric tons), India (1.81 million metric tons), and others (Sadh et al., 2018).

### 2.4 Other Hazardous Waste

Pesticide waste complexity complicates disposal and degradation efforts, exacerbated by equipment rinsing water and spills, particularly prevalent in agriculture where herbicide blends are common. Irrespective of origin, pesticide waste contains concentrated active ingredients and solvents, posing risks of soil and water contamination. Pesticide runoff contaminates surface water and biota, disrupting aquatic ecosystems by hindering top predator populations through growth inhibition and reproductive failure, and posing public health risks through consumption of contaminated fish. Furthermore, windborne transport carries pesticides over vast distances, leading to contamination of aquatic systems far removed from their original application, as evidenced by the presence of tropical/subtropical pesticides in Arctic mammals. Soil pesticide dissipation slows with escalating concentrations, potentially due to microbial toxicity impeding degradation, and heightening risks of runoff and leaching, particularly at higher concentrations. Excessive nutrient discharge, especially phosphorus, leads to eutrophication, causing organoleptic degradation of public water supplies due to increased algal biomass, hypoxia, and potential fish kills in water bodies due to depleted oxygen, and nitrate contamination of groundwater posing public health risks at high levels (Ware et al., 2003; Nagendran et al., 2011).

## 3. IMPORTANCE OF EFFICIENT WASTE MANAGEMENT

Effective waste management is imperative to address environmental, sanitary, and social concerns associated with poorly managed solid waste. Agricultural and industrial residues contain numerous valuable constituents that can be extracted and incorporated into the production of various materials (de Titto & Savino, 2019). Agricultural waste management is the process of collecting, handling, and disposing of waste generated from agricultural activities. Proper agricultural waste management plays a vital role in safeguarding the environment, promoting sustainable agriculture, and ensuring human and animal well-being. Improper waste disposal can lead to water and soil contamination, posing significant health risks to both humans and animals (de Titto, & Savino, 2019). Additionally, agricultural waste contributes to greenhouse gas emissions, particularly methane, when decomposed in landfills. Effective waste management strategies can significantly reduce these emissions and conserve valuable resources. Moreover, sustainable agricultural practices integrate efficient waste management techniques to produce food in an environmentally responsible and economically viable manner (Vaverková et al., 2019).

Organic waste originating from diverse sources, including agricultural operations and municipal waste streams, can be repurposed as organic fertilizers and soil conditioners, thereby enhancing soil quality and fertility. Through the recycling of organic waste materials within agricultural systems, there is a reduction in dependency on synthetic fertilizers, leading to minimized environmental repercussions associated with their manufacture and utilization (Njoku et al., 2019). Furthermore, the integration of organic waste into agricultural practices facilitates improvements in soil structure, moisture retention, and nutrient cycling processes, thereby bolstering long-term soil sustainability and agricultural productivity. Sustainable waste management practices within agricultural contexts not only curtail waste accumulation and environmental contamination but also contribute to the establishment of resilient and sustainable food production systems (Njoku et al., 2019).

While certain countries have been utilizing nutrients derived from waste in agricultural practices for extended periods, research suggests potential environmental concerns associated with the application of these waste materials to soil. These concerns encompass issues such as excessive nitrogen levels, the introduction of pathogens, contamination with heavy metals, soil acidification, and salinization, all of which have the potential to negatively impact the quality of soil and water resources (Ali et al., 2014).

## 3.1 Waste Management Methods

Agricultural waste management is essential for sustainable farming practices and environmental conservation. Various methods can be employed to manage agricultural waste effectively. The primary objective of any waste management system is to optimize the economic value derived from waste resources while ensuring environmental standards are met. Additionally, for practical implementation, the system must be cost-effective and compatible with operational requirements (de Titto & Savino, 2019). Here are some common agricultural waste management methods as illustrated in Fig. 2.

## 3.2 Incineration

Incineration involves the controlled burning of waste to produce inert ash and gases like carbon dioxide and water. Organic compounds, being highly combustible, are preferred for incineration, while inorganic compounds usually transform into oxides during the process. Various types of incinerators, including mass-burn, multiple-hearth, rotary-kiln, fixed-hearth, fluidized-bed, liquid injection, and fume incinerators, cater to different waste types. While the earlier focus was on reducing waste volume, modern incineration also aims at energy generation (Saxena & Jotshi, 1994). Incineration, while offering waste management solutions, presents notable technical and environmental challenges. Combustion processes generate airborne pollutants, including those stemming from incomplete combustion, which pose risks to respiratory health upon inhalation. The complexity of translating the presence of hazardous substances into quantifiable exposure metrics complicates risk assessment methodologies. Factors such as atmospheric

*Figure 2. Classification of common waste management methods*

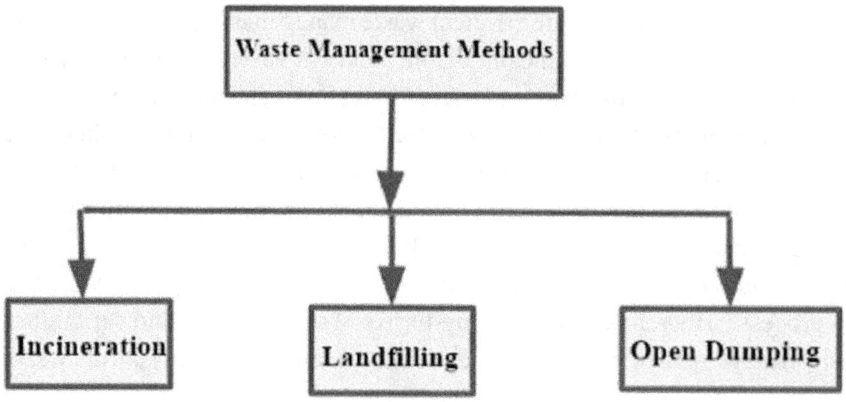

dispersion dynamics, emissions height, and wind patterns significantly influence exposure estimates but are often overlooked or poorly accounted for. The challenge of disentangling the environmental-health relationship in epidemiological studies is further compounded by confounding variables, including demographic factors, socioeconomic status, and lifestyle choices. While advancements in incineration technology have led to reduced emissions in modern facilities compared to older counterparts, concerns persist regarding the long-term health impacts, necessitating thorough investigation and continuous refinement of regulatory frameworks to mitigate potential risks associated with incineration practices (Maheshwari, 2014; Mohan et al., 2024).

## 3.3 Landfilling

Landfilling involves depositing, compacting, and covering waste at suitable sites. Currently, landfilling offers a cost-effective and straightforward method of waste disposal, making it the predominant choice worldwide. Despite a decline in its usage, landfills are expected to remain integral to comprehensive solid waste management systems. Comparative studies of waste management methods, including landfilling, incineration, and composting, indicate that sanitary landfilling or open dumping is widely favored due to its affordability and minimal technical demands across many nations (de Titto & Savino, 2019). However, their environmental impacts are significant and multifaceted. Landfills contribute substantially to anthropogenic greenhouse gas emissions, with the degradation process of deposited waste generating large amounts of methane ($CH_4$) and carbon dioxide ($CO_2$). Additionally, leachate from landfills can contaminate surface and groundwater, posing risks to ecosystems and human health. The emission of noxious gases such as hydrogen sulfide ($H_2S$), volatile organic compounds (VOCs), and acidic gases further exacerbate environmental pollution and can lead to respiratory issues in nearby communities. Landfills also generate various toxic pollutants, including dioxins, polycyclic aromatic hydrocarbons (PAHs), and heavy metals, which pose risks to human health and ecosystems. The presence of bioaerosols and biological agents, such as fungi and bacteria, generated during waste decomposition in landfills, can lead to respiratory diseases and pose occupational health risks to waste handlers and landfill workers. Overall, while landfills offer temporary waste storage solutions, their operation and environmental consequences underscore the urgent need for sustainable waste management practices to mitigate their harmful effects on the environment and public health (Vaverková, 2019).

## 3.4 Open Dumping

In developing nations, the prevalence of open dump sites is widespread, primarily because of limited funding allocated to waste management and a lack of adequately trained personnel (Njoku et al., 2019). In an open dump, waste is deposited directly onto the ground without any attempt to prevent it from infiltrating the soil or reaching underground water sources. Leachate, a harmful mixture of pollutants produced by decomposing waste, contaminates groundwater reserves, making them unsuitable for use. This infiltration of heavy metals, organic contaminants, and dangerous microorganisms puts communities relying on these essential water sources at risk. Moreover, the nearby soil suffers from contamination, affecting agricultural output and disrupting fragile ecosystems. Air quality deteriorates as waste is burned, releasing harmful gases like methane and volatile organic compounds. These not only contribute to climate change but also worsen respiratory illnesses among nearby residents. Additionally, the presence of disease-carrying vectors such as mosquitoes and rodents around open dumps increases the likelihood of epidemics like dengue, malaria, and typhoid fever. The negative effects extend beyond the environment, impacting the socioeconomic dynamics of communities. Property values decline in surrounding areas, impeding local economies (Mohan et al., 2024; Ali et al., 2014).

## 3.5 Sustainable Waste Management Methods

A holistic approach encompassing waste prevention, reduction, reuse, and recycling is imperative. This approach emphasizes converting agricultural waste into valuable resources like compost, biogas, and biochar.

### 3.5.1 Compost in Agriculture

Compost is a cornerstone of sustainable agriculture, serving dual roles as a fertilizer and soil conditioner. It undergoes a controlled, aerobic biodegradation process that transforms raw organic materials into stable, humic-like substances. This process, managed by microbial communities, is characterized by distinct phases dictated by time and temperature. Initially, thermophilic organisms dominate, generating heat through biodegradation, followed by the emergence of mesophilic organisms as the compost cools. The composting process enriches the material with stable organic matter while reducing more degradable components, ultimately fostering carbon sequestration and mitigating $CO_2$ emissions upon soil application. Compost's significance extends beyond soil improvement; it plays a crucial role in combating soil degradation and enhancing carbon storage (Cousins et al., 1997). Compost, being rich in organic matter and nutrients, serves as an effective soil enhancer that bolsters soil structure, fertility, and moisture retention. This fosters optimal soil health, fostering improved plant growth and higher crop yields, consequently boosting agricultural productivity. Composting promotes the aerobic decomposition of organic matter, curbing methane emissions and aiding in climate change mitigation. Additionally, the adoption of composting diminishes the dependency on synthetic fertilizers and pesticides. This leads to cost reductions in production and minimizes detrimental effects on ecosystems and human health (Vatin et al., 2024; Pergola et al., 2018).

Various varieties of compost are available, each possessing unique attributes suited for specific uses and applications. There are two main categories of compost distinguished by the presence or absence of a biologically induced thermophilic phase: traditional compost and vermicompost. Traditional compost is the result of plant and animal residues decomposing at high temperatures (40–70 °C) through the activity of heat-loving microorganisms. This process starts with moderate temperatures (10–40 °C), where easily degradable organic matter is rapidly consumed by mesophilic microorganisms. Then, thermophilic microorganisms raise the temperature to 60°C, breaking down lipids, proteins, and complex carbohydrates. In the final stage, as the material cools down, mesophilic organisms break down the remaining organic matter. Vermicomposting, on the other hand, involves earthworms and mesophilic microorganisms bio-oxidizing and stabilizing organic material without significant heating. Some practitioners combine both methods, initially partially composting at high temperatures before vermicomposting. Vermicomposting stages include acclimatization of worms, hydrolysis of readily degradable organic matter, and curing of more resistant organic matter. Traditional compost selects for adapted thermophilic bacteria during the heating phase, while vermicompost maintains a diverse microbial community throughout the process. Earthworm casts differ from traditional compost due to their mucus cover, which enhances microbial activity. Differences in microbial communities between traditional compost and vermicompost may stem from process disparities or starting materials. Culture-based and molecular studies suggest vermicompost harbors more metabolically diverse microbial populations. Continued advancements in soil microbiology techniques promise deeper insights into compost microbiology and its implications for soil and plant health (Jack & Thies, 2006; Vatin et al., 2024)

### 3.5.2 Biogas as Sustainable Source

Under anaerobic conditions, microorganisms naturally decompose organic matter, producing biogas, a renewable energy source (Scarlat et al., 2018). Anaerobic digestion is a microbial-driven process where organic carbon undergoes successive oxidations and reductions, ultimately oxidized as carbon dioxide, and its most reduced form as methane (Kougias & Angelidaki, 2018). Methane fermentation, a multiphase process, encompasses hydrolysis, acidogenesis, acetogenesis/dehydrogenation, and methanation, each facilitated by diverse microbial consortia under anaerobic conditions. Hydrolytic microorganisms initiate the breakdown of complex organic compounds into acetate, hydrogen, and volatile fatty acids (Angelidaki et al., 1999). Bacteria like Bacteriocides and Clostridia are key players in this phase, along with facultative anaerobes such as Streptococci. Subsequently, obligate hydrogen-producing acetogenic bacteria like Acetobacterium woodii and Clostridium acetium convert volatile fatty acids into acetate and hydrogen. Acetogenic bacteria, crucial for this step, require a low hydrogen partial pressure for optimal function. Methanogens, strict anaerobes like Methanosarcina barkeri, then convert acetate or hydrogen and carbon dioxide into methane (Bagi et al., 2007). Maintaining a balance between the degradation stages is essential for optimal biogas production; disruptions in this balance can lead to pH variations and inhibition of methane production. The kinetics of hydrolysis, particularly of complex substrates like cellulose, rely on factors such as particle size, enzyme production, pH, and temperature (Angelidaki et al., 1999; Gavala et al., 2003). Although the basic metabolic pathways are understood, microbial dynamics within the consortia remain largely unexplored. Molecular techniques like fluorescence in situ hybridization offer insights into microbial communities, aiding in process optimization. Temperature control is critical during digestion, with mesophilic (35–42 °C) and thermophilic (45–60 °C) conditions influencing microbial diversity and reaction rates. Ammonia toxicity, largely affecting thermophilic processes, requires strategies for recovery and mitigation. Optimal pH levels (7.0–8.0) are crucial for methane formation, with volatile fatty acids acting as key intermediates. Nutrient requirements, including macronutrients (carbon, phosphorus, sulfur) and micronutrients (iron, nickel, cobalt, selenium, molybdenum, and tungsten), are essential for microbial growth and activity. Maintaining a balanced nutrient ratio is crucial for stable and efficient biogas production.

### 3.5.3 Biochar

Biochar, derived from organic feedstock through controlled thermal combustion with restricted oxygen, presents a promising avenue for waste utilization. Agricultural residues, municipal solid waste, and even sludge from wastewater treatment, rich in carbon and nutrients, serve as viable sources for biochar production. This carbon-rich material offers distinctive advantages including high carbon content, substantial surface area, and structural stability (Lehmann & Joseph, 2015). Primarily employed for adsorption, biochar's efficacy hinges on its physiochemical properties such as surface area, pore distribution, and functional groups. Production at higher temperatures enhances these properties despite decreasing biochar yields. To optimize biochar's performance, modifications involving acids, alkalis, oxidizing agents, and metal ions are employed to tailor its surface properties (Ahmad et al., 2012). A recent study investigated various biochar conversion techniques, revealing a trade-off between biochar and bio-oil yields dependent on processing parameters. Longer residence times (up to 4 hours) at moderate temperatures reaching up to 500°C favor biochar production (15-35%), while bio-oil production remains moderate (30-50%). Conversely, shorter residence times (up to 2 seconds) prioritize bio-oil, with yields reaching 50-70% (Xie et al., 2015). Thermochemical processes like pyrolysis and carbonization are crucial methods for converting biomass into valuable products. Pyrolysis, conducted in the absence of air at temperatures exceeding 400 °C, converts biomass into biochar, a solid material rich in carbon and other elements. This process can occur through slow, intermediate, or fast modes, with slow pyrolysis typically yielding higher biochar quantities (Kung et al., 2015). For instance, researchers demonstrated the practical application of biochar production with a 15-33% efficiency using a specific type of gasifier for processing rice husk, showcasing its potential for on-farm implementation (Steiner et al., 2018). Carbonization, essentially a slow pyrolysis process specifically aimed at biochar production, has a long history dating back millennia. It involves gradually heating biomass in the absence of air to around 400 °C over an extended period. The heat source for carbonization can be derived directly from the reaction itself, through fuel gases generated from the feedstock, by indirectly heating the reactor walls with hot gases, or through non-gaseous materials like sand heating the reactor walls (Basu, 2018)

## 4. ROLE OF NANOTECHNOLOGY IN AGRICULTURAL WASTE MANAGEMENT

Nanotechnology stands as a transformative force in revolutionizing sustainable agricultural waste management practices, offering innovative solutions to address the multifaceted challenges inherent in traditional waste disposal methods (Ghidan & Antary, 2019). Leveraging nanoscale methodologies and materials, this cutting-edge field enables the efficient extraction and utilization of bioactive compounds embedded within agricultural waste streams, presenting unprecedented avenues for application across various industries (Ghidan & Antary, 2019). Moreover, nanotechnology facilitates the creation of nanostructured materials from agricultural residues, fostering the development of high-value products with enhanced functionalities. These advancements not only elevate the economic value of agricultural waste but also contribute to the sustainable utilization of resources by transforming what was once considered waste into valuable commodities (Elizabath et al., 2019). By embracing nanotechnology-centric strategies, agricultural waste can be repurposed effectively, leading to a significant reduction in environmental impacts associated with waste disposal while propelling the agricultural sector towards a circular economy paradigm. This approach not only addresses the pressing issue of waste management but also underscores the importance of resource efficiency and sustainability in agricultural practices. Additionally, nanotechnology plays a crucial role in developing environmentally friendly alternatives to conventional agricultural inputs, such as pesticides and fertilizers (Elizabath et al., 2019). Nanostructured materials enable more precise delivery and application of agricultural inputs, resulting in targeted treatments that minimize adverse effects on the environment and non-target organisms (Elizabath et al., 2019). In essence, nanotechnology serves as a catalyst for the transition towards a circular economy model in agriculture, where the valorization of agricultural waste streams is central to achieving sustainability goals. Its potential to revolutionize waste management practices while simultaneously enhancing resource efficiency underscores its pivotal role in shaping the future of agriculture towards a more sustainable and environmentally conscious direction (Elizabath et al., 2019).

## 4.1 Nano-Bioremediation for Waste Degradation

Nanotechnology can be employed in bioremediation processes to enhance the degradation of agricultural waste. The various applications of nanomaterials in bioremediation is illustrated in Fig. 3. Nanoparticles can improve the efficiency of microbial degradation by facilitating the transport of microorganisms to targeted waste sites, promoting faster and more effective decomposition. This method employs naturally occurring organisms to convert harmful compounds into less hazardous or benign forms, particularly in water and soil (Hidangmayum et al., 2023). In agricultural settings, nano-bioremediation emerges as a remedial strategy for eliminating a spectrum of pollutants, encompassing inorganic, organic, and emerging contaminants, thereby augmenting soil vitality, stimulating botanical proliferation, and ameliorating degraded terrains (Hidangmayum et al., 2023). Various nanomaterials, including metal oxides and noble metals like titanium dioxide and bimetallic forms, have been developed for environmental remediation purposes. Among these, nanoscale zerovalent iron (nZVI) and its derivatives have emerged as particularly significant in nanoremediation (Shah et al., 2020). They have shown promising results in degrading persistent organic compounds and removing toxic metals such as arsenic and chromium. The preferred method for utilizing these NMs is in situ treatment, which involves injecting them into contaminated porous sites such as sediments, aquifers, and soil due to its cost-effectiveness and scalability (Shah et al., 2020). However, challenges such as the sequential reduction of reactivity and agglomeration of pollutants hinder the widespread use of nZVI in bioremediation (Shah et al., 2020). To tackle these challenges, endeavors have been directed toward enhancing the reactivity, stability, and mobility of nanoscale zerovalent iron (nZVI) by applying surface coatings of organic polymers such as lactate and guar gum. These modifications have yielded improved characteristics, leading to enhanced site decontamination efficacy (Shah et al., 2020). Furthermore, carbon nanotubes have exhibited notably higher dioxin absorption rates compared to activated carbon owing to their increased surface area in nanostructured form. Anionic and hydrophilic carbon-supported zerovalent iron nanoparticles have also been employed for the dehalogenation of chlorinated hydrocarbons present in groundwater and soil environments (Shah et al., 2020). Singh et al. (Singh et al., 2013) documented the integrated remediation of Lindane-contaminated soil utilizing carboxymethyl cellulose-coated Pd/Fe nanoparticles (CMC-Pd/Fe) (Singh et al., 2013). Their findings revealed that suspensions containing 40 mg/L of nanoparticles yielded optimal results for microbial growth. Through the amalgamation of biotic and abiotic systems, they achieved a contaminant degradation rate of 99% within a mere 6-day period, surpassing the efficacy of individual treatments. This study underscores the synergistic effects of nanoparticle-based remediation and microbial activity in enhancing contaminant removal from soil environments (Singh et al., 2013).

## 4.2 Nanobiosensors in Agricultural Waste Management

The Fusion of nanotechnology and biotechnology has led to the evolution of sensors which is illustrated in Fig. 4. In agriculture, a variety of nano-biosensors are employed, such as e-Nose, e-Tongue, electrochemical nanosensors, optical nanosensors, nano-barcode technology, wireless nanosensors, and wireless sensor networks (Hakeem & Pirzadah, 2020). Electrochemical nanosensors, predicated on chemical interactions between nanostructured biomolecules and target analytes, offer heightened sensitivity and rapidity, finding utility in soil analysis, water management, and pesticide application. Optical nanosensors exploit alterations in optical signals, notably fluorescence, for spectroscopic evaluations, enabling dynamic

*Figure 3. The utilization of nanomaterials for bioremediation*

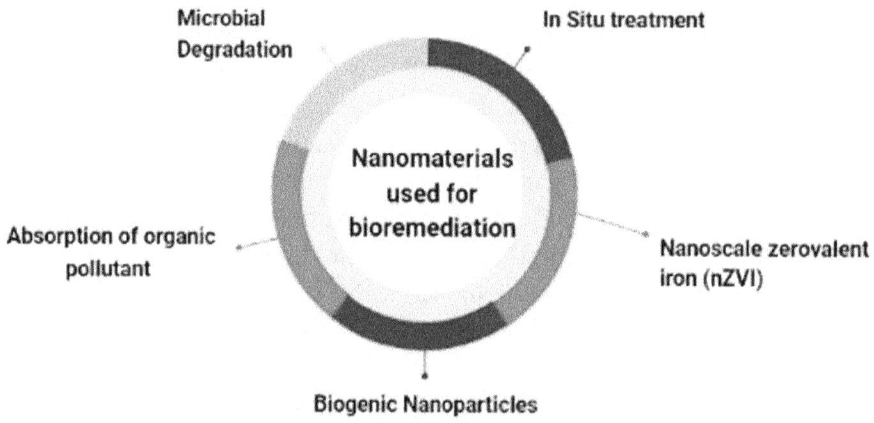


imaging of plant root systems and protein labeling. Nano-barcoding, employing metallic and magnetic nanoparticles, enables swift and dependable pathogen detection in food samples and crop diseases. e-Nose and e-Tongue, emulating human sensory mechanisms, detect gases, odors, and flavors, aiding in precise food quality assessment and plant ailment diagnosis. Wireless nanosensors and sensor networks deliver real-time environmental monitoring, capturing data on soil moisture and temperature, thereby refining precision agriculture and greenhouse control (Hakeem & Pirzadah, 2020). Nanocellulose fibers, for instance, have been successfully produced and hold potential for diverse applications. Similarly, rice husk, a byproduct of rice milling and traditionally considered waste, can be utilized in nano biosensor development, particularly when combined with silica for enhanced functionality. While these materials have shown efficacy in various agricultural applications, their adoption has been hindered by the high costs associated with their design and processing (Thakur et al., 2022). In sustainable agriculture, maximizing the efficiency of water usage is crucial. To address this challenge, nano-biosensors such as nano-hydrogel play a significant role. Nano-hydrogel can absorb and release water molecules, conserving water resources. This technology holds great significance in ensuring the optimal management of water resources for agricultural purposes (Vundavalli et al., 2015). Various nano-biosensors have been created using polymeric, lipid nanoparticles, and inorganic polymeric, via diverse fabrication methods such as emulsification, polymerization, and ionic gelation (Fogel & Limson, 2016). These techniques aim to enhance plant protection and productivity. Through the utilization of intelligent nano-systems derived from these techniques, it becomes viable to mitigate leaching while concurrently enhancing nutrient absorption by plants. Moreover, these systems can contribute to the reduction of eutrophication by restraining the nitrogen transfer to groundwater (Liu & Lal, 2015). Additionally, the employment of nanomaterials holds promise in augmenting the effectiveness of pesticides by improving solubility, fortifying resistance against photodegradation, and hydrolysis, and facilitating a precise and regulated release directed toward specific target organisms (Mishra & Singh, 2015; Grillo et al., 2016; Nuruzzaman et al., 2016). Vamvakaki and Chaniotakis (Vamvakaki & Chaniotakis, 2007) devised nano biosensors employing liposomes for the detection of dichlorvos and paraoxon, prevalent organophosphorus pesticides, along with assessing overall toxicity in drinking water samples (Nath et al., 2023). Employing fluorescence as the transducer mechanism, the biosensors demonstrated exceptional sensitivity, minimal detection thresholds, and a broad dynamic range. Encapsulation of acetylcholinesterase within liposomal nano environments served to significantly enhance enzyme stability, with pesticide-induced inhibition serving as the basis for activity measurement (Mittal et al., 2020).The utilization of nano-sized liposomes provided a conducive milieu for enzyme stabilization, enabling their utility in fluorescence-based biosensing.

## 4.3 Nanostructured Smart Delivery Systems

Nanotechnology enables the development of smart delivery systems for agrochemicals. Nanoformulations can protect active ingredients, control release rates, and enhance the efficacy of pesticides and fertilizers. This not only reduces the amount of chemicals required but also minimizes the environmental impact associated with their application. The utilization of nanotechnology-derived bioactive compound formulations extends across a spectrum of applications, demonstrating heightened efficacy in agricultural pest management, food preservation, and pharmaceuticals. In the agricultural domain, nanoformulations offer innovative solutions for sustainable pest control strategies, mitigating the reliance on conventional chemical pesticides and fostering environmentally friendly agricultural practices (Vamvakaki & Chaniotakis, 2007).

Studies conducted by Rafi et al (Rafi et al., 1997). have shown that plants depend heavily on silica for their growth, and in the absence, they become weaker and more susceptible to various stresses, from both biotic and abiotic factors. Additionally, research by Ahmad et al. (Ahmad et al., 2007) indicated that silica can help plants cope with drought and even increase their growth under stressful conditions. This was particularly evident in maize plants treated with specially designed silica nanoparticles, which performed better than plants treated with regular silica or left untreated altogether (Ahmad et al., 2007). Chitosan stands as a prominent polymer in drug delivery due to its advantageous structural and biological characteristics. Notably, its cationic nature, solubility in acidic aqueous environments, and

*Figure 4. Nanobiosensor for sensing purpose*

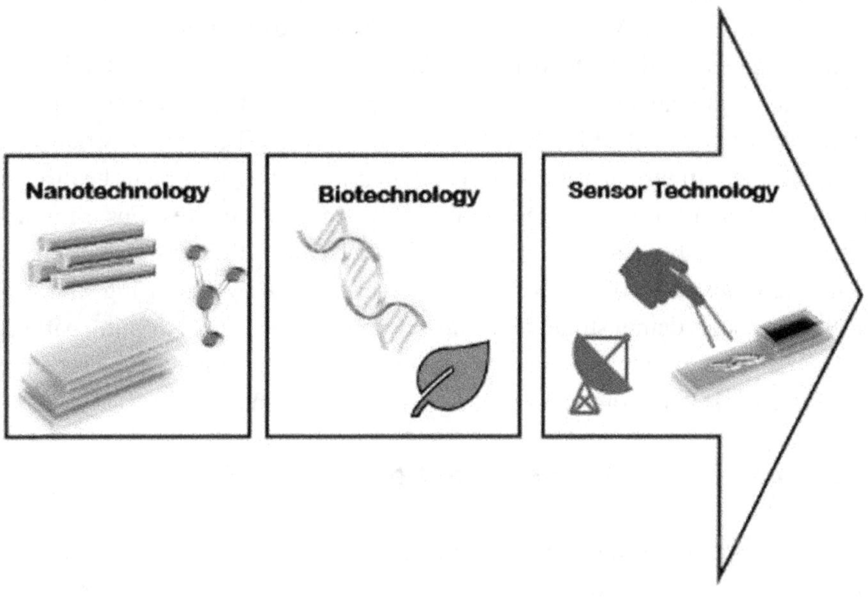

biodegradability render it particularly appealing (Kashyap et al., 2015). Despite its low solubility under physiological conditions with a pH of 7.4, which is attributed to its weakly basic nature (having a pKa range of 6.2–7), chitosan can be synthesized by the removal of the acetate group from chitin through amide hydrolysis under alkaline conditions (using concentrated NaOH) or enzymatic hydrolysis employing chitin deacetylase (Suh & Matthew, 2000). The amine groups present in chitosan readily form complexes with various oppositely charged polymers, thus expanding its application range (Sonia & Sharma, 2011). Additionally, chitosan offers versatility in formulation development, being available in a diverse range of molecular weights (ranging from 500 to 1400 KDa) and degrees of acetylation, while its amine groups can be subject to chemical modifications. Its affinity for plant surfaces enhances its utility in agriculture by extending the interaction time between agrochemicals and the target absorptive surfaces (Tiyaboonchai et al., 2003). Chitosan nanoparticles exhibit enhanced absorption capabilities, facilitating the uptake of active compounds through cell membranes and thereby improving molecular bioavailability. These attributes collectively position chitosan as a promising drug delivery system for sustainable agricultural practices (Kashyap et al., 2015).

## 4.4 Nano-Enabled Plant Disease Management

Nanoparticles can be utilized for the targeted delivery of antifungal agents and other disease-controlling substances to plants. This approach minimizes the use of chemical pesticides, reduces environmental contamination, and enhances the overall health of crops. In the realm of food preservation, nanoencapsulated bioactive compounds exhibit superior antimicrobial properties, prolonging shelf life and enhancing food safety. The integration of nanotechnology paves the way for the development of advanced delivery systems tailored for bioactive compounds. Nanostructured carriers enable controlled release profiles, ensuring optimal dosing and sustained therapeutic effects in pharmaceutical applications (Nath et al., 2023). Furthermore, the precise targeting mechanisms enabled by nanotechnology facilitate the selective delivery of bioactive compounds to specific cellular or tissue targets, thereby augmenting therapeutic outcomes while minimizing off-target effects. The synergistic amalgamation of nanotechnology with agricultural waste valorization harbors significant potential for unlocking the inherent value of bioactive compounds (Noman et al., 2023). Wheat plants were enhanced using phytogenic silica nanoparticles which showed innate defense against Rhizoctonia solani. Similarly, chitosan-coated iron nanoparticles, synthesized by Bacillus aryabhattai RNT7, trigger the expression of PR proteins and antioxidant enzyme genes in rice, combating bacterial leaf blight (Noman et al., 2023). These nanoparticles also act as carriers for essential nutrients, improving nutrient uptake by plants, boosting plant health, and increasing resistance to pathogens. For instance, chemogenic sulfur nanoparticles suppress Fusarium wilt in tomatoes by enhancing sulfur accumulation and plant biomass (Noman et al., 2023). Similarly, chemogenic silica nanoparticles enhance watermelon plant resistance to Fusarium wilt by increasing silicon concentration in plant tissues. Nanoparticles demonstrate direct antimicrobial effects by disrupting pathogen structures and functions, as observed with phytogenic zinc and chitosan nanoparticles, which exhibit significant antibacterial activity against Xanthomonas oryzae pv. Oryzae (Noman et al., 2023).

## 4.5 Green Nanomaterial Synthesis From Agricultural Biomass

Employing innovative nanoengineering methodologies, these agricultural byproducts can be converted into value-added nanoproducts with versatile functionalities (Salama et al., 2021). Nanostructured materials derived from agro-waste demonstrate applicability across diverse sectors, encompassing environmental remediation, food packaging, and agricultural applications. This strategy not only enhances the value of agricultural waste streams but also fosters the development of sustainable nanomaterials with negligible environmental footprints. Agricultural residues, such as crop stalks, husks, and biomass, serve as ample and renewable feedstocks for the production of nanostructured materials. Nanotechnology presents promising opportunities for extracting and utilizing bioactive compounds from agricultural waste streams, thereby creating new avenues for application (Azeez et al., 2017). The integration of nanotechnology with the valorization of agricultural waste has substantial potential for unlocking the value of bioactive compounds and broadening their applications across multiple industries, thereby fostering sustainable waste management practices and promoting resource efficiency (Azeez et al., 2017).

### 4.5.1 Utilization of Nanocellulose From Agro-Waste

Agro-waste presents a valuable resource for the development of nanocellulose, a versatile material with exceptional mechanical properties, biocompatibility, and renewability (Salama et al., 2021). Derived from cellulose-rich agricultural residues like wheat straw or sugarcane bagasse, nanocellulose holds promise for various applications, including reinforcing nanocomposites, biomedical scaffolds, and packaging materials (Salama et al., 2021). Its abundance and eco-friendly nature make it an attractive option for sustainable material development across industries.

### 4.5.2 Production of Biochar Nanoparticles from Agro-Waste

In addition to nanocellulose and nanocomposites, agro-waste can be harnessed for the production of biochar nanoparticles through pyrolysis processes. Biochar, characterized by its high surface area, porosity, and carbon content, exhibits adsorption capabilities beneficial for soil remediation, water purification, and agricultural applications. As carrier systems for agricultural inputs, biochar nanoparticles offer a sustainable solution for delivering nutrients and pesticides effectively while minimizing environmental impact (Nguyen et al., 2017). Velmurugan et al (Velmurugan et al., 2014a; Velmurugan et al., 2014b) undertook the fabrication of Ag and Au nanoparticles (silver and gold) utilizing the shell of cashewnut, showcasing their efficacy in combating fish pathogens. Chicken eggshell membrane (ESM), rich in glycoproteins such as collagen and amino acids, has been investigated for its templating capabilities in fabricating macroporous materials and in applications for heavy metal retrieval and biosensing (Devi et al., 2012). Annona squamosa peel extract was harnessed for AgNPs synthesis, wherein water-soluble ketone and hydroxyl groups played pivotal roles in reduction and stabilization processes (Kumar et al., 2012). Likewise, orange peel aqueous extract (OPE) facilitated the green synthesis of AgNPs, demonstrating stability over 6 months and manifesting antimicrobial and anti-cancer properties against rat glial tumor cells. $Mn_3O_4$ NPs with super capacitive properties were successfully synthesized utilizing banana peel extract, owing to constituents such as pectin, cellulose, hemicellulose, and lignin (Ibrahim et al., 2015). Various agricultural byproducts, such as rice bran, watermelon rind, chicken feathers, cotton fibers, and industrial milk, have also been utilized for metallic nanoparticle synthesis, highlighting a diverse range of applications (Saratale et al., 2018). Leveraging agro-waste for nanostructured material synthesis presents multifaceted benefits, ranging from the development of high-performance materials to environmental remediation and agricultural sustainability (Nguyen et al., 2017).

## 5. FUTURE PROSPECTS

Liposome-mediated gene delivery enables precise genetic modifications in crops, offering the potential for enhanced biodegradation of agricultural waste and optimization of biofuel production, contributing to more sustainable agriculture. In this method, the target gene is introduced into plants via liposomes, intricate structures comprising dual-layered phospholipid cholesterol molecules, diverging from conventional plasmid-based approaches. The process initiates with sonication, facilitating DNA transfer into liposomes, which subsequently exploit an endocytosis mechanism for precise delivery of the genetic material into the plant's nucleus (Hemalatha et al., 2023). Cutting-edge nano-enabled packaging enhances food preservation and prolongs shelf-life post-harvest. Advanced antimicrobial composite films, incorporating silver nanoparticles, zinc oxide nanorods, or graphene, effectively inhibit microbial proliferation on various food items. Oxygen-scavenging nano-polymer films mitigate oxidative deterioration, while nanosensors detect gases indicating food spoilage. These intelligent nanopackaging solutions demonstrate the potential for enhancing food safety and mitigating substantial losses caused by spoilage (Kansotia et al., 2024).

## 6. CONCLUSION

The exponential growth in agricultural production has resulted in a corresponding increase in waste generation, posing significant environmental challenges globally. Effective waste management strategies are crucial to mitigate pollution and environmental degradation associated with agricultural waste, which comprises vegetative residues, animal excreta, and processing byproducts. The adoption of nanotechnology presents promising solutions for agricultural waste management, offering novel approaches to address waste complexities and enhance resource efficiency. Nanostructured materials derived from agricultural residues demonstrate versatile functionalities, with nanotechnology enabling the extraction and utilization of bioactive compounds for innovative applications in pest management, food preservation, and pharmaceuticals. Nano-bioremediation offers efficient solutions for waste degradation, while nano biosensors enable precise monitoring and management of agricultural processes. Additionally, nano-enabled plant disease management strategies offer targeted solutions for crop health, while the synthesis of green nanomaterials from agricultural biomass presents opportunities for sustainable material development. Looking ahead, future prospects include liposome-mediated gene delivery for enhanced biodegradation and nano-enabled packaging for improved food preservation. As the global population continues to increase, coupled with economic expansion and dietary shifts, the agricultural sector faces significant challenges in meeting escalating demand while mitigating environmental impacts. Sustainable waste management practices and the integration of nanotechnology are imperative steps toward achieving food security, environmental sustainability, and economic prosperity in agriculture. Through innovation, collaboration, and regulatory frameworks, stakeholders can pave the way for a resilient and thriving agricultural sector that meets the needs of present and future generations.

## REFERENCES

Ahmad, M., Lee, S. S., Dou, X., Mohan, D., Sung, J. K., Yang, J. E., & Ok, Y. S. (2012). Effects of pyrolysis temperature on soybean stover-and peanut shell-derived biochar properties and TCE adsorption in water. *Bioresource Technology, 118*, 536–544. doi:10.1016/j.biortech.2012.05.042 PMID:22721877

Ali, S. M., Pervaiz, A., Afzal, B., Hamid, N., & Yasmin, A. (2014). Open dumping of municipal solid waste and its hazardous impacts on soil and vegetation diversity at waste dumping sites of Islamabad city. *Journal of King Saud University. Science, 26*(1), 59–65. doi:10.1016/j.jksus.2013.08.003

Angelidaki, I., Ellegaard, L., & Ahring, B. K. (1999). A comprehensive model of anaerobic bioconversion of complex substrates to biogas. *Biotechnology and Bioengineering, 63*(3), 363–372. doi:10.1002/(SICI)1097-0290(19990505)63:3<363::AID-BIT13>3.0.CO;2-Z PMID:10099616

Bagi, Z., Ács, N., Bálint, B., Horváth, L., Dobó, K., Perei, K. R., Rákhely, G., & Kovács, K. L. (2007). Biotechnological intensification of biogas production. *Applied Microbiology and Biotechnology, 76*(2), 473–482. doi:10.1007/s00253-007-1009-6 PMID:17503035

Basu, P. (2018). *Biomass gasification, pyrolysis and torrefaction: practical design and theory*. Academic press.

Bharagava, R. N. (Ed.). (2020). *Emerging eco-friendly green technologies for wastewater treatment. Berlin*. Springer. doi:10.1007/978-981-15-1390-9

. Boechat, C. L., de Santana Arauco, A. M., Duda, R. M., de Sena, A. F. S., de Souza, M. E. L., & Brito, A. C. C. (2017). Solid waste in agricultural soils: an approach based on environmental principles, human health, and food security. *Solid waste management in rural areas*, 81-102.

De Titto, E., & Savino, A. (2019). Environmental and health risks related to waste incineration. *Waste Management & Research, 37*(10), 976–986. doi:10.1177/0734242X19859700 PMID:31319775

Dethier, J. J., & Effenberger, A. (2012). Agriculture and development: A brief review of the literature. *Economic Systems, 36*(2), 175–205. doi:10.1016/j.ecosys.2011.09.003

Duque-Acevedo, M., Belmonte-Ureña, L. J., Cortés-García, F. J., & Camacho-Ferre, F. (2020). Agricultural waste: Review of the evolution, approaches and perspectives on alternative uses. *Global Ecology and Conservation, 22*, e00902. doi:10.1016/j.gecco.2020.e00902

Elizabath, A., Babychan, M., Mathew, A. M., & Syriac, G. M. (2019). Application of nanotechnology in agriculture. *Int. J. Pure Appl. Biosci, 7*(2), 131–139. doi:10.18782/2320-7051.6493

Fogel, R., & Limson, J. (2016). Developing biosensors in developing countries: South Africa as a case study. *Biosensors (Basel), 6*(1), 5. doi:10.3390/bios6010005 PMID:26848700

Food and Agriculture Organization. (2017). The state of food and agriculture 2017: leveraging food systems for inclusive rural transformation (pp. xvii+-160). FAO.

Gavala, H. N., Angelidaki, I., & Ahring, B. K. (2003). Kinetics and modeling of anaerobic digestion process. *Advances in Biochemical Engineering/Biotechnology, I*, 57–93. doi:10.1007/3-540-45839-5_3 PMID:12747561

Ghidan, A. Y., & Al Antary, T. M. (2019). Applications of nanotechnology in agriculture. In *Applications of nanobiotechnology*. IntechOpen.

Grillo, R., Abhilash, P. C., & Fraceto, L. F. (2016). Nanotechnology applied to bio-encapsulation of pesticides. *Journal of Nanoscience and Nanotechnology*, *16*(1), 1231–1234. doi:10.1166/jnn.2016.12332 PMID:27398594

Hakeem, K. R., & Pirzadah, T. B. (2020). *Nanobiotechnology in Agriculture*. Springer International Publishing. doi:10.1007/978-3-030-39978-8

Hemalatha, M. Vinita, Sravanalakshmi, G., Kotibagar, B. C., & Megha. (2023). Nanotechnology for Sustainable Agriculture: Current Trends and Future Prospects. In Modern Nanotechnology: Volume 1: Environmental Sustainability and Remediation (pp. 43-75). Cham: Springer International Publishing

Hidangmayum, A., Debnath, A., Guru, A., Singh, B. N., Upadhyay, S. K., & Dwivedi, P. (2023). Mechanistic and recent updates in nano-bioremediation for developing green technology to alleviate agricultural contaminants. *International Journal of Environmental Science and Technology*, *20*(10), 11693–11718. doi:10.1007/s13762-022-04560-7 PMID:36196301

. Jack, A. L., & Thies, J. E. (2006). Compost and vermicompost as amendments promoting soil health. *Biological approaches to sustainable soil systems*, 453-466.

Kansotia, K., Naresh, R., Sharma, Y., Sekhar, M., Sachan, P., Baral, K., & Pandey, S. K. (2024). Nanotechnology-driven Solutions: Transforming Agriculture for a Sustainable and Productive Future. *Journal of Scientific Research and Reports*, *30*(3), 32–51. doi:10.9734/jsrr/2024/v30i31856

Kashyap, P. L., Xiang, X., & Heiden, P. (2015). Chitosan nanoparticle based delivery systems for sustainable agriculture. *International Journal of Biological Macromolecules*, *77*, 36–51. doi:10.1016/j.ijbiomac.2015.02.039 PMID:25748851

Kougias, P. G., & Angelidaki, I. (2018). Biogas and its opportunities—A review. *Frontiers of Environmental Science & Engineering*, *12*(3), 1–12. doi:10.1007/s11783-018-1037-8

Koul, B., Yakoob, M., & Shah, M. P. (2022). Agricultural waste management strategies for environmental sustainability. *Environmental Research*, *206*, 112285. doi:10.1016/j.envres.2021.112285 PMID:34710442

Kung, C. C., Kong, F., & Choi, Y. (2015). Pyrolysis and biochar potential using crop residues and agricultural wastes in China. *Ecological Indicators*, *51*, 139–145. doi:10.1016/j.ecolind.2014.06.043

Lehmann, J., & Joseph, S. (2015). Biochar for environmental management: an introduction. In *Biochar for environmental management* (pp. 1–13). Routledge.

Liu, R., & Lal, R. (2015). Potentials of engineered nanoparticles as fertilizers for increasing agronomic productions. *The Science of the Total Environment*, *514*, 131–139. doi:10.1016/j.scitotenv.2015.01.104 PMID:25659311

Maheshwari, D. K. (Ed.). (2014). *Composting for sustainable agriculture* (Vol. 3). Springer. doi:10.1007/978-3-319-08004-8

Mishra, S., & Singh, H. B. (2015). Biosynthesized silver nanoparticles as a nanoweapon against phytopathogens: Exploring their scope and potential in agriculture. *Applied Microbiology and Biotechnology*, *99*(3), 1097–1107. doi:10.1007/s00253-014-6296-0 PMID:25547832

Mittal, D., Kaur, G., Singh, P., Yadav, K., & Ali, S. A. (2020). Nanoparticle-based sustainable agriculture and food science: Recent advances and future outlook. *Frontiers in Nanotechnology*, *2*, 579954. doi:10.3389/fnano.2020.579954

Muenchen, D. K., Martinazzo, J., de Cezaro, A. M., Rigo, A. A., Brezolin, A. N., Manzoli, A., & Steffens, J. (2016). Pesticide Detection in Soil Using Biosensors and Nanobiosensors. *Biointerface Research in Applied Chemistry*, *6*(6).

Nagendran, R. (2011, January). Agricultural waste and pollution. In *Waste* (pp. 341–355). Academic Press. doi:10.1016/B978-0-12-381475-3.10024-5

Nath, P. C., Ojha, A., Debnath, S., Sharma, M., Sridhar, K., Nayak, P. K., & Inbaraj, B. S. (2023). Biogeneration of Valuable Nanomaterials from Agro-Wastes: A Comprehensive Review. *Agronomy (Basel)*, *13*(2), 561. doi:10.3390/agronomy13020561

Nguyen, V. T. (2017). *Recovering bioactive compounds from agricultural wastes*. John Wiley & Sons. doi:10.1002/9781119168850

Njoku, P. O., Edokpayi, J. N., & Odiyo, J. O. (2019). Health and environmental risks of residents living close to a landfill: A case study of Thohoyandou Landfill, Limpopo Province, South Africa. *International Journal of Environmental Research and Public Health*, *16*(12), 2125. doi:10.3390/ijerph16122125 PMID:31208082

Noman, M., Ahmed, T., Wang, J., Ijaz, M., Shahid, M., Islam, M. S., Azizullah, Manzoor, I., Li, D., & Song, F. (2023). Nano-enabled crop resilience against pathogens: Potential, mechanisms and strategies. *Crop Health*, *1*(1), 15. doi:10.1007/s44297-023-00015-8

Nuruzzaman, M. D., Rahman, M. M., Liu, Y., & Naidu, R. (2016). Nanoencapsulation, nano-guard for pesticides: A new window for safe application. *Journal of Agricultural and Food Chemistry*, *64*(7), 1447–1483. doi:10.1021/acs.jafc.5b05214 PMID:26730488

Pergola, M., Persiani, A., Palese, A. M., Di Meo, V., Pastore, V., D'Adamo, C., & Celano, G. (2018). Composting: The way for a sustainable agriculture. *Applied Soil Ecology*, *123*, 744–750. doi:10.1016/j.apsoil.2017.10.016

Sadh, P. K., Duhan, S., & Duhan, J. S. (2018). Agro-industrial wastes and their utilization using solid state fermentation: A review. *Bioresources and Bioprocessing*, *5*(1), 1–15. doi:10.1186/s40643-017-0187-z

Salama, D. M., Abd El-Aziz, M. E., El-Naggar, M. E., Shaaban, E. A., & Abd El-Wahed, M. S. (2021). Synthesis of an eco-friendly nanocomposite fertilizer for common bean based on carbon nanoparticles from agricultural waste biochar. *Pedosphere*, *31*(6), 923–933. doi:10.1016/S1002-0160(21)60024-3

Saratale, R. G., Saratale, G. D., Shin, H. S., Jacob, J. M., Pugazhendhi, A., Bhaisare, M., & Kumar, G. (2018). New insights on the green synthesis of metallic nanoparticles using plant and waste biomaterials: Current knowledge, their agricultural and environmental applications. *Environmental Science and Pollution Research International*, *25*(11), 10164–10183. doi:10.1007/s11356-017-9912-6 PMID:28815433

Saxena, S. C., & Jotshi, C. K. (1994). Fluidized-bed incineration of waste materials. *Progress in Energy and Combustion Science*, *20*(4), 281–324. doi:10.1016/0360-1285(94)90012-4

Scarlat, N., Dallemand, J. F., & Fahl, F. (2018). Biogas: Developments and perspectives in Europe. *Renewable Energy*, *129*, 457–472. doi:10.1016/j.renene.2018.03.006

Shah, M. P. (Ed.). (2020). *Microbial bioremediation & biodegradation*. Springer. doi:10.1007/978-981-15-1812-6

Singh, R., Manickam, N., Mudiam, M. K. R., Murthy, R. C., & Misra, V. (2013). An integrated (nano-bio) technique for degradation of γ-HCH contaminated soil. *Journal of Hazardous Materials*, *258*, 35–41. doi:10.1016/j.jhazmat.2013.04.016 PMID:23692681

Steiner, C., Bellwood-Howard, I., Häring, V., Tonkudor, K., Addai, F., Atiah, K., Abubakari, A. H., Kranjac-Berisavljevic, G., Marschner, B., & Buerkert, A. (2018). Participatory trials of on-farm biochar production and use in Tamale, Ghana. *Agronomy for Sustainable Development*, *38*(1), 1–10. doi:10.1007/s13593-017-0486-y

Thakur, M., Wang, B., & Verma, M. L. (2022). Development and applications of nanobiosensors for sustainable agricultural and food industries: Recent developments, challenges and perspectives. *Environmental Technology & Innovation*, *26*, 102371. doi:10.1016/j.eti.2022.102371

United Nations Environment Programme. (2009). *Converting Waste Agricultural Biomass into a Resource-Compendium of Technologies*. UN. https://wedocs.unep.org/20.500.11822/7614

Vatin, N. I., Negi, G. S., Yellanki, S. V., Mohan, C., & Singla, N. (2024). Sustainability Measures: An Experimental Analysis of AI and Big Data Insights in Industry 5.0" *BIO Web of Conferences*. Scopus. doi:10.1051/bioconf/20248601072

Vaverková, M. D. (2019). Landfill impacts on the environment. *Geosciences*, *9*(10), 431. doi:10.3390/geosciences9100431

Vundavalli, R., Vundavalli, S., Nakka, M., & Rao, D. S. (2015). Biodegradable nano-hydrogels in agricultural farming-alternative source for water resources. *Procedia Materials Science*, *10*, 548–554. doi:10.1016/j.mspro.2015.06.005

. Ware, G. W., Albert, L. A., Bro-Rasmussen, F., Crosby, D. G., de Voogt, P., Frehse, H., & & Hamilton, D. J. (2003). *Disposal and degradation of pesticide waste. Reviews of Environmental Contamination and Toxicology: Continuation of Residue Reviews*, 123-200.

# Chapter 15
# Green Blockchain Technology for an Eco-Friendly Environment

**Jonas Yomboi**
https://orcid.org/0000-0002-4897-8388
*St. John's Integrated SHS, Ghana*

**Mohammed Majeed**
https://orcid.org/0000-0001-9804-5335
*Tamale Technical University, Ghana*

**Esther Asiedu**
https://orcid.org/0000-0002-4352-1282
*Ghana Communication Technology University, Ghana*

**Clement Nangpiire**
https://orcid.org/0000-0003-0967-0855
*SDD University of Business and Integrated Development Studies, Ghana*

**Fatawu Alhassan**
https://orcid.org/0000-0001-8257-7409
*Tamale Technical University, Ghana*

**Victoria Manu**
*Sunyani Technical University, Ghana*

## ABSTRACT

*As blockchain technology continues its disruptive influence across various sectors, its environmental implications have raised concerns about long-term viability and global repercussions. This chapter explores the concept of "green blockchain" as a framework for sustainable alternatives. It delves into the transition to energy-efficient consensus mechanisms like proof of stake. The chapter also addresses the regulatory landscape, ethical considerations, and future prospects, including the integration of artificial intelligence to optimize blockchain processes. Despite the potential benefits, challenges such as regulatory uncertainty, scalability concerns, and privacy risks underscore the industry's need to navigate carefully and strike a balance between innovation and environmental responsibility. The chapter advocates for a collective effort to build a culture of sustainability within the blockchain community and highlights successful community-led initiatives and regulatory frameworks as crucial elements in mitigating the environmental impact of blockchain technology.*

DOI: 10.4018/979-8-3693-4264-0.ch015

## 1. INTRODUCTION

Blockchain technology has become a disruptive influence across multiple sectors due to its decentralised and transparent characteristics (Mubarik et al., 2021). Nonetheless, this technological marvel is not devoid of environmental disadvantages, which raises concerns regarding its long-term viability and global repercussions. Fundamentally, blockchain operates as a decentralised and distributed ledger, wherein transactions are documented across an interconnected system of computers (Polas et al., 2022). Every individual transaction is contained within a "block," which is interconnected in a chronological fashion to form a record that is both secure and impervious to tampering. Although the technology rose to prominence alongside the emergence of cryptocurrencies like Bitcoin, its utility transcends digital currencies in a substantial way. Decentralised financing, smart contracts, and supply chain administration are merely a few of the numerous applications of blockchain.

Blockchain is predicated on the tenets of decentralisation, immutability, and transparency. Decentralisation mitigates the potential for manipulation or corruption by obviating the necessity for a central governing body (Joshi et al., 2024; Nygaard & Silkoset, 2023). Immutability guarantees that data entered onto the blockchain is resistant to modification or deletion. By permitting all network participants to observe and authenticate transactions, transparency cultivates a sense of confidence among users. Notwithstanding its capacity for profound change, blockchain technology has encountered censure regarding its ecological ramifications, predominantly attributable to the energy usage inherent in specific consensus mechanisms, including Proof of Work (PoW) (Bradu et al., 2022). In order to validate transactions and append blocks to the blockchain, miners are obligated to solve intricate mathematical riddles using PoW, the consensus algorithm employed by Bitcoin. This procedure necessitates a significant quantity of computational resources and, as a result, energy. Significant environmental challenges are presented by blockchain technology. It has been suggested that the energy usage linked to cryptocurrency mining is comparable in magnitude to the carbon emissions of entire nations. Given the increasing impacts of climate change and the pressing demand for environmentally favourable solutions, it is apparent that a more sustainable approach to blockchain technology is imperative.

In light of the expanding adoption of blockchain technology, it is critical to investigate sustainable alternatives and mitigate its environmental impact (Mohan et al., 2024; Bada et al., 2021). It is imperative to comprehend the distinct environmental obstacles that blockchain presents in order to formulate well-informed resolutions that harmonise technological advancement with ecological accountability. In order to underscore the gravity of the matter, a multitude of studies have emphasised the energy consumption and carbon emissions of widely used blockchain networks. As an illustration, the annual energy consumption of the Bitcoin network in isolation was estimated by Digiconomist to be comparable to that of several minor countries (Alzoubi & Mishra, 2023; Göçen & Kurubacak, 2023; Garg et al., 2024). The urgency for immediate action to enhance the environmental sustainability of blockchain technology is highlighted by these alarming statistics. As a resolution to the environmental issues that conventional blockchain approaches present, the notion of "green blockchain" materialises as a framework focused on finding solutions. The term "green blockchain" pertains to the integration of environmentally sustainable practises, technologies, and consensus mechanisms that aim to reduce energy usage and adverse effects on the environment (Sharma et al., 2020). This chapter seeks to introduce and explore various aspects of green blockchain, including the transition to more energy-efficient consensus mechanisms like Proof of Stake (PoS) and the adoption of sustainable mining practices (Khan et al., 2021). By examining successful case studies and community-led initiatives, the chapter aims to showcase the feasibility and benefits of integrating green principles into blockchain technology.

## Conceptual Framework

The conceptual framework for the study on "Green Blockchain Technology for Eco-Friendly Environment" revolves around a central theme - "Green Blockchain Technology." This central focus is supported by four key pillars: (a) Environmental Challenges in Blockchain, (b) Green Blockchain Solutions, (c) Regulatory and Ethical Considerations, and (d) Future Prospects and Challenges. These pillars signify interconnected aspects influencing and shaping the landscape of environmentally sustainable blockchain practices. The foundational understanding of blockchain technology, its energy consumption implications, and the broader environmental impact sets the stage. Green Blockchain Solutions encompass practical initiatives like green mining and community-led sustainability projects. Regulatory and Ethical Considerations stand as integral components, influencing both the present and future trajectory of the blockchain industry. Education and advocacy play a crucial role, fostering a culture of sustainability. The study also explores emerging technologies, particularly the integration of artificial intelligence, and anticipates future challenges and opportunities in achieving widespread adoption of green blockchain practices.

# 2. LITERATURE REVIEW

## 2.1 Understanding Blockchain Technology

The decentralised and transparent blockchain technology has grown into a revolutionary force that has spread across many sectors (Adeniyi et al., 2022; Joshi et al., 2024). Blockchain technology is built around the principle of decentralisation. The absence of a governing body at the network's core is a hallmark of decentralised systems. Nodes are instead used to describe the people who share data and control. This

*Figure 1. Conceptual framework*

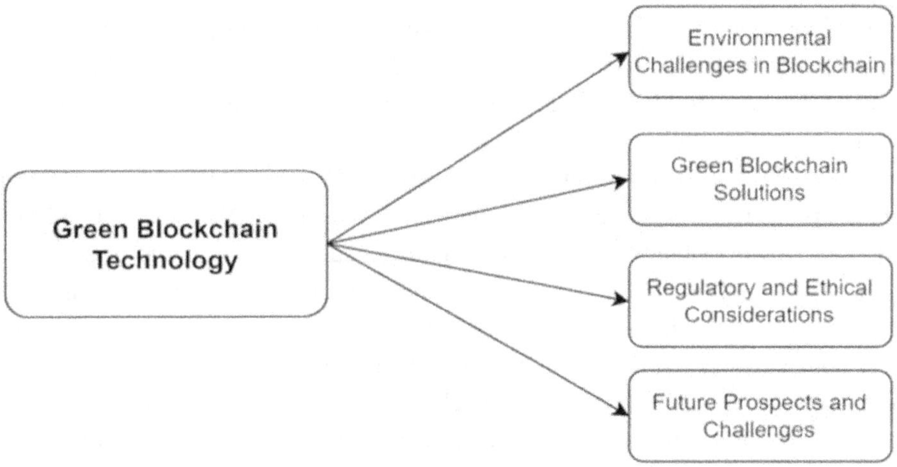

gets rid of middlemen and makes everything more secure by reducing the possibility of a single weak spot. By distributing authority over resources and information, decentralisation improves safety, builds confidence, and gives people agency (Abakah et al., 2023). Data stored on the blockchain is considered immutable once it has been recorded. No one can remove or alter the data contained in a block once it has been added to the chain. The use of cryptographic hashing makes data immutable, which in turn makes transaction records clear and impossible to alter (Miraz et al., 2021). Openness and trust are fostered by transparency, a crucial aspect of the blockchain network. Everyone can see any modifications made to the ledger, and everyone can see the whole history of transactions. Because any fraudulent action would be readily identifiable by the network, this openness aids in preventing fraud. Accountability is aided and user confidence is enhanced (Srivastava et al., 2021; Garg et al., 2024).

## 2.2 Energy Consumption in Traditional Blockchain

The consensus mechanism known as Proof of Work (PoW) is extensively implemented in conventional blockchain networks, including Bitcoin. PoW involves miners engaging in a competitive endeavour to resolve intricate mathematical riddles in order to authenticate transactions and append fresh blocks to the blockchain. According to (Ullah et al., 2021), the initial miner to resolve the riddle is granted the privilege of adding the block and is awarded a newly generated cryptocurrency. Although PoW has demonstrated efficacy in safeguarding blockchain networks, it is accompanied by notable environmental disadvantages. PoW mining entails a significant energy consumption due to the fact that miners vie to solve progressively intricate puzzles. Significant computational resources are necessary for the procedure, resulting in a carbon footprint that is analogous to that of entire nations (Kin et al., 2020; Platt et al., 2021; Xue et al., 2018). There have been concerns expressed regarding the sustainability of blockchain technology due to the environmental impact of PoW. As PoW-based cryptocurrencies gain in prominence, the demand for energy-intensive mining operations increases as well. This has led to demands for alternative consensus mechanisms that are environmentally sustainable and energy-efficient.

## 2.3 Other Consensus Mechanisms and Their Energy Implications

Alternative consensus algorithms have been proposed to decrease energy usage in blockchain networks in response to the environmental issues connected with PoW. One alternative consensus method is Proof of Stake (PoS), which uses the amount of bitcoin a user has and is willing to "stake" as collateral to pick validators to generate new blocks (Sedlmeir et al., 2021). The energy needed to validate blocks is much lower in PoS because, unlike PoW, it does not entail solving complicated problems. As a more energy-efficient alternative, PoS has gained favour. Ethereum plans to migrate to PoS from PoW in its 2.0 release (Ethereum, n.d.). Utilising Directed Acyclic Graphs (DAGs), a need for conventional blocks and miners is rendered obsolete in digital currencies such as IOTA. The graph-like structure of the transactions makes it possible to execute them in parallel and scale them up or down. According to (Schinckus, 2021), this method uses less energy than conventional blockchain designs.

The basic principles of blockchain technology are essential for appreciating its transformative potential across various industries. However, the environmental impact of traditional blockchain, particularly with the energy-intensive PoW consensus mechanism, necessitates exploration of alternative and more sustainable approaches. The shift towards consensus mechanisms like PoS and DAGs reflects the industry's commitment to mitigating blockchain's environmental footprint while maintaining the core principles of decentralization, immutability, and transparency. As the blockchain landscape continues to evolve, striking a balance between innovation and environmental responsibility remains a critical challenge and opportunity for the future.

## 2.4 Environmental Impact of Blockchain

Blockchain technology, while revolutionary in its potential applications, has come under scrutiny for its environmental impact, primarily driven by energy consumption and electronic waste generation.

### 2.4.1 Carbon Footprint of Traditional Blockchain

In the Case Studies on Energy Consumption, Blockchain networks, particularly those employing energy-intensive consensus mechanisms like Proof of Work (PoW), have been associated with significant carbon footprints. Case studies shed light on the energy consumption of major blockchain networks. For instance, a comprehensive analysis by Digiconomist estimates the annual energy consumption of the Bitcoin network. The study reveals that Bitcoin mining alone consumes energy equivalent to that of some small countries (Oudani et al., 2023). The energy-intensive process of solving cryptographic puzzles to validate transactions contributes substantially to the carbon footprint of PoW-based blockchains. Similarly, studies on other PoW-based cryptocurrencies, such as Ethereum and Litecoin, highlight the environmental impact of their energy consumption. These case studies underscore the urgent need to address the carbon footprint associated with traditional blockchain technologies (Chan et al., 2020).

### 2.4.2 Environmental Consequences of High Energy Usage

The high energy usage of traditional blockchain networks results in several environmental consequences. One significant impact is the increased emission of greenhouse gases, contributing to climate change. The carbon intensity of blockchain transactions, especially in PoW-based networks, poses a challenge to the sustainability goals of global communities (Rane et al., 2023). Additionally, the reliance on fossil fuels for energy-intensive mining operations exacerbates the environmental impact. The extraction and burning of fossil fuels release pollutants, further contributing to air and water pollution. As blockchain technology continues to gain prominence, understanding and mitigating these environmental consequences are critical for a sustainable future.

## 2.5 E-WASTE GENERATION

### 2.5.1 Exploration of Electronic Waste Resulting From Blockchain Hardware

The rapid evolution of blockchain technology necessitates powerful and specialized hardware for mining and validating transactions. As a result, the industry witnesses a considerable amount of electronic waste (e-waste) arising from outdated or obsolete blockchain hardware (Oudani et al., 2023). Blockchain hardware, particularly mining rigs used in PoW networks, has a relatively short lifespan due to the constant advancements in technology. Because of this, e-waste accumulates due to the rapid cycle of hardware updates and replacements. In order to investigate blockchain hardware-related e-waste, one must first learn about the various electrical component kinds, their lifespans, and the effects of their disposal on the environment (Hepziba & Ebenesar, 2023).

### 2.5.2 Challenges in Disposing of Blockchain-Related Electronic Components

Disposal of electrical components associated with blockchain technology poses distinct difficulties. Improper management of the hardware used in blockchain networks might result in environmental and health issues due to the presence of dangerous elements including poisonous compounds and heavy metals. Additionally, hardware gets outdated rapidly due to the fast speed of technical progress in the blockchain area, resulting in a flood of abandoned equipment. A further obstacle to effective e-waste management is the absence of defined procedures for recycling blockchain-specific hardware. To reduce the negative effects on the environment caused by blockchain technology's electronic waste, it is essential to use sustainable practises such making hardware that can be recycled and following proper disposal protocols (Suresh & Das, 2023).

## 3. GREEN BLOCKCHAIN SOLUTIONS

There is a rising need to create and execute green blockchain solutions as the environmental consequences of conventional blockchain technology are becoming more and more obvious.

### 3.1 Transition to Proof of Stake (POS) and Other Energy-Efficient Consensus Mechanisms

One of the key concepts behind green blockchain is the replacement of energy-intensive consensus procedures like Proof of Work (PoW) with more sustainable alternatives (Kin et al., 2020). Bitcoin and other cryptocurrencies employ Power-of-Wave (PoW) to solve challenging mathematical puzzles, which requires miners to expend a substantial amount of energy. PoS, on the other hand, is becoming more popular as a more environmentally friendly option. PoS method selects validators to generate new blocks according to their cryptocurrency holdings and their willingness to "stake" (put their holdings up for auction) as collateral. That puts an end to PoW's energy-intensive mining procedure. As part of its 2.0 update, Ethereum, a leading blockchain platform, has begun switching from PoW to PoS in an effort to drastically lower its carbon (Bouraga, 2021; Wagner et al., 2019; Zhang et al., 2019). Potential substitutes for PoW that are energy efficient include Directed Acyclic Graphs (DAGs) and Practical Byzantine Fault Tolerance (PBFT). While decentralised autonomous graphs (DAGs) remove the requirement for conventional blocks and miners, PBFT relies on a voting method among nodes to decrease the necessity for energy-intensive competition (Saad et al., 2021). DAGs are used by cryptocurrencies like IOTA to further reduce energy usage.

### 3.2 Sustainable Mining Practices

Green blockchain solutions extend to the mining process itself, emphasizing the adoption of sustainable practices to minimize energy consumption. This includes:

***Use of Renewable Energy Sources:*** Blockchain mining companies may greatly lessen their impact on the environment by promoting or requiring the usage of renewable energy sources. The use of renewable energy sources including solar, wind, and hydropower by certain mining farms is already underway, bringing blockchain initiatives in line with larger international initiatives to switch to green electricity (Zhang et al., 2019).

***Energy-Efficient Mining Hardware:*** The development and adoption of energy-efficient mining hardware play a pivotal role in reducing the environmental impact of blockchain. Manufacturers are increasingly designing hardware that consumes less power while maintaining the necessary computational capabilities. This transition to more energy-efficient equipment can contribute to a greener blockchain ecosystem (Lasla et al., 2022).

## 3.3 Community-Led Sustainability Projects

Green blockchain solutions are not solely technological but also involve community-driven initiatives that prioritize sustainability (Platt et al., 2021). These projects aim to create awareness, promote responsible practices, and implement eco-friendly measures within the blockchain community. Key components of community-led sustainability projects according to Bourag (2021) include:

***Initiatives to Offset Carbon Emissions:*** Some blockchain projects and communities are taking proactive measures to offset their carbon emissions by investing in carbon offset projects. This involves funding projects that reduce or capture an equivalent amount of greenhouse gases, effectively neutralizing the environmental impact of blockchain activities.

***Education and Advocacy:*** Building a culture of environmental responsibility within the blockchain community is essential. Educational initiatives and advocacy campaigns can raise awareness about the environmental impact of blockchain and promote sustainable practices. Forums, conferences, and online communities are instrumental in fostering a sense of responsibility among developers, users, and industry stakeholders.

## 3.4 Regulatory and Ethical Considerations

Green blockchain solutions also encompass regulatory frameworks and ethical considerations that guide the development and deployment of blockchain technologies (Kalaiarasi & Kirubahari, 2023). Governments and regulatory bodies play a crucial role in shaping the industry's environmental impact through:

***Government Policies:*** The introduction of policies and regulations that incentivize or mandate environmentally friendly practices within the blockchain space can drive industry-wide change. Governments may provide tax incentives for the use of renewable energy in mining operations or impose standards for e-waste management (Platt et al., 2021).

***Ethical Considerations:*** Blockchain developers and users alike bear a responsibility to prioritize ethical considerations, including environmental sustainability. Integrating ethical guidelines into the development process can encourage the creation of eco-friendly blockchain solutions and foster a sense of social responsibility within the industry (Shoker, 2017).

### 3.5 Sustainable Mining Practices

Blockchain's transformative potential comes with environmental challenges, particularly in the energy-intensive process of mining. Sustainable mining practices have become a focal point for mitigating the environmental impact of blockchain technologies (Yadav et al., 2023).

## 4. GREEN MINING INITIATIVES

### 4.1 Use of Renewable Energy Sources

Green mining initiatives prioritize the shift toward renewable energy sources to power blockchain mining operations. Traditional mining, especially in Proof of Work (PoW) consensus mechanisms, relies heavily on energy derived from non-renewable sources, contributing to carbon emissions and environmental degradation. Embracing renewable energy, such as solar, wind, or hydropower, helps reduce the carbon footprint associated with blockchain mining (Khan et al., 2021; Vafaeva et al., 2024). Several blockchain mining farms and projects have started integrating renewable energy into their operations. For example, there are initiatives in regions with abundant renewable resources to establish mining facilities powered entirely by solar or wind energy. This not only makes the mining process more sustainable but also aligns with global efforts to transition to cleaner energy alternatives (Rane et al., 2023).

### 4.2 Energy-Efficient Mining Hardware

Another key aspect of sustainable mining practices involves the development and adoption of energy-efficient mining hardware. Traditional mining rigs are often power-hungry, contributing to excessive energy consumption. The industry is witnessing a shift towards the design and use of hardware that maximizes computational efficiency while minimizing energy requirements. Manufacturers are actively working on creating more energy-efficient Application-Specific Integrated Circuits (ASICs) and Graphics Processing Units (GPUs) for mining. These advancements in hardware technology aim to maintain the necessary computational power for blockchain operations while significantly reducing energy consumption. This transition contributes to a greener and more sustainable approach to blockchain mining (Oudani et al., 2023; Vatin et al., 2024).

### 4.3 Community-Led Sustainability Projects

#### 4.3.1 Initiatives by Blockchain Communities to Reduce Environmental Impact

Blockchain communities are increasingly recognizing their role in addressing the environmental impact of the technology they support. Several initiatives are emerging within these communities to actively reduce their carbon footprint and promote sustainable practices:

**Carbon Offsetting *Programs:*** Some blockchain projects and communities are engaging in carbon offsetting programs. To do this, one must invest in initiatives that neutralise the impact of blockchain operations on the environment by reducing or capturing the same quantity of greenhouse gases. As a result of these efforts, blockchain projects are less harmful to the environment (Hepziba & Ebenesar, 2023).

***Sustainability Guidelines and Best Practices:*** Guidelines and best practises for sustainability are being developed and promoted by blockchain groups. Developers, miners, and users all have a chance to follow these rules and make their activities more environmentally friendly. A culture of sustainability is fostered and ecologically responsible practises are encouraged across the blockchain ecosystem via the creation of a set of standards by the community (Suresh & Das, 2023).

## 4.4 Examples of Successful Community-Driven Sustainability Efforts

Several blockchain communities have successfully implemented sustainability initiatives, setting positive examples for the industry:

***Tezos Foundation:*** The Tezos blockchain has established itself as a pioneer in sustainability efforts. The Tezos Foundation actively supports research and projects focused on reducing the environmental impact of blockchain technology. This includes initiatives related to energy-efficient consensus mechanisms and sustainable development practices within the Tezos ecosystem.

***Ethereum's Move to Proof of Stake:*** Ethereum, one of the most widely used blockchain platforms, has initiated a move from the energy-intensive PoW consensus mechanism to the eco-friendlier PoS. This transition, known as Ethereum 2.0, is a community-driven effort to address the environmental concerns associated with Ethereum's previous energy consumption model (Kalaiarasi & Kirubahari, 2023; Vatin et al., 2024).

## 4.5 Regulatory and Ethical Considerations

Blockchain technology's rapid evolution has prompted a closer examination of its environmental impact, leading to the development of regulatory and ethical frameworks to address sustainability concerns.

### 4.5.1 Government Policies Related to Blockchain and the Environment

The regulatory landscape concerning blockchain and the environment is dynamic and varies across jurisdictions. Governments worldwide are increasingly recognizing the need to address the environmental impact of blockchain technologies and are formulating policies to guide their development. Key aspects of the current regulatory landscape include:

***Energy Consumption Regulations:*** Some governments are actively considering or implementing regulations specifically addressing the energy consumption of blockchain networks. These regulations may include guidelines for using renewable energy sources or incentives for blockchain projects that adopt eco-friendly practices (Xue et al., 2018).

***E-Waste Management:*** Blockchain hardware's electrical components add to e-waste, making e-waste management legislation important. In order to promote a circular economy, governments may establish regulations for the proper disposal and recycling of electrical components associated to blockchain technology (Asif & Hassan, 2023).

***Incentives for Sustainable Practices:*** There are a number of incentives being considered by governments as a means of promoting blockchain sector sustainability. Projects that show they are committed to lowering their environmental effect may be eligible for financial incentives such as grants or tax exemptions (Vafaeva et al., 2024; Yadav et al., 2023).

### 4.5.2 Challenges and Opportunities for Regulatory Frameworks

The international scope of blockchain technology poses a challenge for the regulatory environment as it makes it hard to implement consistent rules. Furthermore, authorities have difficulty in keeping up with the newest innovations in blockchain technology and formulating appropriate laws due to its rapid evolution. Possibilities for governments to cooperate together on global frameworks and standards for sustainable blockchain practises provide opportunities. Environmental problems related to blockchain technology may be better addressed if authorities work together on a global scale. under addition, according to Kalaiarasi and Kirubahari (2023), environmentally responsible innovation may flourish under regulatory frameworks that are well-designed, creating an ideal setting for the creation of green blockchain solutions.

## 4.6 Ethical Considerations in Green Blockchain

### 4.6.1 Social Responsibility of Blockchain Developers and Users

When it comes to creating and implementing green blockchain solutions, ethical concerns are paramount. It is the shared obligation of creators and users to aid in the greater cause of ecological preservation:

***Responsibilities of Developers:*** When it comes to creating and executing solutions that are good for the environment, blockchain developers are indispensable. Optimal environmental performance of blockchain networks, development of sustainable mining algorithms, and prioritisation of energy-efficient consensus processes are all important ethical concerns. According to Kalaiarasi and Kirubahari (2023), developers should make an effort to reduce the amount of carbon footprint and electronic trash that their initiatives generate.

***Users' Responsibility:*** Each user of a blockchain network also has a moral weight to carry. This involves doing things like sponsoring projects that have green initiatives, joining networks that prioritise sustainability, and speaking out against practises that are harmful to the environment. By making decisions that are in line with ecological and ethical standards, users may shape the future of the blockchain sector (Schinckus, 2021).

### 4.6.2 Building a Culture of Sustainability in the Blockchain Community

To effectively address environmental problems in the long run, the blockchain community must foster a culture of sustainability. This involves fostering a collective mindset that values sustainability and integrates ethical considerations into decision-making processes:

***Educational Initiatives:*** Educational initiatives within the blockchain community can raise awareness about the environmental impact of blockchain technology. Workshops, seminars, and online resources can inform developers, users, and stakeholders about the importance of sustainability and provide practical guidance on adopting eco-friendly practices (Platt et al., 2021).

***Community Engagement:*** Active engagement within the blockchain community is vital for building a culture of sustainability. Forums, conferences, and online platforms can serve as spaces for discussing and sharing best practices related to green blockchain initiatives. Collaborative efforts, such as the development of sustainability guidelines, can be instrumental in creating a shared vision for a more environmentally responsible blockchain ecosystem (Platt et al., 2021; Vafaeva et al., 2024).

## 5. FUTURE PROSPECTS AND CHALLENGES

As the evolution of blockchain technology unfolds, its future trajectory becomes intertwined with emerging technologies that hold the potential to redefine its capabilities. However, within these promising opportunities lie challenges and potential impediments that demand careful navigation from the industry. The integration of artificial intelligence (AI) into blockchain technology stands out as a pivotal development, offering substantial promise for process optimization and heightened overall efficiency. The impact of AI on blockchain processes is multifaceted according to Schinckus (2021):

- AI augments the functionality of smart contracts through the introduction of machine learning algorithms. This infusion of intelligence enables smart contracts to dynamically adapt and respond to changing conditions, fostering increased versatility. Automated decision-making within smart contracts benefits from AI-driven predictions and analyses.
- In bolstering the security of blockchain networks, AI technologies, including machine learning algorithms, play a crucial role. They identify and mitigate potential threats, with AI-driven cybersecurity measures capable of detecting anomalies and patterns indicative of malicious activities. This real-time threat intelligence enhances the overall resilience of blockchain systems.
- Addressing scalability challenges within blockchain networks is another realm where AI can make significant contributions. By leveraging machine learning algorithms to optimize network resource allocation, transaction processing, and consensus mechanisms, AI-driven solutions have the potential to enhance scalability. This, in turn, enables blockchain systems to efficiently handle a larger volume of transaction.
- The integration of AI with blockchain also facilitates advanced data analytics and insights. AI algorithms can analyze the extensive datasets stored on the blockchain, extracting valuable information and providing actionable insights. This holds particular significance in sectors like supply chain management and healthcare, where data-driven decision-making plays a critical role.

However, despite the exciting prospects of integrating AI and other emerging technologies with blockchain, achieving widespread adoption is not without its challenges and potential roadblocks:

- Regulatory uncertainty poses a significant hurdle. The evolving regulatory landscape for both blockchain and AI technologies creates an environment of uncertainty, where inconsistent or restrictive regulations can impede innovation and erect barriers for businesses and developers (Hepziba & Ebenesar, 2023).
- Interoperability issues present a formidable challenge. Seamless integration between diverse blockchain networks and AI systems requires addressing interoperability challenges. Ensuring compatibility and efficient communication among different technologies is paramount for unlocking the full potential of their integration ((Hepziba & Ebenesar, 2023).
- Scalability concerns emerge as blockchain networks grow in size and complexity. Integrating AI introduces additional computational demands that may exacerbate existing scalability issues. Striking a balance between increased functionality and maintaining or improving scalability becomes a critical consideration (Rane et al., 2023).
- Privacy and security risks are heightened in the convergence of AI and blockchain. While blockchain is recognized for its robust security features, the introduction of AI may introduce new vul-

nerabilities. Safeguarding the privacy of sensitive data in AI-driven applications on the blockchain necessitates careful consideration and robust security measures (Hepziba & Ebenesar, 2023).

- Technological maturity remains an overarching consideration. Both AI and blockchain technologies are in ongoing stages of evolution, and their levels of maturity can impact widespread adoption. Overcoming technological immaturity, ensuring secure implementations, and addressing potential bugs and vulnerabilities are crucial steps in building trust among users and stakeholders (Hepziba & Ebenesar, 2023).

## 6. CONCLUSION

The trajectory of blockchain technology is marked by a dynamic interplay of innovation, challenges, and evolving considerations for environmental sustainability, regulatory frameworks, and ethical practices. The foundational principles of decentralization, immutability, and transparency have propelled blockchain into various industries, transforming how we manage data, conduct transactions, and envision the future of technology. The environmental impact of blockchain, particularly in terms of energy consumption and e-waste generation, has sparked a growing awareness of the need for sustainable practices. Concepts like "green blockchain" and the transition to energy-efficient consensus mechanisms represent a collective effort to align technological advancements with environmental responsibility. As the industry explores the integration of emerging technologies such as artificial intelligence, the potential for optimization and innovation becomes increasingly evident, promising a more versatile and efficient blockchain ecosystem. Regulatory landscapes around the world are adapting to the challenges and opportunities presented by blockchain. Governments are crafting policies to address energy consumption, e-waste management, and overall sustainability. However, the regulatory journey is marked by complexities, including the need for international collaboration and the continuous evolution of technology, underscoring the importance of agile and adaptable regulatory frameworks. Ethical considerations within the blockchain community play a pivotal role in shaping the technology's future. Developers and users alike bear a responsibility to prioritize sustainability, transparency, and ethical practices. Initiatives by blockchain communities, educational endeavors, and the cultivation of a culture of environmental consciousness contribute to a socially responsible and sustainable blockchain ecosystem. Looking forward, the convergence of blockchain with emerging technologies, notably artificial intelligence, holds immense potential for optimization, enhanced security, and advanced data analytics. Yet, this integration brings forth challenges such as regulatory uncertainty, interoperability issues, scalability concerns, and privacy risks. Overcoming these challenges will be crucial to realizing the full potential of blockchain and AI synergy. The future of blockchain hinges on a delicate balance—between technological innovation and environmental responsibility, regulatory frameworks and industry collaboration, and ethical considerations that prioritize sustainability. As the industry continues to evolve, addressing these considerations collectively will be essential in shaping a blockchain ecosystem that not only thrives on innovation but also contributes positively to the global landscape, fostering a sustainable, secure, and ethically-driven future.

## REFERENCES


Abakah, E. J. A., Wali Ullah, G. M., Adekoya, O. B., Osei Bonsu, C., & Abdullah, M. (2023). Blockchain market and eco-friendly financial assets: Dynamic price correlation, connectedness and spillovers with portfolio implications. *International Review of Economics & Finance*, *87*, 218–243. doi:10.1016/j.iref.2023.04.028

. Adeniyi, E. A., Ogundokun, R. O., Misra, S., Awotunde, J. B., & Abiodun, K. M. (2022). Enhanced Security and Privacy Issue in Multi-Tenant Environment of Green Computing Using Blockchain Technology. *EAI/Springer Innovations in Communication and Computing*, 65–83. doi:10.1007/978-3-030-89546-4_4

Alzoubi, Y. I., & Mishra, A. (2023). Green blockchain – A move towards sustainability. In *Journal of Cleaner Production* (Vol. 430, p. 139541). Elsevier. doi:10.1016/j.jclepro.2023.139541

Asif, R., & Hassan, S. R. (2023). Shaping the future of Ethereum: Exploring energy consumption in Proof-of-Work and Proof-of-Stake consensus. *Frontiers in Blockchain*, *6*, 1151724. doi:10.3389/fbloc.2023.1151724

Bada, A. O., Damianou, A., Angelopoulos, C. M., & Katos, V. (2021). Towards a Green Blockchain: A Review of Consensus Mechanisms and their Energy Consumption. *Proceedings - 17th Annual International Conference on Distributed Computing in Sensor Systems, DCOS 2021*, 503–511. 10.1109/DCOSS52077.2021.00083

Bouraga, S. (2021). A taxonomy of blockchain consensus protocols: A survey and classification framework. *Expert Systems with Applications*, *168*, 114384. doi:10.1016/j.eswa.2020.114384

Bradu, P., Biswas, A., Nair, C., Sreevalsakumar, S., Patil, M., Kannampuzha, S., Mukherjee, A. G., Wanjari, U. R., Renu, K., Vellingiri, B., & Gopalakrishnan, A. V. (2022). Recent advances in green technology and Industrial Revolution 4.0 for a sustainable future. In *Environmental Science and Pollution Research*. Springer., doi:10.1007/s11356-022-20024-4

. Göçen, S., & Kurubacak, G. (2023). From black to green: Eco-friendly learning with blockchain technology. *Glocal Policy and Strategies for Blockchain: Building Ecosystems and Sustainability*, 80–99. doi:10.4018/978-1-6684-4153-4.ch005

. Hepziba, G. R., & Ebenesar, A. B. J. (2023). Eco-friendly blockchain for smart cities. *Green Blockchain Technology for Sustainable Smart Cities*, 65–96. doi:10.1016/B978-0-323-95407-5.00015-3

Joshi, P. B. (2024). *Full blown green metrics*. Elsevier Publishing. doi:10.1016/B978-0-443-18959-3.00013-6

. Kalaiarasi, H., & Kirubahari, S. (2023). Green finance for sustainable development using blockchain technology. *Green Blockchain Technology for Sustainable Smart Cities*, 167–185. doi:10.1016/B978-0-323-95407-5.00003-7

Khan, S. A. R., Godil, D. I., Jabbour, C. J. C., Shujaat, S., Razzaq, A., & Yu, Z. (2021). Green data analytics, blockchain technology for sustainable development, and sustainable supply chain practices: Evidence from small and medium enterprises. *Annals of Operations Research*. doi:10.1007/s10479-021-04275-x

Kin, C. W., Zhang, R., & Chan, W. K. (2020). Evaluation of Energy Consumption in Block-Chains with Proof of Work and Proof of Stake. *Journal of Physics: Conference Series*, *1584*(1), 012023. doi:10.1088/1742-6596/1584/1/012023

Lasla, N., Al-Sahan, L., Abdallah, M., & Younis, M. (2022). Green-PoW: An energy-efficient blockchain Proof-of-Work consensus algorithm. *Computer Networks*. https://www.sciencedirect.com/science/article/pii/S1389128622002432

Miraz, M. H., Excell, P. S., & Sobayel, K. (2021). Evaluation of green alternatives for blockchain proof-of-work (Pow) approach. *Annals of Emerging Technologies in Computing*, *5*(4), 54–59. doi:10.33166/AETiC.2021.04.005

Mohan, C., Kumari, N., & Sushma, A. (2024). *Introduction to environmental and green chemistry*. Elsevier Publishing., doi:10.1016/B978-0-443-18959-3.00005-7

Mohan, C., Robinson, J., Vodwal, L., & Kumari, N. (2024). *Sustainable Development Goals for addressing environmental challenges*. Elsevier Publishing. doi:10.1016/B978-0-443-18959-3.00007-0

Mubarik, M., Raja Mohd Rasi, R. Z., Mubarak, M. F., & Ashraf, R. (2021). Impact of blockchain technology on green supply chain practices: Evidence from emerging economy. *Management of Environmental Quality*, *32*(5), 1023–1039. doi:10.1108/MEQ-11-2020-0277

Nygaard, A., & Silkoset, R. (2023). Sustainable development and greenwashing: How blockchain technology information can empower green consumers. *Business Strategy and the Environment*, *32*(6), 3801–3813. doi:10.1002/bse.3338

Oudani, M., Sebbar, A., Zkik, K., El Harraki, I., & Belhadi, A. (2023). Green Blockchain based IoT for secured supply chain of hazardous materials. *Computers & Industrial Engineering*, *175*, 108814. Advance online publication. doi:10.1016/j.cie.2022.108814

Platt, M., Sedlmeir, J., Platt, D., Xu, J., Tasca, P., Vadgama, N., & Ibanez, J. I. (2021). The Energy Footprint of Blockchain Consensus Mechanisms Beyond Proof-of-Work. *Proceedings - 2021 21st International Conference on Software Quality, Reliability and Security Companion, QRS-C 2021*, (pp. 1135–1144). IEEE. 10.1109/QRS-C55045.2021.00168

Polas, M. R. H., Kabir, A. I., Sohel-Uz-zaman, A. S. M., Karim, R., & Tabash, M. I. (2022). Blockchain Technology as a Game Changer for Green Innovation: Green Entrepreneurship as a Roadmap to Green Economic Sustainability in Peru. In Journal of Open Innovation: Technology, Market, and Complexity, 8(2). Elsevier. doi:10.3390/joitmc8020062

Rane, S. B., Potdar, P. R., & Aware, S. (2023). *Strategies for development of smart and green products using Blockchain-IoT integrated architecture*. Operations Management Research., doi:10.1007/s12063-023-00398-5

Saad, M., Qin, Z., Ren, K., Nyang, D. H., & Mohaisen, D. (2021). E-PoS: Making Proof-of-Stake Decentralized and Fair. *IEEE Transactions on Parallel and Distributed Systems*, *32*(8), 1961–1973. doi:10.1109/TPDS.2020.3048853

Schinckus, C. (2021). Proof-of-work based blockchain technology and Anthropocene: An undermined situation? *Renewable & Sustainable Energy Reviews*, *152*, 111682. Advance online publication. doi:10.1016/j.rser.2021.111682

Sedlmeir, J., Buhl, H. U., Fridgen, G., & Keller, R. (2021). Recent Developments in Blockchain Technology and their Impact on Energy Consumption. *ArXiv Preprint ArXiv:2102.07886.* https://doi.org/doi:10.1007/s00287-020-01321-z

Sharma, P. K., Kumar, N., & Park, J. H. (2020). Blockchain technology toward green IoT: Opportunities and challenges. *IEEE Network*, *34*(4), 263–269. doi:10.1109/MNET.001.1900526

Shoker, A. (2017). Sustainable blockchain through proof of exercise. *2017 IEEE 16th International Symposium on Network Computing and Applications*. IEEE. 10.1109/NCA.2017.8171383

Srivastava, A., Dubey, V., & Hazela, B. (2021). Delivering Green Supply Chain Using Blockchain Technology for Sustainable Environment: A Survey. *Emerging Technologies in Data Mining*. Springer. doi:10.1007/978-981-15-9927-9_74

Suresh, B. C. V., & Das, S. (2023). Impact of blockchain technology on the stock market. *Revolutionizing Financial Services and Markets Through FinTech and Blockchain*, 44–59. doi:10.4018/978-1-6684-8624-5.ch004

Ullah, S., Arjun, K. P., & Sureshbabu, A. P. (2021). Green Computing and Blockchain Fundamentals. In *Handbook of Green Computing and Blockchain Technologies* (pp. 1–16). CRC Press. doi:10.1201/9781003107507-1

Vafaeva K. M., Chhetri A., Sudan P., Mishra M., Pakkiraiah B., & Mohan C. (2024). Polymer Matrix Nanocomposites for Sustainable Packaging: A Green Approach. *E3S Web of Conferences, 511.* doi:10.1051/e3sconf/202451101008

Vatin, N. I., Negi, G. S., Yellanki, S. V., Mohan, C., & Singla, N. (2024). Sustainability Measures: An Experimental Analysis of AI and Big Data Insights in Industry 5.0. *BIO Web of Conferences, 86.* 10.1051/bioconf/20248601072

Wagner, K., Keller, T., & Seiler, R. (2019). a Comparative Analysis of Cryptocurrency Consensus Algorithms. In *Proceedings of the 16th …* (pp. 217–225). researchgate.net. 10.33965/ac2019_201912L026

Xue, T., Yuan, Y., Ahmed, Z., Moniz, K., Cao, G., & Wang, C. (2018). Proof of Contribution: A Modification of Proof of Work to Increase Mining Efficiency. *Proceedings - International Computer Software and Applications Conference,* (pp. 636–644). IEEE. 10.1109/COMPSAC.2018.00096

Yadav, A. K., Singh, K., Amin, A. H., Almutairi, L., Alsenani, T. R., & Ahmadian, A. (2023). A comparative study on consensus mechanism with security threats and future scopes: Blockchain. *Computer Communications*, *201*, 102–115. doi:10.1016/j.comcom.2023.01.018

Zhang, P., Schmidt, D. C., White, J., & Dubey, A. (2019). Consensus mechanisms and information security technologies. *Advances in Computers*, *115*, 181–209. doi:10.1016/bs.adcom.2019.05.001

# Compilation of References

. Adeniyi, E. A., Ogundokun, R. O., Misra, S., Awotunde, J. B., & Abiodun, K. M. (2022). Enhanced Security and Privacy Issue in Multi-Tenant Environment of Green Computing Using Blockchain Technology. *EAI/Springer Innovations in Communication and Computing*, 65–83. doi:10.1007/978-3-030-89546-4_4

. Atser, G., Tahirou, A., Dixon, A., Ekeleme, F., Alloh, J., & Akande, A. (2019). *Transforming Oyo State to an Agribusiness Hub:(Our vision, Our Plan) 2019-2023.* Oyo State.

. Boechat, C. L., de Santana Arauco, A. M., Duda, R. M., de Sena, A. F. S., de Souza, M. E. L., & Brito, A. C. C. (2017). Solid waste in agricultural soils: an approach based on environmental principles, human health, and food security. *Solid waste management in rural areas*, 81-102.

. Buckler, C., & Creech, H. (2014). *Shaping the Future We Want: UN Decade of Education for Sustainable Development (2005-2014*).

. Delina, L. L., Perez, O. A., Afable, S. B., & Steuer, B. (2023). Do global public health crises change people's behaviors towards sustainability? Evidence of the COVID-19 and sustainability nexus from Hong Kong. *Cleaner and responsible consumption, 10*, 100132.

. Ghodke, P. K., Anthony, C., & Sharma, A. K. (2024). Bioprocessing of Agricultural and Forest Waste. *Sustainable Production Innovations: Bioremediation and Other Biotechnologies*, 367-393.

. Gigauri, I., Vasilev, V., & Mushkudiani. Z. (2022). *In pursuit of sustainability: Towards sustainable future through education. 1*(37). doi:10.31435/rsglobal_ijite/30032022/7798

. Göçen, S., & Kurubacak, G. (2023). From black to green: Eco-friendly learning with blockchain technology. *Glocal Policy and Strategies for Blockchain: Building Ecosystems and Sustainability*, 80–99. doi:10.4018/978-1-6684-4153-4.ch005

. Hepziba, G. R., & Ebenesar, A. B. J. (2023). Eco-friendly blockchain for smart cities. *Green Blockchain Technology for Sustainable Smart Cities*, 65–96. doi:10.1016/B978-0-323-95407-5.00015-3

. Jack, A. L., & Thies, J. E. (2006). Compost and vermicompost as amendments promoting soil health. *Biological approaches to sustainable soil systems*, 453-466.

. Kalaiarasi, H., & Kirubahari, S. (2023). Green finance for sustainable development using blockchain technology. *Green Blockchain Technology for Sustainable Smart Cities*, 167–185. doi:10.1016/B978-0-323-95407-5.00003-7

. Katuwal, S., & Afzal, M. (2022). *A framework for managing food wastage: Strategies to promote recycling and sustainability in the hotel industry.*

. Kumar, P., & Joshi, L. (2013). Pollution caused by agricultural waste burning and possible alternate uses of crop stubble: a case study of Punjab. *Knowledge systems of societies for adaptation and mitigation of impacts of climate change*, 367-385.

. Meena, H. N., Singh, S. K., Meena, M. S., Narayan, R., & Sen, B. (2022). *Crop residue: waste or wealth?*

. Mohan, C., Robinson, J., Lal, C., Jammala, A. P. N., Meena, P. L. (2024). Sustainable Energy Solutions for Environmental Pollution Control. *E3S Web of Conferences, 511*, 01023.

. Pilapitiya, P. N. T., & Ratnayake, A. S. (2024). The world of plastic waste: a review. *Cleaner Materials, 100220.*

. Puniya, R., Bazaya, B. R., Kumar, A., Sharma, B. C., Nesar, N. A., Botchily, R. S. (2023). Effect of residue and weed management practices on weed flora, yield, energetics, carbon footprint, economics and soil quality of zero tillage wheat. *National library of medicine, 13*, 19311.

. Toivonen, A. (2020). *Sustainable space tourism: an introduction* (*Vol. 3*). Channel view publications.

. Vital-Grappin, A. D., Ariza-Tarazona, M. C., Sandoval, I., Rojas-Guerrero, C. A., Villarreal-Chiu, J. F., & Cedillo-González, E. I. (2023). Design of Green TiO2-Based Semiconductors for Microplastic and Nanoplastic Photocatalytic Degradation.

. Voorhoeve, J. (2021). Fighting Poverty and Violence: How Basic Human Needs Could Be Met.

. Ware, G. W., Albert, L. A., Bro-Rasmussen, F., Crosby, D. G., de Voogt, P., Frehse, H., & & Hamilton, D. J. (2003). *Disposal and degradation of pesticide waste. Reviews of Environmental Contamination and Toxicology: Continuation of Residue Reviews*, 123-200.

United Nations Environment Programme. (2019). *Textiles and the Environment in a Circular Economy*. UN.

Abakah, E. J. A., Wali Ullah, G. M., Adekoya, O. B., Osei Bonsu, C., & Abdullah, M. (2023). Blockchain market and eco-friendly financial assets: Dynamic price correlation, connectedness and spillovers with portfolio implications. *International Review of Economics & Finance*, *87*, 218–243. doi:10.1016/j.iref.2023.04.028

Abatan, A., Adeyinka, M. A., Sodiya, E. O., Jacks, B. S., Ugwuanyi, E. D., Daraojimba, O. H., & Lottu, O. A. (2024). *The role of IT in sustainable environmental management: A global perspective review.*

Abitbol, T., Rivkin, A., Cao, Y., Nevo, Y., Abraham, E., Ben-Shalom, T., Lapidot, S., & Shoseyov, O. (2016). Nanocellulose, a tiny fiber with huge applications. *Current Opinion in Biotechnology*, *39*, 76–88. doi:10.1016/j.copbio.2016.01.002 PMID:26930621

Abo-Shama, U. H., El-Gendy, H., Mousa, W. S., Hamouda, R. A., Yousuf, W. E., Hetta, H. F., & Abdeen, E. E. (2020). Synergistic and antagonistic effects of metal nanoparticles in combination with antibiotics against some reference strains of pathogenic microorganisms. *Infection and Drug Resistance*, *13*, 351–362. doi:10.2147/IDR.S234425 PMID:32104007

Abu Jadayil, W., Qureshi, M. R. N. M., Ajaj, R., Aqil, E., Shawahin, G., Anver, H., & Aljeawi, S. (2022). An Empirical Investigation on Plastic Waste Issues and Plastic Disposal Strategies to Protect the Environment: A UAE Perspective. *Sustainability (Basel)*, *14*(24), 16719. doi:10.3390/su142416719

Abubakar, I., Maniruzzaman, K., Dano, U., AlShihri, F., AlShammari, M., Ahmed, S., Al-Gehlani, W., & Alrawaf, T. (2022). Environmental Sustainability Impacts of Solid Waste Management Practices in the Global South. *International Journal of Environmental Research and Public Health*, *19*(19), 12717. doi:10.3390/ijerph191912717 PMID:36232017

Achmon, Y., Fernández, J. D., Hernandez, K., & McCurry, D. G. (2017). Weed seed inactivation in soil mesocosms via bio solarization with mature compost and tomato processing waste amendments. *Pest Management Science*, *73*(5), 862–873. doi:10.1002/ps.4354 PMID:27391139

Aditya, A., & Kurniawati, K. (2023). Food Waste Management Challenges and Strategies in The Hotel Industry in Jakarta. *Journal of Social Research*, *2*(9), 2983–2989. doi:10.55324/josr.v2i9.1295

Advanced Manufacturing & Industrial Decarbonization. (2021). *Department of energy, in partnership with remade institute, awards over $16 million for 23 projects that will reduce carbon emissions across the manufacturing sector.* Office of Energy Efficiency & Renewable Energy.

Agamuthu, P. (2009, November). Challenges and opportunities in agro-waste management: An Asian perspective. In *Inaugural meeting of first regional 3R forum in Asia, 11*(12). University of Malasiya.

Agrawal, A., Sharma, R., & Singh, G. (2021). Sustainable Leather: A Review on Innovative Materials and Production Processes. *Journal of Cleaner Production, 293*, 126084. doi:10.1016/j.jclepro.2021.126084

Ahmad, M., Lee, S. S., Dou, X., Mohan, D., Sung, J. K., Yang, J. E., & Ok, Y. S. (2012). Effects of pyrolysis temperature on soybean stover-and peanut shell-derived biochar properties and TCE adsorption in water. *Bioresource Technology, 118*, 536–544. doi:10.1016/j.biortech.2012.05.042 PMID:22721877

Aimara, M. S., & Olanrewaju, O. S. (2020). Waste Management through Composting: Challenges and Potentials. *Sustainability (Basel), 12*(11), 4456. doi:10.3390/su12114456

Akmal, T., & Jamil, F. (2021). Health impact of solid waste management practices on household: The case of metropolitans of Islamabad-Rawalpindi, Pakistan. *Heliyon*, 7.

Akter, N., & Hossain, M. I. (2023). Role of informal and formal actors in e-waste management. *The Business Standard.* https://www.tbsnews.net/thoughts/role-informal-and-formal-actors-e-waste-management-746346

Alaerts, L., Augustinus, M., & Van Acker, K. (2018). Impact of bio-based plastics on current recycling of plastics. *Sustainability (Basel), 10*(5), 1487. doi:10.3390/su10051487

Alblooshi, B. G. K. M., Ahmad, S. Z., Hussain, M., & Singh, S. K. (2022). Sustainability management of electronic waste: Empirical evidences from a stakeholders' perspective. *Business Strategy and the Environment, 21*(4), 1856–1874. doi:10.1002/bse.2987

Al-Dailami, A., Ahmad, I., Kamyab, H., Abdullah, N., Koji, I., Ashokkumar, V., & Zabara, B. (2022). Sustainable solid waste management in Yemen: Environmental, social aspects, and challenges. *Biomass Conversion and Biorefinery*, 1–27. doi:10.1007/s13399-022-02871-w

Algozeeb, W. A., Savas, P. E., Luong, D. X., Chen, W., Kittrell, C., Bhat, M., Shahsavari, R., & Tour, J. M. (2020). Flash graphene from plastic waste. *ACS Nano, 14*(11), 15595–15604. doi:10.1021/acsnano.0c06328 PMID:33119255

Aliasghar, A. (2024). *A history of the concept of sustainable development: Literature, review.* Research Gate. https://www.researchgate.net/publication/242219096_A_history_of_the_concept_of_sustainable_development_Literature_review

Ali, R., Irfan, S., & Hamid, S. (2024). Emerging themes in food tourism: A systematic literature review and research agenda. *British Food Journal, 126*(1), 372–408. doi:10.1108/BFJ-11-2022-0939

Ali, S. M., Pervaiz, A., Afzal, B., Hamid, N., & Yasmin, A. (2014). Open dumping of municipal solid waste and its hazardous impacts on soil and vegetation diversity at waste dumping sites of Islamabad city. *Journal of King Saud University. Science, 26*(1), 59–65. doi:10.1016/j.jksus.2013.08.003

Alves, B. (2023, April 17). *Consumer behaviour with old electronics worldwide 2021.* Statista. https://www.statista.com/statistics/1337224/e-waste-consumer-behavior-worldwide/

Alves, B. (2023, December 18). *Global e-wastes statistics & facts.* Statista. https://www.statista.com/topics/3409/electronic-waste-worldwide/#topicOverview

Alves, B. (2023, November 7). *E-waste flow in the European Union 2021, by method.* Statista. https://www.statista.com/statistics/1337536/electronic-waste-flow-european-union-by-operation/

Alves, B. (2023, November 7). *E-waste per capital collection in the European Union 2021, by country.* Statista. https://www.statista.com/statistics/1337511/electronic-waste-collection-european-union-per-capita/

Alzoubi, Y. I., & Mishra, A. (2023). Green blockchain – A move towards sustainability. In *Journal of Cleaner Production* (Vol. 430, p. 139541). Elsevier. doi:10.1016/j.jclepro.2023.139541

Amicarelli, V., & Bux, C. (2021). Food waste measurement toward a fair, healthy and environmental-friendly food system: A critical review. *British Food Journal, 123*(8), 2907–2935. doi:10.1108/BFJ-07-2020-0658

Amoabediny, G. H., Naderi, A., Malakootikhah, J., Koohi, M. K., Mortazavi, S. A., Naderi, M., & Rashedi, H. (2009). Guidelines for safe handling, use and disposal of nanoparticles. *Journal of Physics: Conference Series, 170*, 012037. doi:10.1088/1742-6596/170/1/012037

Andrady, A. L. (2011). Microplastics in the marine environment. *Marine Pollution Bulletin, 62*(8), 1596–1605. doi:10.1016/j.marpolbul.2011.05.030 PMID:21742351

Angelidaki, I., Ellegaard, L., & Ahring, B. K. (1999). A comprehensive model of anaerobic bioconversion of complex substrates to biogas. *Biotechnology and Bioengineering, 63*(3), 363–372. doi:10.1002/(SICI)1097-0290(19990505)63:3<363::AID-BIT13>3.0.CO;2-Z PMID:10099616

Araujo, C. M. L. D. (2023). *The effects of retail sustainable strategies on the sustainability of the food value chain* [Doctoral dissertation].

Aromar, R., Amir, B., Chandni, S., & Prathigna, P. (2021). *The Author Team Poonam Batra University of Delhi Indian Institute for Human Settlements.* IIHS. https://iihs.co.in/knowledge-gateway/wp-content/uploads/2021/01/TESF-India-Background-paper.pdf

Asif, R., & Hassan, S. R. (2023). Shaping the future of Ethereum: Exploring energy consumption in Proof-of-Work and Proof-of-Stake consensus. *Frontiers in Blockchain, 6*, 1151724. doi:10.3389/fbloc.2023.1151724

Ayangbenro, A. S. S., & Chukwuemeka, C. F. (2022). Harnessing the Rhizosphere Soil Microbiome of Organically Amended Soil for Plant Productivity. *Agronomy (Basel), 12*(12), 3179. doi:10.3390/agronomy12123179

Aziz, S., Mahmood, M., & Rehman, Z. (2018). Implementation of CIPP Model for Quality Evaluation at School Level: A Case Study. *Journal of Education and Educational Development, 5*(1), 189. doi:10.22555/joeed.v5i1.1553

Bada, A. O., Damianou, A., Angelopoulos, C. M., & Katos, V. (2021). Towards a Green Blockchain: A Review of Consensus Mechanisms and their Energy Consumption. *Proceedings - 17th Annual International Conference on Distributed Computing in Sensor Systems, DCOS 2021*, 503–511. 10.1109/DCOSS52077.2021.00083

Bagi, Z., Ács, N., Bálint, B., Horváth, L., Dobó, K., Perei, K. R., Rákhely, G., & Kovács, K. L. (2007). Biotechnological intensification of biogas production. *Applied Microbiology and Biotechnology, 76*(2), 473–482. doi:10.1007/s00253-007-1009-6 PMID:17503035

Banik, S. (2018). Small scale industries in India: Opportunities and challenges. *International Journal of Creative Research Thoughts, 6*(1), 337–341.

Barrett, M., Marino, M., Brkic, F., & Pratesi, C. A. (2023). *How to Create a Sustainable Food Industry: A Practical Guide to Perfect Food.* Taylor & Francis. doi:10.4324/9781003449744

Barth, M., Godemann, J., Rieckmann, M., & Stoltenberg, U. (2007). Developing key competencies for sustainable development in higher education. *International Journal of Sustainability in Higher Education*, *8*(4), 416–430. doi:10.1108/14676370710823582

Bartzas, G., & Komnitsas, K. (2020). An integrated multi-criteria analysis for assessing sustainability of agricultural production at regional level. *Information Processing in Agriculture*, *7*(2), 223–232. doi:10.1016/j.inpa.2019.09.005

Barua, C., Halai, P., Patwary, P., Bora, M., Barua, A. G., & Bordoloi, M. J. (2015). Evaluation of anti-amnestic activity of few medicinal plants against scopolamine-induced amnesia. *Indian Journal of Traditional Knowledge*, *14*(4), 581–589.

Basu, P. (2018). *Biomass gasification, pyrolysis and torrefaction: practical design and theory.* Academic press.

Baumann, E. (2018). *The Circular Economy: A New Sustainability Paradigm.* Springer.

Bennett, C. F. (n.d.). *Analyzing impacts of extension programs.*

Besseling, E., Wang, B., & Lübken, M. (2020). Coconut Leather: A Sustainable Alternative for Leather Fashion Products. *Sustainability*, *12*(8), 3201. doi:10.3390/su12083201

Bharagava, R. N. (Ed.). (2020). *Emerging eco-friendly green technologies for wastewater treatment. Berlin.* Springer. doi:10.1007/978-981-15-1390-9

Bhardwaj, V., & Fairhurst, A. (2010). Fast Fashion: Response to Changes in the Fashion Industry. *International Review of Retail, Distribution and Consumer Research*, *20*(1), 165–173. doi:10.1080/09593960903498300

Bhattacharyya, P., Bisen, J., Bhaduri, D., Priyadarsini, S., Munda, S., Chakraborti, M., & Nimbrayan, P. (2021). Turn the wheel from waste to wealth: Economic and environmental gain of sustainable rice straw management practices over field burning in reference to India. *The Science of the Total Environment*, *775*, 145896. doi:10.1016/j.scitotenv.2021.145896

Bhuvaneshwari, S., Hettiarachchi, H., & Meegoda, J. N. (2019). Crop residue burning in India: Policy challenges and potential solutions. *International Journal of Environmental Research and Public Health*, *16*(5), 832. doi:10.3390/ijerph16050832 PMID:30866483

Birgitta, N. & Helen, A. (2021). Global Learning For Sustainable Development. *Sustainability*, *13*(6), 3451. doi:10.3390/su13063451

Birla, D., Singh, D., Yadav, B., Kumar, A., & Rai, S. (2020). Modern Strategies for Crop Residue Management. *Agriculture and Food: E-Newsletter*, *2*(11), 172–175.

Birnbaum, L. S. (2013). Designing safer chemicals. *Environmental Health Perspectives*, *121*(1), a9–a9. doi:10.1289/ehp.1206349 PMID:23287533

Bisaria, P. (2022). *Environmental issues and corporate social responsibility.*

Black, J., & Harben, A. M. (2015). The Environmental Impact of Traditional vs. Digital Printing in the Textile Industry. *Journal of Textile and Apparel. Technology and Management*, *10*(3), 1–13.

Black, S. (2019). Sustainable Fashion: Past, Present, and Future. In T. Lang, P. J. Goodman, & H. Molesworth (Eds.), *The Sage Handbook of Nature* (pp. 670–684). Sage.

Bouraga, S. (2021). A taxonomy of blockchain consensus protocols: A survey and classification framework. *Expert Systems with Applications*, *168*, 114384. doi:10.1016/j.eswa.2020.114384

Bradu, P., Biswas, A., Nair, C., Sreevalsakumar, S., Patil, M., Kannampuzha, S., Mukherjee, A. G., Wanjari, U. R., Renu, K., Vellingiri, B., & Gopalakrishnan, A. V. (2022). Recent advances in green technology and Industrial Revolution 4.0 for a sustainable future. In *Environmental Science and Pollution Research*. Springer., doi:10.1007/s11356-022-20024-4

Brown, H., & Jones, S. (2024). *Upcycling as a Sustainable Solution: Case Studies and Best Practices*. Sustainable Development Journal.

Brown, H., & Martinez, P. (2022). Economic Benefits of Recycling Programs: Cost-Benefit Analysis and Financial Implications. *Journal of Economics and Finance*.

Brown, K. (2017). *Sustainable Fashion and Textiles: Design Journeys*. Routledge.

Buhalis, D., O'Connor, P., & Leung, R. (2023). Smart hospitality: From smart cities and smart tourism towards agile business ecosystems in networked destinations. *International Journal of Contemporary Hospitality Management*, *35*(1), 369–393. doi:10.1108/IJCHM-04-2022-0497

Cai, Z., Li, M., Zhu, Z., Wang, X., Huang, Y., Li, T., Gong, H., & Yan, M. (2023). Biological degradation of plastics and microplastics: A recent perspective on associated mechanisms and influencing factors. *Microorganisms*, *11*(7), 1661. doi:10.3390/microorganisms11071661 PMID:37512834

Camilleri, M. A. (2021). Sustainable production and consumption of food. Mise-en-place circular economy policies and waste management practices in tourism cities. *Sustainability (Basel)*, *13*(17), 9986. doi:10.3390/su13179986

Cannon, T. (Ed.). (2015). *Global waste management outlook*. UNEP. https://www.unep.org/resources/report/global-waste-management-outlook

Carew, A., & Mitchell, C. (2008). Teaching sustainability as a contested concept: Capitalizing on variation in engineering educators' conceptions of environmental, social and economic sustainability. *Production*, *16*, 105–115.

Ceballos, D., & Dong, Z. (2016). Formal e-recycling: The complexity of solving the e-waste problem worldwide. *Harvard T. H. Chan*. https://www.hsph.harvard.edu/hoffman-program/2016/08/26/formal-e-recycling-the-complexity-of-solving-the-e-waste-problem-worldwide/#:~:text=Developed%20countries%20generally%20use%20formal,such%20as%20plastic%20and%20metal

Ceballos, D. M., & Dong, Z. (2016). The formal electronic recycling industry: Challenges and opportunities in occupational and environmental health research. *Environment International*, *95*, 157–166. doi:10.1016/j.envint.2016.07.010 PMID:27568575

Chalak, A., Abou-Daher, C., & Abiad, M. G. (2018). Generation of food waste in the hospitality and food retail and wholesale sectors: Lessons from developed economies. *Food Security*, *10*(5), 1279–1290. doi:10.1007/s12571-018-0841-0

Chandran, R. R., Thomson, B. I., Natishah, A. J., Mary, J., & Nachiyar, V. (2023). Nanotechnology in Plastic Degradation. *Biosciences Biotechnology Research Asia*, *20*(1), 53–68. doi:10.13005/bbra/3068

Chaudhary, C., Yadav, D. B., Yadav, A., Chaudhary, A., & Hooda, V. S. (2024). Influence of crop residue and nitrogen management on nutrient uptake, yield, and economics of rice-wheat cropping system. *Journal of Plant Nutrition*, *47*(3), 376–391. doi:10.1080/01904167.2023.2277399

Choi, T. M., & Cheng, T. C. E. (2015). Socially Responsible Supplier Selection: Literature Review and Future Research. *International Journal of Production Economics*, *159*, 1–17. doi:10.1016/j.ijpe.2014.09.026

Choi, T. M., & Li, Y. (2019). The Impact of Transparency on Green Supply Chain Management: A Review. *Sustainability*, *11*(15), 4085. doi:10.3390/su11154085

Chopra, D., Gulati, K., & Ivanovski, S. (2021). Understanding and optimizing the antibacterial functions of anodized nano-engineered titanium implants. *Acta Biomaterialia*, *127*, 80–101. doi:10.1016/j.actbio.2021.03.027 PMID:33744499

Circular by Design. (2019). *Products in the Circular Economy*. European Environment Agency.

Circularity Indicators. (2020). *An Approach to Measuring Circularity*. Ellen MacArthur Foundation.

Closing the Loop. (2021). *Commission Adopts New Circular Economy Action Plan*. European Commission.

Cointreau, S. J. (2006). *Occupational and environmental health issues of solid waste management: Special emphasis on middle- and lower-income countries*. World Bank.

Colvin, V. L. (2003). The potential environmental impact of engineered nanomaterials. *Nature Biotechnology*, *21*(10), 1166–1170. doi:10.1038/nbt875 PMID:14520401

Coulibaly, T. P., Du, J., & Diakité, D. (2021). Sustainable agricultural practices adoption. *Agriculture (Pol'nohospodárstvo)*, *67*(4), 166–176. doi:10.2478/agri-2021-0015

Dalia, K. (2020). Green garbage: A state comparison of Morison of marijuana packaging and waste management. *Golden Gate University Environmental Law Journal*, *12*, 175.

Danesh, P., Niaparast, P., Ghorbannezhad, P., & Ali, I. (2023). Biochar production: Recent developments, applications, and challenges. *Fuel*, *337*, 126889. doi:10.1016/j.fuel.2022.126889

Daniel, W. E., Koza, M. B., Parker, S. D., & Long, M. D. (2012). Traditional RCRA and Subpart K: UNC-CH's foray into the waste management weeds. *Journal of Chemical Health and Safety*, *19*(4), 25–30. doi:10.1016/j.jchas.2012.01.011

Das, S., Biswas, S., Ramakrishnan, B., Das, T. K., Purakayastha, T. J., Gawade, B. H., & Sinha, K. (2024). Biological soil health with conventional and qPCR-based indicators under conservation agriculture based rice-wheat cropping system in Indo-Gangetic Plain. *Applied Soil Ecology*, *193*, 105128. doi:10.1016/j.apsoil.2023.105128

David, M., Prakash, L., Sangeetha, J., Naik, J., Thangadurai, D., & Thimmappa, S. C. (2020). Degradation of Plastics Using Nanomaterials. In O. Kharissova, L. Martínez, & B. Kharisov (Eds.), *Handbook of Nanomaterials and Nanocomposites for Energy and Environmental Applications*. Springer. doi:10.1007/978-3-030-11155-7_74-1

De Bernardi, P., Bertello, A., & Forliano, C. (2023). Circularity of food systems: A review and research agenda. *British Food Journal*, *125*(3), 1094–1129. doi:10.1108/BFJ-05-2021-0576

De Titto, E., & Savino, A. (2019). Environmental and health risks related to waste incineration. *Waste Management & Research*, *37*(10), 976–986. doi:10.1177/0734242X19859700 PMID:31319775

de Visser-Amundson, A. (2022). A multi-stakeholder partnership to fight food waste in the hospitality industry: A contribution to the United Nations Sustainable Development Goals 12 and 17. *Journal of Sustainable Tourism*, *30*(10), 2448–2475. doi:10.1080/09669582.2020.1849232

Despoudi, S., Bucatariu, C., Otles, S., & Kartal, C. (2021). Food waste management, valorization, and sustainability in the food industry. In *Food waste recovery* (pp. 3–19). Academic Press. doi:10.1016/B978-0-12-820563-1.00008-1

Dethier, J. J., & Effenberger, A. (2012). Agriculture and development: A brief review of the literature. *Economic Systems*, *36*(2), 175–205. doi:10.1016/j.ecosys.2011.09.003

Deus, R., Mele, F., Bezerra, B., & Battistelle, R. (2020). A municipal solid waste indicator for environmental impact: Assessment and identification of best management practices. *Journal of Cleaner Production*, *242*, 118433. doi:10.1016/j.jclepro.2019.118433

Dhillon, J., & Sandhu, V. (2017). Integrating the activities of informal and formal sector in e-waste management: Potential benefits and challenges. *International Journal of Latest Technology in Engineering Management & Applied Science, 6*(7), 133–136. https://www.ijltemas.in/DigitalLibrary/Vol.6Issue7/133-136.pdf

Dhir, A., Talwar, S., Kaur, P., & Malibari, A. (2020). Food waste in hospitality and food services: A systematic literature review and framework development approach. *Journal of Cleaner Production, 270*, 122861. doi:10.1016/j.jclepro.2020.122861

Dhote, S., & Dixit, S. (2009). Water quality improvement through macrophytes - A review. *Environmental Monitoring and Assessment, 152*(1-4), 149–153. doi:10.1007/s10661-008-0303-9 PMID:18537050

Dianati, K., Schäfer, L., Milner, J., Gómez-Sanabria, A., Gitau, H., Hale, J., Langmaack, H., Kiesewetter, G., Muindi, K., Mberu, B., Zimmermann, N., Michie, S., Wilkinson, P., & Davies, M. (2021). A system dynamics-based scenario analysis of residential solid waste management in Kisumu, Kenya. *The Science of the Total Environment, 777*, 146200. doi:10.1016/j.scitotenv.2021.146200

Dias, S. (2011). *Integrating informal workers into selective waste collection: the case of Belo Horizonte, Brazil.* Wiego policy brief (urban policies) Number 4.

Donzelot, J. (1977). *The +Policing of the Families.* The Johns Hopkins University Press.

Dubé, M. A., & Salehpour, S. (2014). Applying the principles of green chemistry to polymer production technology. *Macromolecular Reaction Engineering, 8*(1), 7–28. doi:10.1002/mren.201300103

Dunn, R. A., & Welden, N. A. (2023). Management of Environmental Plastic Pollution: A Comparison of Existing Strategies and Emerging Solutions from Nature. *Water, Air, and Soil Pollution, 234*(3), 201. doi:10.1007/s11270-023-06190-2

Duque-Acevedo, M., Belmonte-Ureña, L. J., Cortés-García, F. J., & Camacho-Ferre, F. (2020). Agricultural waste: Review of the evolution, approaches and perspectives on alternative uses. *Global Ecology and Conservation, 22*, e00902. doi:10.1016/j.gecco.2020.e00902

Dutta, D., & Goel, S. (2021). Understanding the gap between formal and informal e-waste recycling facilities in India. *Waste Management (New York, N.Y.), 125*, 163–171. doi:10.1016/j.wasman.2021.02.045 PMID:33706255

Education for Sustainable Development. (2020). United Nations Educational, Scientific and Cultural Organization.

El Janati, M., Akkal-Corfini, N., Bouaziz, A., Oukarroum, A., Robin, P., Sabri, A., Chikhaoui, M., & Thomas, Z. (2021). Benefits of Circular Agriculture for Cropping Systems and Soil Fertility in Oases. *Sustainability (Basel), 13*(9), 4713. doi:10.3390/su13094713

El Zein, B., Elrashidi, A., Dahlan, M., Al Jarwan, A., & Jabbour, G. (2024). *Nano and Society 5.0: Advancing the Human-Centric Revolution.*

Elander, M., & Nylund, P. (2018). Ethical Consumption of Fast Fashion: Literature Review and Research Agenda. *Journal of Retailing and Consumer Services, 41*, 308–318. doi:10.1016/j.jretconser.2017.11.014

Elizabath, A., Babychan, M., Mathew, A. M., & Syriac, G. M. (2019). Application of nanotechnology in agriculture. *Int. J. Pure Appl. Biosci, 7*(2), 131–139. doi:10.18782/2320-7051.6493

Ellen MacArthur Foundation. (2017). *A New Textiles Economy.* Redesigning Fashion's.

Elsheekh, K., Kamel, R., Elsherif, D., & Shalaby, A. (2018). Achieving sustainable development goals from the perspective of solid waste management plans. *Journal of Engineering and Applied Sciences (Asian Research Publishing Network), 68*(9), 1–15.

Evode, N., Qamar, S. A., Bilal, M., Barceló, D., & Iqbal, H. M. (2021). Plastic waste and its management strategies for environmental sustainability. *Case Studies in Chemical and Environmental Engineering*, *4*, 100142. doi:10.1016/j.cscee.2021.100142

Fang, B., Yu, J., Chen, Z., Osman, A. I., Farghali, M., Ihara, I., Hamza, E. G., Rooney, D. W., & Yap, P. (2023). Artificial intelligence for waste management in smart cities: A review. *Environmental Chemistry Letters*, *21*(4), 1959–1989. doi:10.1007/s10311-023-01604-3 PMID:37362015

FAO (Ed.). (2017). *Towards zero hunger 1945-2030*. Food and Agriculture Organization of the United Nations.

Fashion on Climate. (2019). *The Relationship between Fashion and Climate Change*. World Resources Institute.

Ferreira, J., Ryan, L., & Tilbury, D. (2006). *Whole-School Approaches to Sustainability*. Aries. http://www.aries.mq.edu.au/projects/preservice/

Filimonau, V., Coşkun, A., Derqui, B., & Matute, J. (2022). Restaurant management and food waste reduction: Factors affecting attitudes and intentions in restaurants of Spain. *International Journal of Contemporary Hospitality Management*, *34*(3), 1177–1203. doi:10.1108/IJCHM-07-2021-0899

Filimonau, V., Krivcova, M., & Pettit, F. (2019). An exploratory study of managerial approaches to food waste mitigation in coffee shops. *International Journal of Hospitality Management*, *76*, 48–57. doi:10.1016/j.ijhm.2018.04.010

Filimonau, V., Matute, J., Kubal-Czerwińska, M., Krzesiwo, K., & Mika, M. (2020). The determinants of consumer engagement in restaurant food waste mitigation in Poland: An exploratory study. *Journal of Cleaner Production*, *247*, 119105. doi:10.1016/j.jclepro.2019.119105

Filimonau, V., & Uddin, R. (2021). Food waste management in chain-affiliated and independent consumers' places: A preliminary and exploratory study. *Journal of Cleaner Production*, *319*, 128721. doi:10.1016/j.jclepro.2021.128721

Fitchuk, C. (2021). *Creating a Sustainable Business Strategy for a SME*.

Fletcher, K. (2018). *Fashioned from Nature: How Sustainable Materials Have Changed the Fashion Industry*. Victoria and Albert Museum.

Fletcher, K., & Grose, L. (2012). *Fashion and Sustainability: Design for Change*. Laurence King Publishing.

Fletcher, K., & Tham, M. (2019). *The Handbook of Sustainability and Fashion*. Routledge.

Fogel, R., & Limson, J. (2016). Developing biosensors in developing countries: South Africa as a case study. *Biosensors (Basel)*, *6*(1), 5. doi:10.3390/bios6010005 PMID:26848700

Food and Agriculture Organization. (2017). The state of food and agriculture 2017: leveraging food systems for inclusive rural transformation (pp. xvii+-160). FAO.

Food and agriculture: Driving action across the 2030 Agenda for Sustainable Development. (n.d.). *FOOD AND AGRICULTURE*.

Forbes, J. (2020). *Design for Sustainability: A Practical Approach for Textiles and Fashion*. Bloomsbury Visual Arts.

Forti, V., Balde, C. P., Kuehr, R., & Bel, G. (2020). *The global e-waste monitor 2020* (1st ed.). Unitar Scycle., https://ewastemonitor.info/wp-content/uploads/2020/11/GEM_2020_def_july1_low.pdf

Gao, J. Q., Li, D., Qiao, G. H., Jia, Q. R., Li, S. R., & Gao, H. L. (2024). Circular economy strategies in supply chains, enhancing resource efficiency and sustainable development goals. *Environmental Science and Pollution Research International*, *31*(6), 8751–8767. doi:10.1007/s11356-023-31551-z PMID:38180660

Garofalo, E., Di Maio, L., Scarfato, P., Di Gregorio, F., & Incarnato, L. (2019). Nanotechnology-based strategy to upgrade the performances of plastic flexible film waste. *Polymers, 11*(5), 830. doi:10.3390/polym11050830 PMID:31071960

Gaur, M., Misra, C., Yadav, A. B., Swaroop, S., Maolmhuaidh, F. Ó., Bechelany, M., & Barhoum, A. (2021). Biomedical Applications of Carbon Nanomaterials: Fullerenes, Quantum Dots, Nanotubes, Nanofibers, and Graphene. *Materials (Basel), 14*(20), 5978. doi:10.3390/ma14205978 PMID:34683568

Gavala, H. N., Angelidaki, I., & Ahring, B. K. (2003). Kinetics and modeling of anaerobic digestion process. *Advances in Biochemical Engineering/Biotechnology, I*, 57–93. doi:10.1007/3-540-45839-5_3 PMID:12747561

Geyer, R., Jambeck, J. R., & Law, K. L. (2017). Production, use, and fate of all plastics ever made. *Science Advances, 3*(7), e1700782. doi:10.1126/sciadv.1700782 PMID:28776036

Ghidan, A. Y., & Al Antary, T. M. (2019). Applications of nanotechnology in agriculture. In *Applications of nanobiotechnology*. IntechOpen.

Ghulam, S. T., & Abushammala, H. (2023). Challenges and opportunities in the management of electronic waste and its impact on human health and environment. *Sustainability (Basel), 15*(3), 1837–1859. doi:10.3390/su15031837

Gkountela, C. I., & Vouyiouka, S. N. (2022). Enzymatic polymerization as a green approach to synthesizing bio-based polyesters. *Macromol, 2*(1), 30–57. doi:10.3390/macromol2010003

Glebova, E., Gerke, A., & Book, R. (2023). The transformational role of technology in sports events. In *Sports Management in an Uncertain Environment* (pp. 169–187). Springer Nature Singapore. doi:10.1007/978-981-19-7010-8_8

Goh, E., & Jie, F. (2019). To waste or not to waste: Exploring motivational factors of Generation Z hospitality employees towards food wastage in the hospitality industry. *International Journal of Hospitality Management, 80*, 126–135. doi:10.1016/j.ijhm.2019.02.005

Goh, E., Okumus, B., Jie, F., Djajadikerta, H. G., & Lemy, D. M. (2022). Managing food wastage in hotels: Discrepancies between injunctive and descriptive norms amongst hotel food and beverage managers. *British Food Journal, 124*(12), 4666–4685. doi:10.1108/BFJ-05-2021-0513

Goldsworthy, M. (2018). *Designing for the Circular Economy*. Bloomsbury Visual Arts.

Goswami, S. B., Mondal, R., & Mandi, S. K. (2020). Crop residue management options in rice–rice system: A review. *Archives of Agronomy and Soil Science, 66*(9), 1218–1234. doi:10.1080/03650340.2019.1661994

Green, L., & White, M. (2024). Sustainable Materials in Fashion Design: A Review of Alternatives to Virgin Fibers. *Journal of Sustainable Fashion*.

Green, L., & Smith, J. (2024). *The Future of Recycling: Emerging Trends and Technologies*. Sustainability Review.

Green, L., & Wilson, D. (2023). *Educational Initiatives for Recycling Awareness: Strategies for Effective Outreach*. Environmental Education Journal.

Green, L., & Wilson, D. (2024). *Incentivizing Recycling Behavior: Strategies for Effective Policy Implementation*. Environmental Policy Journal.

Grillo, R., Abhilash, P. C., & Fraceto, L. F. (2016). Nanotechnology applied to bio-encapsulation of pesticides. *Journal of Nanoscience and Nanotechnology, 16*(1), 1231–1234. doi:10.1166/jnn.2016.12332 PMID:27398594

Gunamantha, M., & Sarto. (2012). Life cycle assessment of municipal solid waste treatment to energy options: Case study of Kartamantul region, Yogyakarta. *Renewable Energy, 41*, 277–284. doi:10.1016/j.renene.2011.11.008

Gupta, M., Choudhary, B., & Sharma, D. (2023). Emersion and Immersion of Technology in the Development of Smart Cities: A Bibliometric Analysis. In *Enabling Technologies for Effective Planning and Management in Sustainable Smart Cities* (pp. 303–318). Springer International Publishing. doi:10.1007/978-3-031-22922-0_12

Gupta, R., Hirani, H., & Shankar, R. (2023). Sustainable solid waste management system using technology-enabled end-of-pipe strategies. *Journal of Environmental Management*, *347*, 119122. doi:10.1016/j.jenvman.2023.119122 PMID:37793298

Gwilt, A., & Rissanen, T. (2011). *Shaping Sustainable Fashion: Changing the Way We Make and Use Clothes*. Earthscan.

Gwilt, A., & Rissanen, T. (2019). *Zero Waste Fashion Design*. Bloomsbury Visual Arts.

Hadden, M., Martinez-Martin, D., Yong, K.-T., Ramaswamy, Y., & Singh, G. (2022). Recent Advancements in the Fabrication of Functional Nanoporous Materials and Their Biomedical Applications. *Materials (Basel)*, *15*(6), 2111. doi:10.3390/ma15062111 PMID:35329563

Hahn, T., Figge, F., Pinkse, J., & Preuss, L. (2018). A Paradox Perspective on Corporate Sustainability: Descriptive, Instrumental, and Normative Aspects. *Journal of Business Ethics*, *148*(2), 235–248. doi:10.1007/s10551-017-3587-2

Hakeem, K. R., & Pirzadah, T. B. (2020). *Nanobiotechnology in Agriculture*. Springer International Publishing. doi:10.1007/978-3-030-39978-8

Hald Albertsen, C., Kulkarni, J. A., Witzigmann, D., Lind, M., Petersson, K., & Simonsen, J. B. (2022). The role of lipid components in lipid nanoparticles for vaccines and gene therapy. *Advanced Drug Delivery Reviews*, *188*, 114416. doi:10.1016/j.addr.2022.114416 PMID:35787388

Hannon, J. (2020). Exploring and illustrating the (inter-) disciplinarity of waste and zero waste management. *Urban Science (Basel, Switzerland)*, *4*(4), 73. doi:10.3390/urbansci4040073

Harper-Harris, J., Kant, K., & Singh, G. (2021). Oleic Acid-Assisted Synthesis of Tunable High-Aspect-Ratio Multiply-Twinned Gold Nanorods for Bioimaging. *ACS Applied Nano Materials*, *4*(4), 3325–3330. doi:10.1021/acsanm.1c00360

Hassoun, A., Aït-Kaddour, A., Abu-Mahfouz, A. M., Rathod, N. B., Bader, F., Barba, F. J., Biancolillo, A., Cropotova, J., Galanakis, C. M., Jambrak, A. R., Lorenzo, J. M., Måge, I., Ozogul, F., & Regenstein, J. (2023). The fourth industrial revolution in the food industry—Part I: Industry 4.0 technologies. *Critical Reviews in Food Science and Nutrition*, *63*(23), 6547–6563. doi:10.1080/10408398.2022.2034735 PMID:35114860

Häußler, M., Eck, M., Rothauer, D., & Mecking, S. (2021). Closed-loop recycling of polyethylene-like materials. *Nature*, *590*(7846), 423–427. doi:10.1038/s41586-020-03149-9 PMID:33597754

Hayes, G., Laurel, M., MacKinnon, D., Zhao, T., Houck, H. A., & Becer, C. R. (2022). Polymers without petrochemicals: Sustainable routes to conventional monomers. *Chemical Reviews*, *123*(5), 2609–2734. doi:10.1021/acs.chemrev.2c00354 PMID:36227737

Helal, E., Shaban, M., Abd El-Karim, E., & Ellayen, A. (2021). A comprehensive management model to maximize the benefits of agriculture drainage water reuse. *Water Environment Research*, *93*(9), 1722–1733. doi:10.1002/wer.1559 PMID:33725400

Hemalatha, M. Vinita, Sravanalakshmi, G., Kotibagar, B. C., & Megha. (2023). Nanotechnology for Sustainable Agriculture: Current Trends and Future Prospects. In Modern Nanotechnology: Volume 1: Environmental Sustainability and Remediation (pp. 43-75). Cham: Springer International Publishing

He, Y., Deng, X., Jiang, L., Hao, L., Shi, Y., Lyu, M., & Wang, S. (2023). Current advances, challenges and strategies for enhancing the biodegradation of plastic waste. *The Science of the Total Environment*, 167850. PMID:37844647

Hidangmayum, A., Debnath, A., Guru, A., Singh, B. N., Upadhyay, S. K., & Dwivedi, P. (2023). Mechanistic and recent updates in nano-bioremediation for developing green technology to alleviate agricultural contaminants. *International Journal of Environmental Science and Technology*, *20*(10), 11693–11718. doi:10.1007/s13762-022-04560-7 PMID:36196301

Higg Materials Sustainability Index (MSI), *Sustainable Apparel Coalition*, (2020).

Höjer, M., & Wangel, J. (2015). Smart Sustainable Cities: Definition and Challenges. In M. Höjer & J. Wangel (Eds.), *Smart Cities: Foundations, Principles, and Applications* (pp. 3–19). John Wiley & Sons.

Holfelder, A. K. (2019). Towards a sustainable future with education. *Sustainability Science*, *14*(4), 943–952. doi:10.1007/s11625-019-00682-z

Hong, J., Jaegler, A., & Gergaud, O. (2024). Mobile applications to reduce food waste in supply chains: A systematic literature review. *British Food Journal*, *126*(2), 509–530. doi:10.1108/BFJ-09-2022-0742

Hoornweg, D., & Bhada-Tata, P. (2012). *What a Waste: A Global Review of Solid Waste Management. Urban Development Series Knowledge Papers*. World Bank.

Hopewell, J., Dvorak, R., & Kosior, E. (2009). Plastics recycling: Challenges and opportunities. *Philosophical Transactions of the Royal Society of London. Series B, Biological Sciences*, *364*(1526), 2115–2126. doi:10.1098/rstb.2008.0311 PMID:19528059

Horejs, C. (2021). Lasting impact of lipid nanoparticles. *Nature Reviews. Materials*, *6*(12), 1071. doi:10.1038/s41578-021-00392-y

Horton, A. A. (2022). Plastic pollution: When do we know enough? *Journal of Hazardous Materials*, *422*, 126885. doi:10.1016/j.jhazmat.2021.126885 PMID:34418830

*How disposable tech is feeding an e-waste crisis*. (2022). UNEP. https://www.unep.org/news-and-stories/story/how-disposable-tech-feeding-e-waste-crisis

Huang, J., Veksha, A., Chan, W. P., Giannis, A., & Lisak, G. (2022). Chemical recycling of plastic waste for sustainable material management: A prospective review on catalysts and processes. *Renewable & Sustainable Energy Reviews*, *154*, 111866. doi:10.1016/j.rser.2021.111866

Huang, Y., Ma, E., & Wang, D. (2021). Message framing strategies, food waste prevention, and diners' repatronage intentions: The mediating role of corporate social responsibility. *Journal of Sustainable Tourism*, *29*(10), 1694–1715. doi:10.1080/09669582.2020.1867151

Hunter, D., Monville-Oro, E., Burgos, B., Roel, C. N., Calub, B. M., Gonsalves, J., & Lauridsen, N. (Eds.). (2020). *Agrobiodiversity, school gardens and healthy diets: Promoting biodiversity, food and sustainable nutrition*. Routledge. doi:10.4324/9780429053788

Iacovidou, E., Ohandja, D., Gronow, J., & Voulvoulis, N. (2012). The household use of food waste disposal units as a waste management option: A review. *Critical Reviews in Environmental Science and Technology*, *42*(14), 1485–1508. doi:10.1080/10643389.2011.556897

Ignatowicz, K. (2009). Selection of sorbent for removing pesticides during water treatment. *Journal of Hazardous Materials*, *169*(1-3), 953–957. doi:10.1016/j.jhazmat.2009.04.061 PMID:19446393

Islam, J. U., Nazir, O., & Rahman, Z. (2023). Sustainably engaging employees in food wastage reduction: A conscious capitalism perspective. *Journal of Cleaner Production*, *389*, 136091. doi:10.1016/j.jclepro.2023.136091

IslamM.NayemA. I.BristyU. A.EshaS. A. (2023). Approaches to e-waste reduction through 4R, public awareness, and design for sustainability. *IN4OBE Global Virtual Summit 2022 Transforming Education and Empowering Learners.* [REMOVED HYPERLINK FIELD] doi:10.2139/ssrn.4458682

Jain, M. S., & Kalamdhad, A. S. (2018). Efficacy of batch mode rotary drum composter for management of aquatic weed. *Journal of Environmental Management*, *1*(221), 20–27. doi:10.1016/j.jenvman.2018.05.055 PMID:29787969

Jambeck, J., Geyer, R., Wilcox, C., Siegler, T., Perryman, M., Andrady, A., Narayan, R., & Law, L. (2015). Plastic waste inputs from land into the ocean. *Science*, *347*(6223), 768–771. doi:10.1126/science.1260352 PMID:25678662

Jauhar, S., Pratap, S., Lakshay, Paul, S., & Gunasekaran, A. (2023). Internet of things based innovative solutions and emerging research clusters in circular economy. *Operations Management Research : Advancing Practice Through Research*, *16*(4), 1968–1988. doi:10.1007/s12063-023-00421-9

Jayathilaka, W. A. D. M., Qi, K., Qin, Y., Chinnappan, A., Serrano-García, W., Baskar, C., Wang, H., He, J., Cui, S., Thomas, S. W., & Ramakrishna, S. (2019). Significance of Nanomaterials in Wearables: A Review on Wearable Actuators and Sensors. *Advanced Materials*, *31*(7), 1805921. doi:10.1002/adma.201805921 PMID:30589117

Jiao, H., Ali, S. S., Alsharbaty, M. H. M., Elsamahy, T., Abdelkarim, E., Schagerl, M., Al-Tohamy, R., & Sun, J. (2024). A critical review on plastic waste life cycle assessment and management: Challenges, research gaps, and future perspectives. *Ecotoxicology and Environmental Safety*, *271*, 115942. doi:10.1016/j.ecoenv.2024.115942 PMID:38218104

Jirapong, K., Cagarman, K., & von Arnim, L. (2021). Road to Sustainability: University–Start-Up Collaboration. *Sustainability (Basel)*, *13*(11), 6131. doi:10.3390/su13116131

Johnson, R., & Jones, S. (2023). The Role of Government in Promoting Recycling: Policy Recommendations for Sustainable Development. *Journal of Public Policy*.

Johnson, R., & Martinez, P. (2022). *The Environmental Impact of Fashion Waste: Challenges and Solutions*. Environmental Management Review.

Jones, R. (2023). Collaborative Initiatives for Environmental Impact. *Sustainable Fashion Forum*. (2023).

Jones, S., & Martinez, P. (2024). Consumer Behavior and Recycling: Insights from Market Research. *Journal of Consumer Studies*.

Jones, S., & Brown, A. (2024). *Circular Design Principles in Fashion: Strategies for Sustainable Product Development*. Design Journal.

Joshi, N., Singh, V. P., Dhyani, V. C., Singh, S. P., & Pratap, T. (2015). Effect of plant geometry and weed management practices on weed growth and yield of direct seeded rice. *Journal of Environment and Bioscience*, *29*(1), 23–28.

Joshi, P. B. (2024). *Full blown green metrics*. Elsevier Publishing.

Joshi, P. B. (2024). *N. C. Durvel and C. Mohan, "Full blown green metrics"*. Elsevier Publishing., doi:10.1016/B978-0-443-18959-3.00013-6

Jyrki, R., & Anita, P. (2024). *Education for Sustainable Development Research from the Researchers' Point of View*. Research Gate. www.researchgate.net/publication/258157579 p-9

Kabera, T., Wilson, D., & Nishimwe, H. (2019). Benchmarking performance of solid waste management and recycling systems in East Africa: Comparing Kigali Rwanda with other major cities. *Waste Management & Research*, *37*(1_suppl), 58–72. doi:10.1177/0734242X18819752 PMID:30761955

Kandpal, A., Singh, A., Jain, D., Upadhyay, S. K., Siddiqui, A. J., Sharma, K., & Saxena, J. (2023). Managing plastic waste with nanotechnology: Current sustainability prospects. *Nanotechnology for Environmental Engineering, 8*(4), 1015–1031. doi:10.1007/s41204-023-00346-5

Kannan, H., Rodriguez, R. V., Paprika, Z. Z., & Ade-Ibijola, A. (Eds.). (2023). *Exploring Ethical Dimensions of Environmental Sustainability and Use of AI*. IGI Global. doi:10.4018/979-8-3693-0892-9

Kansotia, K., Naresh, R., Sharma, Y., Sekhar, M., Sachan, P., Baral, K., & Pandey, S. K. (2024). Nanotechnology-driven Solutions: Transforming Agriculture for a Sustainable and Productive Future. *Journal of Scientific Research and Reports, 30*(3), 32–51. doi:10.9734/jsrr/2024/v30i31856

Kasavan, S., Mohamed, A. F., & Halim, S. A. (2019). Drivers of food waste generation: Case study of island-based hotels in Langkawi, Malaysia. *Waste Management (New York, N.Y.), 91*, 72–79. doi:10.1016/j.wasman.2019.04.055 PMID:31203944

Kasavan, S., Siron, R., Yusoff, S., & Fakri, M. F. R. (2022). Drivers of food waste generation and best practice towards sustainable food waste management in the hotel sector: A systematic review. *Environmental Science and Pollution Research International, 29*(32), 48152–48167. doi:10.1007/s11356-022-19984-4 PMID:35579831

Kashyap, P. L., Xiang, X., & Heiden, P. (2015). Chitosan nanoparticle based delivery systems for sustainable agriculture. *International Journal of Biological Macromolecules, 77*, 36–51. doi:10.1016/j.ijbiomac.2015.02.039 PMID:25748851

Kauser, H., Pal, S., Haq, I., & Khwairakpam, M. (2020). Evaluation of rotary drum composting for the management of invasive weed Mikania micrantha Kunth and its toxicity assessment. *Bioresource Technology, 313*, 123678. doi:10.1016/j.biortech.2020.123678 PMID:32570078

Kaza, S., Yao, L., Congyuan, B., Perinaz, W., Frank, M., Thierry, R., Serrona, K., & Roy, T., & (2021). *What a Waste 2.0: A global snapshot of solid waste management to 2050. Urban development series Washington*. World Bank Group.

Khadka, R. B., Velazquez, A. D. S., & Cardina, J. (2022). Evaluation of Anaerobic Soil Disinfestation for Environmentally Sustainable Weed Management. *Agronomy (Basel), 2*(12), 3147. doi:10.3390/agronomy12123147

Khan, S. A. R., Godil, D. I., Jabbour, C. J. C., Shujaat, S., Razzaq, A., & Yu, Z. (2021). Green data analytics, blockchain technology for sustainable development, and sustainable supply chain practices: Evidence from small and medium enterprises. *Annals of Operations Research*. doi:10.1007/s10479-021-04275-x

Kibria, G., Masuk, N., Safayet, R., Nguyen, H., & Mourshed, M. (2023). Plastic waste: Challenges and opportunities to mitigate pollution and effective management. *International Journal of Environmental Research, 17*(1), 20. doi:10.1007/s41742-023-00507-z PMID:36711426

Kijo-Kleczkowska, A., & Gnatowski, A. (2022). Recycling of plastic waste, with particular emphasis on thermal methods. *Energies, 15*(6), 2114. doi:10.3390/en15062114

Kilibarda, N., Djokovic, F., & Suzic, R. (2019). Food waste management–reducing and managing food waste in hospitality. *Scientific journal. Metals Technology, 60*(2), 134–142.

Kim, H., & Kim, W. (2018). Sustainable Operations Management in the Fashion Industry: A Review. *Sustainability, 10*(10), 3577. doi:10.3390/su10103577

Kin, C. W., Zhang, R., & Chan, W. K. (2020). Evaluation of Energy Consumption in Block-Chains with Proof of Work and Proof of Stake. *Journal of Physics: Conference Series, 1584*(1), 012023. doi:10.1088/1742-6596/1584/1/012023

Koskela-Huotari, K., Svärd, K., Williams, H., Trischler, J., & Wikström, F. (2024). Drivers and hinderers of (un) sustainable service: A systems view. *Journal of Service Research, 27*(1), 106–123. doi:10.1177/10946705231176071

Kougias, P. G., & Angelidaki, I. (2018). Biogas and its opportunities—A review. *Frontiers of Environmental Science & Engineering*, *12*(3), 1–12. doi:10.1007/s11783-018-1037-8

Koul, B., Yakoob, M., & Shah, M. P. (2022). Agricultural waste management strategies for environmental sustainability. *Environmental Research*, *206*, 112285. doi:10.1016/j.envres.2021.112285 PMID:34710442

Kristanto, G., & Koven, W. (2019). Estimating greenhouse gas emission from municipal solid waste management in Depok, Indonesia. *City and Environment Interactions*, *4*, 100027. doi:10.1016/j.cacint.2020.100027

Kumar, P., Kumar, S., Joshi, L., Kumar, P., Kumar, S., & Joshi, L. (2015). Policies for restricting the agriculture residue burning in Punjab. Socioeconomic and Environmental Implications of Agricultural Residue Burning: A Case Study of Punjab, India, 117-131.

Kumari, N., Mohan, C., & Negi, A. (2023). An Investigative Study on the Thermal and Mechanical Properties of Clay PVC Polymer Composite Films. *Polymers*, *15*(8), 1922. doi:10.3390/polym15081922 PMID:37112069

Kumari, P., Kumari, N., Mohan, C., Chinglenthoiba, C., & Amesho, K. T. T. (2023). Environmentally Benign Approach to Formulate Nanoclay/Starch hydrogel for Controlled Release of Zinc and Its Application in Seed Coating of Oryza Sativa Plant. *International Journal of Biological Macromolecules*, *257*, 128278. doi:10.1016/j.ijbiomac.2023.128278 PMID:38029920

Kumar, M., Mohan, C., Kumar, S., Epifantsev, K., Singh, V., Dixit, S., & Singh, R. (2022). Coordination behavior of Schiff base copper complexes and structural characterization. *MRS Advances*, *7*(31), 939–943. doi:10.1557/s43580-022-00348-6

Kumar, R. (2016). Crop residue generation and management in Punjab state. *Indian Journal of Economics and Development*, *12*(1a), 477–483. doi:10.5958/2322-0430.2016.00109.8

Kung, C. C., Kong, F., & Choi, Y. (2015). Pyrolysis and biochar potential using crop residues and agricultural wastes in China. *Ecological Indicators*, *51*, 139–145. doi:10.1016/j.ecolind.2014.06.043

Labadi, S., Giliberto, F., Rosetti, I., Shetabi, L., & Yildirim, E. (2021). Heritage and the sustainable development goals: Policy guidance for heritage and development actors. *International Journal of Heritage Studies*.

Labelling, M., & Thörn, C. (2018). Ethical and Sustainable Fashion: An Overview. In E. Shultz & K. W. Hassan (Eds.), *Handbook of Research on the Global Impacts and Roles of Immersive Media* (pp. 1–25). IGI Global., doi:10.4018/978-1-5225-5449-5.ch001

Lagioia, G., Amicarelli, V., Strippoli, R., Bux, C., & Gallucci, T. (2023). Sustainable and circular practices in the hotel industry in Southern Italy: Opportunities, barriers and trends in food waste management. *British Food Journal*.

Lange, J. P. (2021). Managing plastic waste— Sorting, recycling, disposal, and product redesign. *ACS Sustainable Chemistry & Engineering*, *9*(47), 15722–15738. doi:10.1021/acssuschemeng.1c05013

Lange, J.-P. (2021). Managing Plastic Waste—Sorting, Recycling, Disposal, and Product Redesign ACS Sustainable. *Chemical Engineering (Albany, N.Y.)*, *9*(47), 15722–15738.

Larney, F. J., & Hao, X. (2007). A review of composting as a management alternative for beef cattle feedlot manure in southern Alberta. *Bioresource Technology*, *98*(17), 3221–3227. doi:10.1016/j.biortech.2006.07.005 PMID:17276674

Lasla, N., Al-Sahan, L., Abdallah, M., & Younis, M. (2022). Green-PoW: An energy-efficient blockchain Proof-of-Work consensus algorithm. *Computer Networks*. https://www.sciencedirect.com/science/article/pii/S1389128622002432

Leal Filho,W., Barbir, J., Abubakar, I., Paço, A., Stasiskiene, Z., Hornbogen, M., Christin Fendt, M., Voronova, V., & Klõga, M. (2022). *Consumer attitudes and concerns with bioplastics use: An international study.*

Lee, H. H., & Huang, P. Y. (2023). Food Waste and Environmental Sustainability of the Hotel Industry in Taiwan. *Sustainability (Basel), 15*(21), 15459. doi:10.3390/su152115459

Lee, J. E., Kim, Y., Kim, K., Lee, S., & Kim, S. (2018). Eco-Friendly Dyeing of Cotton Fabric Using Various Plant Extracts. *Fibers and Polymers, 19*(6), 1244–1253.

Legrand, W., Chen, J. S., & Laeis, G. C. (2022). *Sustainability in the hospitality industry: Principles of sustainable operations*. Taylor & Francis.

Legrand, W., Kuokkanen, H., & Day, J. (Eds.). (2023). *Critical Questions in Sustainability and Hospitality*. Taylor & Francis. doi:10.4324/9781003218425

Lehmann, J., & Joseph, S. (2015). Biochar for environmental management: an introduction. In *Biochar for environmental management* (pp. 1–13). Routledge.

Li, J., Wang, C., Chen, X., Zhang, Y., Zhang, Y., Fan, K., Zong, L., & Wang, L. (2023). Flash synthesis of ultrafine and active NiRu alloy nanoparticles on N-rich carbon nanotubes via joule heating for efficient hydrogen and oxygen evolution reaction. *Journal of Alloys and Compounds, 959*, 170571. doi:10.1016/j.jallcom.2023.170571

Lisbona, P., Pascual, S., & Perez, V. (2023). Waste to energy: Trends and perspectives. *Chemical Engineering Journal Advances, 14*, 100494. doi:10.1016/j.ceja.2023.100494

Liu, K., Tan, Q., Yu, J., & Wang, M. (2023). A global perspective on e-waste recycling. *Circular Economy, 2*(1), 1–16. doi:10.1016/j.cec.2023.100028

Liu, R., & Lal, R. (2015). Potentials of engineered nanoparticles as fertilizers for increasing agronomic productions. *The Science of the Total Environment, 514*, 131–139. doi:10.1016/j.scitotenv.2015.01.104 PMID:25659311

Lohan, S. K., Jat, H. S., Yadav, A. K., Sidhu, H. S., Jat, M. L., Choudhary, M., Peter, J. K., & Sharma, P. C. (2018). Burning issues of paddy residue management in north-west states of India. *Renewable & Sustainable Energy Reviews, 81*, 693–706. doi:10.1016/j.rser.2017.08.057

Lombardi, P., & Todella, E. (2023). Multi-Criteria Decision Analysis to Evaluate Sustainability and Circularity in Agricultural Waste Management. *Sustainability (Basel), 15*(20), 14878. doi:10.3390/su152014878

Lorenzo, P., Guilherm, R., & Barbosa, S. (2022). Agri-Food Waste as a Method for Weed Control and Soil Amendment in Crops. *Agronomy (Basel), 12*(5), 1184. doi:10.3390/agronomy12051184

Luu, T. T. (2020). Reducing food waste behavior among hospitality employees through communication: Dual mediation paths. *International Journal of Contemporary Hospitality Management, 32*(5), 1881–1904. doi:10.1108/IJCHM-09-2019-0779

Lu, Y., Li, L., & Xu, L. (2020). How Do Transparency and Information Disclosure Affect Environmental Protection in Fashion Supply Chains? Evidence from China. *Sustainability, 12*(7), 2975. doi:10.3390/su12072975

Ma, Y., Xiu, Q., Shao, L., & Yao, H. (2022). Promoting the Sustainable Improvement of Educational Empirical Research Quality. *Sustainability*. doi:10.3390/su14063380

Machado, M., & Hatakeyama, K. (2023, July). Digital Transformation as an Enabler of Sustainability in Supply Chain and Logistics: Evidence from the Field. In *2023 Portland International Conference on Management of Engineering and Technology (PICMET)* (pp. 1-6). IEEE. 10.23919/PICMET59654.2023.10216881

Maheshwari, D. K. (Ed.). (2014). *Composting for sustainable agriculture* (Vol. 3). Springer. doi:10.1007/978-3-319-08004-8

Marble, S. C., Steed, S. T., Saha, D., & Khamare, Y. (2019). On-farm evaluations of wood-derived, waste paper, and plastic mulch materials for weed control in Florida container nurseries. *HortTechnology*, *29*(6), 866–873. doi:10.21273/HORTTECH04437-19

Marcotullio, P., & Sorensen, A. (2023). Editorial: future urban worlds: theories, models, scenarios, and observation of urban spatial expansion. *Frontiers in Built Environment*, *9*, 1194813. doi:10.3389/fbuil.2023.1194813

Marouli, C. (2021). Sustainability Education for the Future? Challenges and Implications for Education and Pedagogy in the 21st Century. *Sustainability (Basel)*, *13*(5), 2021. doi:10.3390/su13052901

Martin-Rios, C., Demen Meier, C., & Pasamar, S. (2022). Sustainable waste management solutions for the foodservice industry: A Delphi study. *Waste Management & Research*, *40*(9), 1412–1423. doi:10.1177/0734242X221079306 PMID:35176923

Maturi, K. C., Haq, I., & Kalamdhad, A. S. (2022). Biodegradation of an intrusive weed Parthenium heterophorias through in-vessel composting technique: Toxicity assessment and spectroscopic study. *Environmental Science and Pollution Research International*, *29*(56), 84600–84615. doi:10.1007/s11356-022-21816-4 PMID:35788476

Mazhandu, Z., Muzenda, E., Mamvura, T., Belaid, M., & Nhubu, T. (2020). Integrated and consolidated review of plastic waste management and bio-based biodegradable plastics: Challenges and opportunities. *Sustainability (Basel)*, *12*(20), 8360. doi:10.3390/su12208360

McCance, K. R., Topliceanu, A. M., Echeverria, D., McAlexander, S. L., Blanchard, M. R., & Venditti, R. A. (2023). Fluffy, Fluffier, and Fluffiest: Creating and Testing Biodegradable Starch Packing Peanuts. *Journal of Chemical Education*, *100*(10), 4031–4039. doi:10.1021/acs.jchemed.3c00510

McKinsey & Company. (2016). Style that's Sustainable: A New Fast Fashion Formula. McKinsey.

Meena, R. S. (2015). A review of weed management in India: The need of new direction for sustainable agriculture. *The Bioscan*, *10*, 253–263.

Meier, T., von Borstel, T., Welte, B., Hogan, B., Finn, S. M., Bonaventura, M., Friedrich, S., Weber, K., & Dräger de Teran, T. (2021). Food waste in healthcare, business and hospitality catering: Composition, environmental impacts and reduction potential on company and national levels. *Sustainability (Basel)*, *13*(6), 3288. doi:10.3390/su13063288

Menikpura, S., Gheewala, S., Bonnet, S., & Chiemchaisri, C. (2013). Evaluation of the effect of recycling on sustainability of municipal solid waste management in Thailand. *Waste and Biomass Valorization*, *4*(2), 237–257. doi:10.1007/s12649-012-9119-5

Menzies, J., Sabert, B., Hassan, R., & Mensah, P. K. (2024). Artificial intelligence for international business: Its use, challenges, and suggestions for future research and practice. *Thunderbird International Business Review*, *66*(2), 185–200. doi:10.1002/tie.22370

Miconi, M. (2022). *The approach to sustainability policies in family firms: how this phenomenon has impacted management remuneration.*

Miraz, M. H., Excell, P. S., & Sobayel, K. (2021). Evaluation of green alternatives for blockchain proof-of-work (Pow) approach. *Annals of Emerging Technologies in Computing*, *5*(4), 54–59. doi:10.33166/AETiC.2021.04.005

Mishra, P., & Singh, G. (2023). Introduction: Importance of Sustainable Smart City. In *Sustainable Smart Cities: Enabling Technologies, Energy Trends and Potential Applications* (pp. 1–37). Springer International Publishing.

Mishra, P., & Singh, G. (2023). *Sustainable Smart Cities: Enabling Technologies, Energy Trends and Potential Applications*. Springer Nature. doi:10.1007/978-3-031-33354-5

Mishra, S., & Singh, H. B. (2015). Biosynthesized silver nanoparticles as a nanoweapon against phytopathogens: Exploring their scope and potential in agriculture. *Applied Microbiology and Biotechnology*, *99*(3), 1097–1107. doi:10.1007/s00253-014-6296-0 PMID:25547832

Mittal, D., Kaur, G., Singh, P., Yadav, K., & Ali, S. A. (2020). Nanoparticle-based sustainable agriculture and food science: Recent advances and future outlook. *Frontiers in Nanotechnology*, *2*, 579954. doi:10.3389/fnano.2020.579954

Mittal, S., & Lai, J.-C. (2022). Wireless, closed-loop, smart bandage with integrated sensors and stimulators for advanced wound care and accelerated healing. *Nature Biotechnology*. PMID:36424488

Modak, P., Pariatamby, A., Seadon, J., Tata, B. T., Borongan, G., Thawan, N. S., & Lim, M. B. (2020). *Highlights Asia waste management outlook summary for decision-makers*. UNEP. https://wedocs.unep.org/bitstream/handle/20.500.11822/27293/AsiaWMO_Sum.pdf?sequence=1&isAllowed=y

Mohan C., Malik D., Pandey S. & Kumar V. (2020). Effect of COVID-19 on India, International *Journal of Advanced Educational Research.*

Mohan C., Robinson J., Lal C., Jammala A.P.N. & Meena P. L. (2024). Sustainable Energy Solutions for Environmental Pollution Control. *E3S Web of Conferences*, *511*.

Mohan, C., Chauhan, S., Shekhar, S., & Kaushik, P. (2020). Madhu, Sharma M., Kumar V., A Brief Study on rate of morphological and genomic similarities between SARS and COVID-19. *International Journal of Pharma Research and Health Sciences*, 8.

Mohan, C., Chauhan, S., Shekhar, S., Kaushik, P., Madhu, M., Sharma, M., & Kumar, V. (2020). A Brief Study on rate of morphological and genomic similarities between SARS and COVID-19. *International Journal of Pharma Research and Health Sciences*, *8*(3), 3143. doi:10.21276/ijprhs.2020.03.03

Mohan, C., Chugh, V., & Pandey, G. (2022). Heterocyclic compounds of thiazoles as important material in medicinal chemistry. *Materials Today: Proceedings*, *69*, 478–481. doi:10.1016/j.matpr.2022.09.150

Mohan, C., & Kumari, N. (2022). Effect of various types of naturally occurring clay minerals on mechanical and thermal properties of PMMA polymer composite films. *MRS Advances*, *7*, 933–938. doi:10.1557/s43580-022-00347-7

Mohan, C., & Kumari, N. (2022). Synthesis of Acid Activated and Silicotungstic Acid (STA) Intercalated Montmorillonite Clay as Green Catalyst. *Materials Today: Proceedings*, *56*, 971–975. doi:10.1016/j.matpr.2022.03.093

Mohan, C., Kumari, N., Jindal, R., & Gautam, R. (2022). *Application of Efficient Naturally Occurring Clay Mineral for Fuchsin Basic Dye Removal, Advances in Functional and Smart Materials, Lecture Notes in Mechanical Engineering*. Springer.

Mohan, C., Kumari, N., Jindal, R., & Gautam, R. (2022). *Application of Efficient Naturally Occurring Clay Mineral for Fuchsin Basic Dye Removal*. Springer Germany.

Mohan, C., Kumari, N., & Robinson, J. (2022). Sustainable and environmental friendly energy materials. *Materials Today: Proceedings*, *69*, 494–498. doi:10.1016/j.matpr.2022.09.187

Mohan, C., Kumari, N., & Sushma, A. (2024). *Introduction to environmental and green chemistry*. Elsevier Publishing., doi:10.1016/B978-0-443-18959-3.00005-7

Mohan, C., Kumari, N., & Varma, R. (2023). Noval Approach for the *Synthesis of Hybrid-Clay based Nano Pigments and their Application as Coloring Agents. Clean Technologies and Environmental Policy*, *25*(1), 310–322. doi:10.1007/s10098-023-02468-5

Mohan, C., & Kumar, V. (2021). A Comparative Study of SARS, MERS with COVID-19. *Coronaviruses*, *2*(3), 379–383. doi:10.2174/2666796701999200905093233

Mohan, C., & Robinson, J. (2024). *"Life cycle analysis of bio-based lubricants", Biolubricants: Feedstocks*. Catalysts, and Nanotechnology., doi:10.1515/9783111145136-011

Mohan, C., Robinson, J., & Negi, A. (2023). Ion-Selective Electrode (ISE) Based on Polyvinyl Chloride Membrane Formed from Heterocyclic Quinazoline Compounds as Ionophore material. *Engineering Proceedings*, *48*, 10. doi:10.3390/CSAC2023-14914

Mohan, C., Robinson, J., Vodwal, L., & Kumari, N. (2024). *Sustainable Development Goals for addressing environmental challenges*. Elsevier Publishing., doi:10.1016/B978-0-443-18959-3.00007-0

Mohan, C., Yadav, S., & Kumari, N. (2022). Interaction of Indigo Carmine with naturally occuring clay minerals: An approach for the synthesis of Nano pigments. *Materials Today: Proceedings*, *69*, 82–86. doi:10.1016/j.matpr.2022.08.081

Moldan, B., Billharz, S., & Matravers, R. (Eds.). (1997). *Sustainability indicators: A report on the project on indicators of sustainable development*. Wiley.

Morais, J., Corder, G., Golev, A., Lawson, L., & Ali, S. (2022). Global review of human waste-picking and its contribution to poverty alleviation and a circular economy. *Environmental Research Letters*, *17*(6), 063002. doi:10.1088/1748-9326/ac6b49

Morán, J. I., Alvarez, V. A., Cyras, V. P., & Vázquez, A. (2008). Extraction of cellulose and preparation of nanocellulose from sisal fibers. *Cellulose (London, England)*, *15*(1), 149–159. doi:10.1007/s10570-007-9145-9

Mor, S., Santosh, B., & Narsi, R. B. (2022). Crop residue burning: Impacts, management techniques and policies. *Annals of Biology*, *35*(2), 245–251.

Mubarik, M., Raja Mohd Rasi, R. Z., Mubarak, M. F., & Ashraf, R. (2021). Impact of blockchain technology on green supply chain practices: Evidence from emerging economy. *Management of Environmental Quality*, *32*(5), 1023–1039. doi:10.1108/MEQ-11-2020-0277

Muenchen, D. K., Martinazzo, J., de Cezaro, A. M., Rigo, A. A., Brezolin, A. N., Manzoli, A., & Steffens, J. (2016). Pesticide Detection in Soil Using Biosensors and Nanobiosensors. *Biointerface Research in Applied Chemistry*, *6*(6).

Muheirwe, F., Kombe, W., & Kihila, J. (2022). The paradox of solid waste management: A regulatory discourse from Sub-Saharan Africa. *Habitat International*, *119*, 102491. doi:10.1016/j.habitatint.2021.102491

Muiruri, J., Wahome, R., & Karatu, K. (2020). Assessment of methods practiced in the disposal of solid waste in Eastleigh Nairobi County, Kenya. *AIMS Environmental Science*, *7*(5), 434–448. doi:10.3934/environsci.2020028

Munir, K. (2022). Sustainable food waste management strategies by applying practice theory in hospitality and food services-a systematic literature review. *Journal of Cleaner Production*, *331*, 129991. doi:10.1016/j.jclepro.2021.129991

Muralia, G., Haridharanb, M. K., Mohan, C., & Khera, G. S. (2023). Compressive strength and impact strength of pre-placed aggregate fibre-reinforced concrete. *Materials Today: Proceedings*. Advance online publication. doi:10.1016/j.matpr.2023.07.300

Muthu, S. S. (2015). *Assessing the Environmental Impact of Textiles and the Clothing Supply Chain*. Woodhead Publishing. doi:10.1016/b978-1-78242-373-6.00001-2

Muthu, S. S., Li, Y., Hu, J. Y., Mok, P. Y., & Liu, Y. (2020). Sustainable Leather Production and Processing: An Overview. *Sustainability*, *12*(11), 4601. doi:10.3390/su12114601

Muzamil, M., Rasool, S., & Masoodi, U. H. (2022). In Situ and Ex Situ Agricultural Waste Management System. Agricultural Waste - New Insights. IntechOpen. doi:10.5772/intechopen.108239

Nagendran, R. (2011, January). Agricultural waste and pollution. In *Waste* (pp. 341–355). Academic Press. doi:10.1016/B978-0-12-381475-3.10024-5

Naresh, R. K., Dhaliwal, S. S., Chandra, M. S., Malhotra, S. K., Harish, J., Singh, P. K., Kumar, V., Baliyan, A., & Gawdiya, S. (2021). Alternative uses, in and off-field managements to reduce adverse impact of crop residue burning on environment: A review. *International Journal of Environment and Climate Change*, *11*(1), 100–118. doi:10.9734/ijecc/2021/v11i130347

Nath, P. C., Ojha, A., Debnath, S., Sharma, M., Sridhar, K., Nayak, P. K., & Inbaraj, B. S. (2023). Biogeneration of Valuable Nanomaterials from Agro-Wastes: A Comprehensive Review. *Agronomy (Basel)*, *13*(2), 561. doi:10.3390/agronomy13020561

Ng, K., & Yang, A. (2023). Development of a system model to predict flows and performance of regional waste management planning: A case study of England. *Journal of Environmental Management*, *325*, 116585. doi:10.1016/j.jenvman.2022.116585 PMID:36326526

Nguyen, V. T. (2017). *Recovering bioactive compounds from agricultural wastes.* John Wiley & Sons. doi:10.1002/9781119168850

Niinimäki, K. (2018). *Sustainable Fashion: New Approaches*. Routledge. doi:10.4324/9781315162243

Njoku, P. O., Edokpayi, J. N., & Odiyo, J. O. (2019). Health and environmental risks of residents living close to a landfill: A case study of Thohoyandou Landfill, Limpopo Province, South Africa. *International Journal of Environmental Research and Public Health*, *16*(12), 2125. doi:10.3390/ijerph16122125 PMID:31208082

Noman, M., Ahmed, T., Wang, J., Ijaz, M., Shahid, M., Islam, M. S., Azizullah, Manzoor, I., Li, D., & Song, F. (2023). Nano-enabled crop resilience against pathogens: Potential, mechanisms and strategies. *Crop Health*, *1*(1), 15. doi:10.1007/s44297-023-00015-8

Nuruzzaman, M. D., Rahman, M. M., Liu, Y., & Naidu, R. (2016). Nanoencapsulation, nano-guard for pesticides: A new window for safe application. *Journal of Agricultural and Food Chemistry*, *64*(7), 1447–1483. doi:10.1021/acs.jafc.5b05214 PMID:26730488

Nwokolo, S. C., Eyime, E. E., Obiwulu, A. U., & Ogbulezie, J. C. (2023). Africa's Path to Sustainability: Harnessing Technology, Policy, and Collaboration. *Trends in Renewable Energy*, *10*(1), 98–131. doi:10.17737/tre.2024.10.1.00166

Nygaard, A., & Silkoset, R. (2023). Sustainable development and greenwashing: How blockchain technology information can empower green consumers. *Business Strategy and the Environment*, *32*(6), 3801–3813. doi:10.1002/bse.3338

Nyika, J., & Dinka, M. (2022a). Converting solid waste materials to energy: A review. *Materials Today: Proceedings*, *57*(2), 964–968. doi:10.1016/j.matpr.2022.03.240

Nyika, J., & Dinka, M. (2022b). Sustainable management of textile solid waste materials: The progress and prospects. *Materials Today: Proceedings*, *62*(6), 3320–3324. doi:10.1016/j.matpr.2022.04.241

Nyika, J., & Dinka, M. (2022c). Recycling plastic waste materials for building and construction materials: A minireview. *Materials Today: Proceedings*, *62*, 3257–3262. doi:10.1016/j.matpr.2022.04.226

Nyika, J., & Dinka, M. (2023). Comparative urban waste management in developing countries—case studies of Nairobi and Johannesburg cities of Africa. In U. Chatterjee, N. Bandyopadhyay, M. D. Setiawati, & S. Sarkar (Eds.), *Urban commons, future smart cities, and sustainability. Springer Geography*. Springer. doi:10.1007/978-3-031-24767-5_27

Nyika, J., Onyari, E., Dinka, M., & Shivani, B. (2019). Heavy metal pollution and mobility in soils within a landfill vicinity: A South African case study. *Oriental Journal of Chemistry*, *35*(4), 1286–1297. doi:10.13005/ojc/350406

Nyika, J., Onyari, E., Dinka, M., & Shivani, B. (2020). Waste Management in South Africa. In A. Pariatamby, F. Hamid, & M. Bhatti (Eds.), *Sustainable Waste Management Challenges in Developing Countries*. IGI Global Publishers. doi:10.4018/978-1-7998-0198-6.ch014

Obi, F., Ugwuishiwu, B., & Nwakaire, J. (2016). AGRICULTURAL WASTE CONCEPT, GENERATION, UTILIZATION AND MANAGEMENT. *Nigerian Journal of Technology*, *35*(4), 957. doi:10.4314/njt.v35i4.34

Ogola, J., Chimuka, L., & Tshivhase, S. (2011). Management of municipal solid wastes: a case study in Limpopo province, South Africa. In S. Kumar (Ed.), *Integrated waste management*.

Ogunseitan, O. O. (2023). The environmental justice agenda for e-waste management. *Environment*, *65*(2), 15–25. doi:10.1080/00139157.2023.2167457

Okumus, B., Taheri, B., Giritlioglu, I., & Gannon, M. J. (2020). Tackling food waste in all-inclusive resort hotels. *International Journal of Hospitality Management*, *88*, 102543. doi:10.1016/j.ijhm.2020.102543

OmarI.HassanH.JayaramanR.SalahK.OmarM. (2023). Using Blockchain Technology to Achieve Sustainability in the Hospitality Industry by Reducing Food Waste. *Available at* SSRN 4530352. doi:10.2139/ssrn.4530352

Ordóñez de Pablos, P., Almunawar, M. N., & Anshari, M. (Eds.). (2023). *Perspectives on the Transition Toward Green and Climate Neutral Economies in Asia*. IGI Global. doi:10.4018/978-1-6684-8613-9

Oriekhoe, O. I., Ashiwaju, B. I., Ihemereze, K. C., Ikwue, U., & Udeh, C. A. (2024). Review Of Technological Advancements In Food Supply Chain Management: A Comparative Study Between The Us And Africa. *International Journal of Management & Entrepreneurship Research*, *6*(1), 132–149. doi:10.51594/ijmer.v6i1.713

Orlando, M., Molla, G., Castellani, P., Pirillo, V., Torretta, V., & Ferronato, N. (2023). Microbial enzyme biotechnology to reach plastic waste circularity: Current status, problems and perspectives. *International Journal of Molecular Sciences*, *24*(4), 3877. doi:10.3390/ijms24043877 PMID:36835289

Oudani, M., Sebbar, A., Zkik, K., El Harraki, I., & Belhadi, A. (2023). Green Blockchain based IoT for secured supply chain of hazardous materials. *Computers & Industrial Engineering*, *175*, 108814. Advance online publication. doi:10.1016/j.cie.2022.108814

Ouyang, C., Zhang, S., Xue, C., Yu, X., Xu, H., Wang, Z., Lu, Y., & Wu, Z.-S. (2020). Precision-Guided Missile-Like DNA Nanostructure Containing Warhead and Guidance Control for Aptamer-Based Targeted Drug Delivery into Cancer Cells in Vitro and in Vivo. *Journal of the American Chemical Society*, *142*(3), 1265–1277. doi:10.1021/jacs.9b09782 PMID:31895985

Ozdemir, G., & Güçer, E. (2018). Food waste management within sustainability perspective: A study on five star chain hotels. *Journal of Tourism & Gastronomy Studies*, *6*(1), 280–299. doi:10.21325/jotags.2018.187

Padmakumar, V., Ravi, D., Rupam, B., Kanwar, S., Ramakrishna, R., Prasad, K. V. S. V., & Jones, C. S. (2024). Genetic variation for grain yield and straw quality traits in 55 diverse rice varieties in Assam. *Animal Nutrition and Feed Technology*, *24*(1), 93–103. doi:10.5958/0974-181X.2024.00007.6

Pandey, A. (2021). Food wastage: Causes, impacts and solutions. *Science Heritage Journal*, *5*(1), 17–20. doi:10.26480/gws.01.2021.17.20

Parajuly, K., Kuehr, R., Awasthi, A. K., Fitzpatrick, C., Lepawsky, J., Smith, E., Widmer, R., & Zeng, X. (2019). Future e-waste scenarios. *UNEP*. https://wedocs.unep.org/bitstream/handle/20.500.11822/30809/FutEWSc.pdf?sequence=1&isAllowed=y

Parajuly, K., Ternald, D., & Kuehr, R. (2022). *The future of electric vehicles and material resources* (1st ed.). UNEP. https://wedocs.unep.org/bitstream/handle/20.500.11822/34225/ElecVe.pdf?sequence=1&isAllowed=y

Paul, R. K., Ghosh, S., Paul, N. K., Ghosh, M. K., & Kundu, A. (2019). Sustainable Textile Dyeing and Finishing. In S. S. Muthu (Ed.), *Sustainable Textiles: Life Cycle and Environmental Impact* (pp. 95–121). Springer.

Pereira, T., & Fernandino, G. (2019). Evaluation of solid waste management sustainability of a coastal municipality from northeastern Brazil. *Ocean and Coastal Management*, *179*, 104839. doi:10.1016/j.ocecoaman.2019.104839

Pergola, M., Persiani, A., Palese, A. M., Di Meo, V., Pastore, V., D'Adamo, C., & Celano, G. (2018). Composting: The way for a sustainable agriculture. *Applied Soil Ecology*, *123*, 744–750. doi:10.1016/j.apsoil.2017.10.016

Platt, M., Sedlmeir, J., Platt, D., Xu, J., Tasca, P., Vadgama, N., & Ibanez, J. I. (2021). The Energy Footprint of Blockchain Consensus Mechanisms Beyond Proof-of-Work. *Proceedings - 2021 21st International Conference on Software Quality, Reliability and Security Companion, QRS-C 2021*, (pp. 1135–1144). IEEE. 10.1109/QRS-C55045.2021.00168

Polas, M. R. H., Kabir, A. I., Sohel-Uz-zaman, A. S. M., Karim, R., & Tabash, M. I. (2022). Blockchain Technology as a Game Changer for Green Innovation: Green Entrepreneurship as a Roadmap to Green Economic Sustainability in Peru. In Journal of Open Innovation: Technology, Market, and Complexity, 8(2). Elsevier. doi:10.3390/joitmc8020062

Pookulangara, S., & Shephard, A. (2013). Slow Fashion Movement: Understanding Consumer Perceptions and Practices. *Journal of Retailing and Consumer Services*, *20*(2), 200–206. doi:10.1016/j.jretconser.2012.12.002

Poon, K., Lu, Z., De Deene, Y., Ramaswamy, Y., Zreiqat, H., & Singh, G. (2021). Tuneable manganese oxide nanoparticle based theranostic agents for potential diagnosis and drug delivery. *Nanoscale Advances*, *3*(14), 4052–4061. doi:10.1039/D0NA00991A PMID:36132835

Principato, L., Pratesi, C. A., & Secondi, L. (2018). Towards zero waste: An exploratory study on restaurant managers. *International Journal of Hospitality Management*, *74*, 130–137. doi:10.1016/j.ijhm.2018.02.022

Pujara, Y., Pathak, P., Sharma, A., & Govani, J. (2019). Review on Indian municipal solid waste management practices for reduction of environmental impacts to achieve sustainable development goals. *Journal of Environmental Management*, *248*, 109238. doi:10.1016/j.jenvman.2019.07.009 PMID:31319199

PwC. (2020). *Reinventing Fashion: A Circular Approach to Redesigning Fashion's Future*. PwC.

Qiu, J., Souza, M. C. D., Aguilar, A. A. R., & Ghysels, S. (2023). Improving biochar properties by co-pyrolysis of pig manure with bio-invasive weed for use as the soil amendment. *Chemosphere*, *312*, 137229. doi:10.1016/j.chemosphere.2022.137229 PMID:36372342

Raghupathy, L., Kruger, C., Chaturvedi, A., Arora, R., & Henzler, M. P. (2018*). E-waste recycling in India - bridging the gap between the informal and formal sector*. Greene. https://greene.gov.in/wp-content/uploads/2018/01/Krueger.pdf

Rahimi-Midani, A. (2023). *Deep Technology for Sustainable Fisheries and Aquaculture*.

Ramadass, R., Sathyanarayana, N., & Fathima, Y. (2023). Improving citizen lifestyle in green smart city. In *Green Blockchain Technology for Sustainable Smart Cities* (pp. 29–63). Elsevier. doi:10.1016/B978-0-323-95407-5.00007-4

RaneN.ChoudharyS.RaneJ. (2023). Metaverse as a cutting-edge platform for attaining Sustainable Development Goals (SDGs). SSRN 4644035. doi:10.2139/ssrn.4644035

Rane, S. B., Potdar, P. R., & Aware, S. (2023). *Strategies for development of smart and green products using Blockchain-IoT integrated architecture*. Operations Management Research., doi:10.1007/s12063-023-00398-5

Ribeiro, A. I., Dias, A. M., & Zille, A. (2022). Synergistic effects between metal nanoparticles and commercial antimicrobial agents: A review. *ACS Applied Nano Materials*, *5*(3), 3030–3064. doi:10.1021/acsanm.1c03891 PMID:36568315

Ribeiro, A. I., Vieira, B., Dantas, D., Silva, B., Pinto, E., Cerqueira, F., Silva, R., Remião, F., Padrão, J., Dias, A. M., & Zille, A. (2023). Synergistic Antimicrobial Activity of Silver Nanoparticles with an Emergent Class of Azoimidazoles. *Pharmaceutics*, *15*(3), 926. doi:10.3390/pharmaceutics15030926 PMID:36986787

Rivera-Ferre, M., Ortega-Cerdà, M., & Baumgärtner, J. (2013). Rethinking Study and Management of Agricultural Systems for Policy Design. *Sustainability (Basel)*, *5*(9), 3858–3875. doi:10.3390/su5093858

Robertson, K. J., Brar, R., Randhawa, P., Stark, C., & Baroutian, S. (2023). Opportunities and challenges in waste management within the medicinal cannabis sector. *Industrial Crops and Products*, *197*, 116639. doi:10.1016/j.indcrop.2023.116639

Rodríguez-Espinosa, T., Papamichael, I., Voukkali, I., Gimeno, A. P., Candel, M. B. A., Navarro-Pedreño, J., Zorpas, A. A., & Lucas, I. G. (2023). Nitrogen management in farming systems under the use of agricultural wastes and circular economy. *The Science of the Total Environment*, *876*, 162666. doi:10.1016/j.scitotenv.2023.162666 PMID:36894085

Roman, B., & Brennan, R. A. (2021). Coupling ecological wastewater treatment with the production of livestock feed and irrigation water provides net benefits to human health and the environment: A life cycle assessment. *Journal of Environmental Management*, *288*, 112361. doi:10.1016/j.jenvman.2021.112361 PMID:33756385

Ruggeri, M., Bianchi, E., Rossi, S., Vigani, B., Bonferoni, M. C., Caramella, C., Sandri, G., & Ferrari, F. (2020). Nanotechnology-Based Medical Devices for the Treatment of Chronic Skin Lesions, *Research to the Clinic. Pharmaceutics*, *12*(9), 815. doi:10.3390/pharmaceutics12090815 PMID:32867241

Ruiz Molina, M. E., Belda-Miquel, S., Hytti, A., & Gil-Saura, I. (2022). Addressing sustainable food management in hotels: Proposing a framework and examining hotel groups. *British Food Journal*, *124*(2), 462–492. doi:10.1108/BFJ-12-2020-1171

Saad, M., Qin, Z., Ren, K., Nyang, D. H., & Mohaisen, D. (2021). E-PoS: Making Proof-of-Stake Decentralized and Fair. *IEEE Transactions on Parallel and Distributed Systems*, *32*(8), 1961–1973. doi:10.1109/TPDS.2020.3048853

Sadh, P. K., Duhan, S., & Duhan, J. S. (2018). Agro-industrial wastes and their utilization using solid state fermentation: A review. *Bioresources and Bioprocessing*, *5*(1), 1–15. doi:10.1186/s40643-017-0187-z

Said, Z., Sharma, P., Nhuong, Q. T. B., Bora, B. J., Lichtfouse, E., Khalid, H. M., & Hoang, A. T. (2023). Intelligent approaches for sustainable management and valorisation of food waste. *Bioresource Technology*, *377*, 128952. doi:10.1016/j.biortech.2023.128952 PMID:36965587

Sala-Garrido, R., Mocholi-Arce, M., Molinos-Senante, M., & Maziotis, A. (2023). Monetary valuation of unsorted waste: A shadow price approach. *Journal of Environmental Management*, *325*, 116668. doi:10.1016/j.jenvman.2022.116668 PMID:36343396

Salama, D. M., Abd El-Aziz, M. E., El-Naggar, M. E., Shaaban, E. A., & Abd El-Wahed, M. S. (2021). Synthesis of an eco-friendly nanocomposite fertilizer for common bean based on carbon nanoparticles from agricultural waste biochar. *Pedosphere*, *31*(6), 923–933. doi:10.1016/S1002-0160(21)60024-3

Saras, P. K., & Patel, B. D. (2014). Weed management by cultural and chemical practices and their effect on phytotoxicity and growth, quality and yield parameters of cluster bean. *Trends in Biosciences*, *7*(24), 4196–4199.

Saratale, R. G., Saratale, G. D., Shin, H. S., Jacob, J. M., Pugazhendhi, A., Bhaisare, M., & Kumar, G. (2018). New insights on the green synthesis of metallic nanoparticles using plant and waste biomaterials: Current knowledge, their agricultural and environmental applications. *Environmental Science and Pollution Research International*, *25*(11), 10164–10183. doi:10.1007/s11356-017-9912-6 PMID:28815433

Sarkis, J. (2018). A brave new world of waste: Exploring the environmental and resource implications of China's ban on waste imports and the global solid waste trade. *Resources, Conservation and Recycling*, *2018*(135), 1–3. doi:10.1016/j.resconrec.2017.12.022

Sartor, D. (2021). *An Exploratory Study of Woolworths Consumer's Post-Covid Food Waste Behaviour.*

Saxena, S. C., & Jotshi, C. K. (1994). Fluidized-bed incineration of waste materials. *Progress in Energy and Combustion Science*, *20*(4), 281–324. doi:10.1016/0360-1285(94)90012-4

Scarlat, N., Dallemand, J. F., & Fahl, F. (2018). Biogas: Developments and perspectives in Europe. *Renewable Energy*, *129*, 457–472. doi:10.1016/j.renene.2018.03.006

Schinckus, C. (2021). Proof-of-work based blockchain technology and Anthropocene: An undermined situation? *Renewable & Sustainable Energy Reviews*, *152*, 111682. Advance online publication. doi:10.1016/j.rser.2021.111682

Scholz, P., Linderová, I., & Konečná, K. (2020). Green Management Tools as a Way to Sustainable Behaviour in the Hotel Industry: Case Study from Czechia. *Sustainability (Basel)*, *12*(23), 10027. doi:10.3390/su122310027

Sedlmeir, J., Buhl, H. U., Fridgen, G., & Keller, R. (2021). Recent Developments in Blockchain Technology and their Impact on Energy Consumption. *ArXiv Preprint ArXiv:2102.07886*. https://doi.org/ doi:10.1007/s00287-020-01321-z

Seif, R., Salem, F. Z., & Allam, N. K. (2023). E-waste recycled materials as efficient catalysts for renewable energy technologies and better environmental sustainability. *Environment, Development and Sustainability*, *26*(3), 5473–5508. doi:10.1007/s10668-023-02925-7 PMID:36691418

Seo, G., Lee, G., Kim, M. J., Baek, S.-H., Choi, M., Ku, K. B., Lee, C.-S., Jun, S., Park, D., Kim, H. G., Kim, S.-J., Lee, J.-O., Kim, B. T., Park, E. C., & Kim, S. I. (2020). Rapid Detection of COVID-19 Causative Virus (SARS-CoV-2) in Human Nasopharyngeal Swab Specimens Using Field-Effect Transistor-Based Biosensor. *ACS Nano*, *14*(4), 5135–5142. doi:10.1021/acsnano.0c02823 PMID:32293168

Shahin, M. A. K., Azadi, H., & Hojjati, S. M. (2017). Environmental Impact of Textile Dyeing. In J. Uddin (Ed.), *Dyeing: Processes, Techniques and Applications* (pp. 125–144). Nova Science Publishers.

Shah, M. P. (Ed.). (2020). *Microbial bioremediation & biodegradation*. Springer. doi:10.1007/978-981-15-1812-6

Shamsi, R., Mir Mohamad Sadeghi, G., Vahabi, H., Seyfi, J., Sheibani, R., Zarrintaj, P., Laoutid, F., & Saeb, M. R. (2019). Hopes beyond PET recycling: Environmentally clean and engineeringly applicable. *Journal of Polymers and the Environment*, *27*(11), 2490–2508. doi:10.1007/s10924-019-01507-x

Shams, S. R., Vrontis, D., Weber, Y., Tsoukatos, E., & Iaia, L. (Eds.). (2023). *Technology, Business and Sustainable Development: Advances for People, Planet and Profit*. Taylor & Francis. doi:10.4324/9781003293187

Sharma, A., & Singh, B. (2022). Measuring Impact of E-commerce on Small Scale Business: A Systematic Review. *Journal of Corporate Governance and International Business Law*, *5*(1).

Sharma, P. K., Kumar, N., & Park, J. H. (2020). Blockchain technology toward green IoT: Opportunities and challenges. *IEEE Network*, *34*(4), 263–269. doi:10.1109/MNET.001.1900526

Sharma, T. (2020). What a waste: confronting consumer food waste behavior in hospitality settings. In *Advances in hospitality and leisure* (pp. 167–176). Emerald Publishing Limited.

Sharma, Y., Sijariya, R., & Gupta, P. (2023). How Deep Learning Can Help in Regulating the Subscription Economy to Ensure Sustainable Consumption and Production Patterns (12th Goal of SDGs). In *Deep Learning Technologies for the Sustainable Development Goals: Issues and Solutions in the Post-COVID Era* (pp. 1–20). Springer Nature Singapore. doi:10.1007/978-981-19-5723-9_1

Shoker, A. (2017). Sustainable blockchain through proof of exercise. *2017 IEEE 16th International Symposium on Network Computing and Applications*. IEEE. 10.1109/NCA.2017.8171383

Shreyas, M. A. S., Rajaraman, R., Harini, S., & Kiliroor, C. C. (2022). Application of artificial intelligence to enhance collection of e-waste: A potential solution for household WEEE collection and segregation in India. *Waste Management & Research*, *40*(7), 1047–1053. doi:10.1177/0734242X211052846 PMID:34726090

Shyamsundar, P., Springer, N. P., Tallis, H., Polasky, S., Jat, M. L., Sidhu, H. S., & Somanathan, R. (2019). Fields on fire: Alternatives to crop residue burning in India. *Science*, *365*(6453), 536–538. doi:10.1126/science.aaw4085 PMID:31395767

Singh, B. (2023). Blockchain Technology in Renovating Healthcare: Legal and Future Perspectives. In Revolutionizing Healthcare Through Artificial Intelligence and Internet of Things Applications (pp. 177-186). IGI Global.

Singh, A., Singh, K., & Singh, D. V. (2008). Suitability of organic mulch (distillation waste) and herbicides for weed management in perennial aromatic grasses. *Tropical Pest Management*, *37*(2), 162–165. doi:10.1080/09670879109371567

Singh, B. (2019). Profiling Public Healthcare: A Comparative Analysis Based on the Multidimensional Healthcare Management and Legal Approach. *Indian Journal of Health and Medical Law*, *2*(2), 1–5.

Singh, B. (2022). COVID-19 Pandemic and Public Healthcare: Endless Downward Spiral or Solution via Rapid Legal and Health Services Implementation with Patient Monitoring Program. *Justice and Law Bulletin*, *1*(1), 1–7.

Singh, B. (2022). Relevance of Agriculture-Nutrition Linkage for Human Healthcare: A Conceptual Legal Framework of Implication and Pathways. *Justice and Law Bulletin*, *1*(1), 44–49.

Singh, B. (2023). Federated Learning for Envision Future Trajectory Smart Transport System for Climate Preservation and Smart Green Planet: Insights into Global Governance and SDG-9 (Industry, Innovation and Infrastructure). *National Journal of Environmental Law*, *6*(2), 6–17.

Singh, B., & Sharma, N. (2008). Mechanistic implications of plastic degradation. *Polymer Degradation & Stability*, *93*(3), 561–584. doi:10.1016/j.polymdegradstab.2007.11.008

Singh, C., Tiwari, S., Rai, P. K., & Singh, J. S. (2021). Sustainable management of paddy crop residues: Effects on methanotrophs diversity and value for soil health restoration. *Land Degradation & Development*, *32*(14), 4121–4131. doi:10.1002/ldr.3982

Singh, G., & Arya, S. K. (2021). A review on management of rice straw by use of cleaner technologies: Abundant opportunities and expectations for Indian farming. *Journal of Cleaner Production*, *291*, 125278. doi:10.1016/j.jclepro.2020.125278

Singh, R., Manickam, N., Mudiam, M. K. R., Murthy, R. C., & Misra, V. (2013). An integrated (nano-bio) technique for degradation of γ-HCH contaminated soil. *Journal of Hazardous Materials*, *258*, 35–41. doi:10.1016/j.jhazmat.2013.04.016 PMID:23692681

Singh, S., Ranguwal, S., & Guleria, A. (2023). Determinants of crop residue burning practices in Punjab. *Agricultural Research Journal*, *60*(2), 289–294. doi:10.5958/2395-146X.2023.00043.1

Single-Use Plastics. (2018). *A Roadmap for Sustainability.* United Nations Environment Programme.

Smieja, J. (2023, March 24). *The enormous opportunity of e-waste recycling.* Weforum. https://www.weforum.org/agenda/2023/03/the-enormous-opportunity-of-e-waste-recycling/

Smith, J., & Brown, A. (2023). *Circular Economy Strategies for the Fashion Industry.* Sustainability Journal.

Smith, J., & Taylor, K. (2022). *Circular Economy Business Models: Case Studies in Fashion Industry.* Sustainability Review.

Soni, V. K., Singh, G., Vijayan, B. K., Chopra, A., Kapur, G. S., & Ramakumar, S. S. V. (2021). Thermochemical recycling of waste plastics by pyrolysis: A review. *Energy & Fuels, 35*(16), 12763–12808. doi:10.1021/acs.energyfuels.1c01292

Spassky, C. E., Regnier, E. E. S., Harrison, S. K., Bennett, M. A., & Metzger, J. D. (2016). Weed Suppression in Pumpkin by Mulches Composed of Organic Municipal Waste Materials. *Horticultural Science (Prague), 51*, 720–726.

Spiess-Knafl, W. (2022). *Artificial Intelligence and Blockchain for Social Impact: Social Business Models and Impact Finance.* Taylor & Francis. doi:10.4324/9781003218913

Srivastava, A., Dubey, V., & Hazela, B. (2021). Delivering Green Supply Chain Using Blockchain Technology for Sustainable Environment: A Survey. *Emerging Technologies in Data Mining.* Springer. doi:10.1007/978-981-15-9927-9_74

Steiner, C., Bellwood-Howard, I., Häring, V., Tonkudor, K., Addai, F., Atiah, K., Abubakari, A. H., Kranjac-Berisavljevic, G., Marschner, B., & Buerkert, A. (2018). Participatory trials of on-farm biochar production and use in Tamale, Ghana. *Agronomy for Sustainable Development, 38*(1), 1–10. doi:10.1007/s13593-017-0486-y

Sternberg, J., Sequerth, O., & Pilla, S. (2021). Green chemistry design in polymers derived from lignin: Review and perspective. *Progress in Polymer Science, 113*, 101344. doi:10.1016/j.progpolymsci.2020.101344

Stewart, J., & Gorman, C. (2024). Authentic Assessment and Embedding Sustainable Literacy and Technology Skills in Tourism Education. *Irish Journal of Academic Practice, 11*(2), 10.

Sumiyati, S., Samadikun, B. P., Widiyanti, A., Budihardjo, M. A., Al Qadar, S., & Puspita, A. S. (2024). Life cycle assessment of agricultural waste recycling for sustainable environmental impact. *Global Journal of Environmental Science and Management.* doi:10.22034/gjesm.2024.02.30

Surender, M., Mohan, C., & Kumar, R. (2023). Novel modification of activated charcoal sheet with N-methylpolypyrrole and silver nanoparticles for removal of hexavalent chromium in water treatment processes. *Materials Protection, 64*(4), 503–511. doi:10.5937/zasmat2304503S

Suresh, B. C. V., & Das, S. (2023). Impact of blockchain technology on the stock market. *Revolutionizing Financial Services and Markets Through FinTech and Blockchain*, 44–59. doi:10.4018/978-1-6684-8624-5.ch004

Sustainable Apparel Coalition. (2021). *Higg Index.* SAC.

Suvedi, M., Den Biggelaar, C., & Morford, S. (2003). Conceptual Framework for Evaluating Sustainable Agriculture. *Journal of Crop Production, 9*(1–2), 433–454. doi:10.1300/J144v09n01_05

Synani, K., Abeliotis, K., & Lasaridi, K. (2021). *Food waste in the catering and hospitality sector: generation factors and innovative practices.*

Taneja, A., Sharma, R., Khetrapal, S., Sharma, A., Nagraik, R., Venkidasamy, B., Ghate, M. N., Azizov, S., Sharma, S., & Kumar, D. (2023). Value Addition Employing Waste Bio-Materials in Environmental Remedies and Food Sector. *Metabolites, 13*(5), 624. doi:10.3390/metabo13050624 PMID:37233665

Tapo, A. A., Traore, A., Danioko, S., & Tembine, H. (2024). Machine Intelligence in Africa: a survey. *arXiv preprint arXiv:2402.02218*.

Taylor, K., & Brown, A. (2022). Social Impacts of Recycling Programs: Empowering Communities and Promoting Equity. *Journal of Social Responsibility*.

Taylor, K., & Smith, J. (2024). The Future of Recycling and Upcycling: Trends and Opportunities in Sustainable Fashion. *Journal of Sustainable Development*.

Taylor, K., & Wilson, D. (2023). The Economics of Recycling Infrastructure: Investment Strategies for Sustainable Growth, *Journal of Environmental Economics*.

Taylor, K., & Smith, J. (2022). Technological Innovations in Recycling: Advancements and Challenges. *Journal of Technology & Innovation*.

Tchobanoglous, G., Theisen, H., & Vigil, S. A. (1993). *Integrated Solid Waste Management: Engineering Principles and Management Issues*. McGraw-Hill Education.

Thakur, M., Wang, B., & Verma, M. L. (2022). Development and applications of nanobiosensors for sustainable agricultural and food industries: Recent developments, challenges and perspectives. *Environmental Technology & Innovation*, *26*, 102371. doi:10.1016/j.eti.2022.102371

The Future of Sustainable Fashion. (2020). *A Collaboration with Good on You*. Business for Social Responsibility.

Thirtle, C. (1995). Monitoring and evaluating agricultural research: A sourcebook. D. Horton, P. Ballantyne, W. Peterson, B. Uribe, D. Gapasin and K. Sheridan. CAB International in association with the International Service for National Agricultural Research, Wallingford, Oxon OX10 8DE, 1993, 219 pp. *Public Administration and Development*, *15*(2), 185–185. doi:10.1002/pad.4230150210

Thunman, H., Vilches, T. B., Seemann, M., Maric, J., Vela, I. C., Pissot, S., & Nguyen, H. N. (2019). Circular use of plastics-transformation of existing petrochemical clusters into thermochemical recycling plants with 100% plastics recovery. *Sustainable Materials and Technologies*, *22*, e00124. doi:10.1016/j.susmat.2019.e00124

Tice, P. H. (2024). *The Race to Zero: How ESG Investing will Crater the Global Financial System*. Encounter Books.

Tirkolaee, E. B., Goli, A., Gütmen, S., Weber, G. W., & Szwedzka, K. (2023). A novel model for sustainable waste collection arc routing problem: Pareto-based algorithms. *Annals of Operations Research*, *324*(1-2), 189–214. doi:10.1007/s10479-021-04486-2 PMID:35068644

Tiseo, I. (2023, August 29). *E-waste management market value worldwide 2022-2029*. Statista. https://www.statista.com/statistics/1154804/global-e-waste-management-market-value/

Tiseo, I. (2024, January 9). *Outlook on global e-waste generation 2019-2030*. Statista. https://www.statista.com/statistics/1067081/generation-electronic-waste-globally-forecast/

Tomaszewska, M., Bilska, B., Tul-Krzyszczuk, A., & Kołożyn-Krajewska, D. (2021). Estimation of the scale of food waste in hotel food services—A case study. *Sustainability (Basel)*, *13*(1), 421. doi:10.3390/su13010421

Tóth, F. A., Magyar, T., Tamás, J., & Nagy, P. T. (2024). Improving the Nutrient Management of an Apple Orchard by Using Organic-Based Composites Derived from Agricultural Waste. *Horticulturae*, *10*(2), 172. doi:10.3390/horticulturae10020172

Tournier, V., Duquesne, S., Guillamot, F., Cramail, H., Taton, D., Marty, A., & André, I. (2023). Enzymes' power for plastics degradation. *Chemical Reviews*, *123*(9), 5612–5701. doi:10.1021/acs.chemrev.2c00644 PMID:36916764

Tran, H.-V., Ngo, N. M., Medhi, R., Srinoi, P., Liu, T., Rittikulsittichai, S., & Lee, T. R. (2022). Multifunctional Iron Oxide Magnetic Nanoparticles for Biomedical Applications: A Review. *Materials (Basel), 15*(2), 503. doi:10.3390/ma15020503 PMID:35057223

Trivedy, R. K., & Thomas, S. (2004). Review of performance data of few pilot and full-scale aquatic weed-based waste treatment plants from India. *Asian Journal of Microbiology, Biotechnology & Environmental Sciences : AJMBES, 6*(1), 157–162.

Tsachidou, B., Hissler, C., Noo, A., Lemaigre, S., Daigneux, B., Gennen, J., Pacaud, S., George, I. F., & Delfosse, P. (2021). Biogas residues in the battle for terrestrial carbon sequestration: A comparative decomposition study in the grassland soils of the Greater Region. *Journal of Environmental Management, 286*, 112272. doi:10.1016/j.jenvman.2021.112272 PMID:33677337

Tsai, F., Bui, T., Tseng, M., Lim, M., & Tan, R. (2021). Sustainable solid-waste management in Vietnam's coastal and marine tourism cities: A hierarchical-level approach. *Resources, Conservation and Recycling, 168*, 105266. doi:10.1016/j.resconrec.2020.105266

Tsydenova, O., & Bengtsson, M. (2011). Chemical hazards associated with treatment of waste electrical and electronic equipment. *Waste Management (New York, N.Y.), 31*(1), 45–58. doi:10.1016/j.wasman.2010.08.014 PMID:20869229

Ullah, S., Arjun, K. P., & Sureshbabu, A. P. (2021). Green Computing and Blockchain Fundamentals. In *Handbook of Green Computing and Blockchain Technologies* (pp. 1–16). CRC Press. doi:10.1201/9781003107507-1

United Nations Environment Programme. (2009). *Converting Waste Agricultural Biomass into a Resource-Compendium of Technologies*. UN. https://wedocs.unep.org/20.500.11822/7614

United Nations. (2015). *Transforming our World: The 2030 Agenda for Sustainable Development*. United Nations.

Vafaeva K. M., Chhetri A., Sudan P., Mishra M., Pakkiraiah B., & Mohan C. (2024). Polymer Matrix Nanocomposites for Sustainable Packaging: A Green Approach. *E3S Web of Conferences, 511*. doi:10.1051/e3sconf/202451101008

Vafaeva, K. M., Chhetri, A., Sudan, P., Mishra, M., Pakkiraiah, B., & Mohan, C. (2024). Polymer Matrix Nanocomposites for Sustainable Packaging: A Green Approach. E3S Web of Conferences, 511.

Vargas, D. B., Campos, L. M. S., & Luna, M. M. M. (2024). Brazil's formal e-waste recycling system: From disposal to reverse manufacturing. *Sustainability (Basel), 16*(1), 1–16. doi:10.3390/su16010066

Vargas-Sanchez, A. (2023). Toward a circular tourism industry: The importance of a start-up ecosystem. *Worldwide Hospitality and Tourism Themes, 15*(6), 625–632. doi:10.1108/WHATT-09-2023-0111

Vatin, N. I., Negi, G. S., Yellanki, S. V., Mohan, C., & Singla, N. (2024). Sustainability Measures: An Experimental Analysis of AI and Big Data Insights in Industry 5.0. *BIO Web of Conferences, 86*.

Vatin, N. I., Negi, G. S., Yellanki, S. V., Mohan, C., & Singla, N. (2024). Sustainability Measures: An Experimental Analysis of AI and Big Data Insights in Industry 5.0" *BIO Web of Conferences*. Scopus. doi:10.1051/bioconf/20248601072

Vatin, N. I., Negi, G. S., Yellanki, S. V., Mohan, C., & Singla, N. (2024). Sustainability Measures: An Experimental Analysis of AI and Big Data Insights in Industry. *Bio Web of Conferences, 86*.

Vaverková, M. D. (2019). Landfill impacts on the environment. *Geosciences, 9*(10), 431. doi:10.3390/geosciences9100431

Verma, V., Joshi, A., Kalra, S., Singh, A., Singh, A., & Mohan, C. (2024). Novel Nano catalysts for Sustainable Hydrogen Production from Renewable Resources. E3S Web of Conferences.

Verma, V., Joshi, A., Kalra, S., Singh, A., Singh, A., & Mohan, C. (2024). Novel Nanocatalysts for Sustainable Hydrogen Production from Renewable Resources. *E3S Web of Conferences, 511.* SCOPUS. doi:10.1051/e3sconf/202451101006

Verma, A., Grobbelaar, W., & Shukla, V. K. (2023). Smart Innovation in the Hospitality and Tourism Industry. In *Computational Intelligence in Urban Infrastructure* (pp. 94–116). CRC Press. doi:10.1201/9781003218715-9

Vieira, O., Ribeiro, R. S., de Tuesta, J. L. D., Gomes, H. T., & Silva, A. M. (2022). A systematic literature review on the conversion of plastic wastes into valuable 2D graphene-based materials. *Chemical Engineering Journal*, *428*, 131399. doi:10.1016/j.cej.2021.131399

Vinci, G., Prencipe, S. A., Pucinischi, L., Perrotta, F., & Ruggeri, M. (2023). Sustainability assessment of waste and wastewater recovery for edible mushroom production through an integrated nexus. A case study in Lazio. *The Science of the Total Environment*, *903*, 166044. doi:10.1016/j.scitotenv.2023.166044 PMID:37572921

Vind, D. L. (2023). *Danish Design Heritage and Global Sustainability*. Taylor & Francis. doi:10.4324/9781003261018

Vundavalli, R., Vundavalli, S., Nakka, M., & Rao, D. S. (2015). Biodegradable nano-hydrogels in agricultural farming-alternative source for water resources. *Procedia Materials Science*, *10*, 548–554. doi:10.1016/j.mspro.2015.06.005

Wagner, K., Keller, T., & Seiler, R. (2019). a Comparative Analysis of Cryptocurrency Consensus Algorithms. In *Proceedings of the 16th …* (pp. 217–225). researchgate.net. 10.33965/ac2019_201912L026

Wamsler, C. (2020). Education for sustainability: Fostering a more conscious society and transformation towards sustainability. *International Journal of Sustainability in Higher Education*, *21*(1), 112–130. doi:10.1108/IJSHE-04-2019-0152

Wang, J., Tan, Z., Peng, J., Qiu, Q., & Li, M. (2016). The behaviors of microplastics in the marine environment. *Marine Environmental Research*, *113*, 7–17. doi:10.1016/j.marenvres.2015.10.014 PMID:26559150

Wang, M., Wang, H., Gu, Y., Zhu, M., Kumar, M., Liang, J., Tie, Z., Ma, J., & Jin, Z. (2023). In Situ Generation of Flash Graphene Supported Spherical Bismuth Nanoparticles in Less than 200 ms for Highly Selective Carbon Dioxide Electroreduction. *ACS Materials Letters*, *6*(1), 100–108. doi:10.1021/acsmaterialslett.3c01183

Wang, Z., Miao, R., Ning, P., He, L., & Guan, Q. (2021). From wastes to functions: A paper mill sludge-based calcium-containing porous biochar adsorbent for phosphorus removal. *Journal of Colloid and Interface Science*, *593*, 434–446. doi:10.1016/j.jcis.2021.02.118 PMID:33765625

Wani, N. R., Rather, R. A., Farooq, A., Padder, S. A., Baba, T. R., Sharma, S., Mubarak, N. M., Khan, A. H., Singh, P., & Ara, S. (2023). New insights in food security and environmental sustainability through waste food management. *Environmental Science and Pollution Research International*, *31*(12), 17835–17857. doi:10.1007/s11356-023-26462-y PMID:36988800

Wanyama, J., Bwambale, E., Kiraga, S., Katimbo, A., Nakawuka, P., Kabenge, I., & Oluk, I. (2024). A systematic review of fourth industrial revolution technologies in smart irrigation: Constraints, opportunities, and future prospects for sub-Saharan Africa. *Smart Agricultural Technology*, *7*, 100412. doi:10.1016/j.atech.2024.100412

Webb, H. K., Arnott, J., Crawford, R. J., & Ivanova, E. P. (2013). Plastic Degradation and Its Environmental Implications with Special Reference to Poly(ethylene terephthalate). *Polymers*, *5*(1), 1–18. doi:10.3390/polym5010001

Wilson, D. C., Rodic, L., Cowing, M. J., Velis, C., Whiteman, A. D., Scheinberg, A., Vilches, R., Masterson, D., Stretz, J., & Oelz, B. (2015). Wasteaware' benchmark indicators for integrated sustainable waste management in cities. *Waste Management (New York, N.Y.)*, *35*, 329–342. doi:10.1016/j.wasman.2014.10.006 PMID:25458855

Wojnowska-Baryła, I., Bernat, K., & Zaborowska, M. (2022). Plastic waste degradation in landfill conditions: The problem with microplastics, and their direct and indirect environmental effects. *International Journal of Environmental Research and Public Health*, *19*(20), 13223. doi:10.3390/ijerph192013223 PMID:36293805

Workie, E., Kumar, V., Bhatnagar, A., He, Y., Dai, Y., Tong, Y. W., & Fu, C. (2023). Advancing the bioconversion process of food waste into methane: A systematic review. *Waste Management (New York, N.Y.)*, *156*, 187–197. doi:10.1016/j.wasman.2022.11.030 PMID:36493662

Xin, C., Zhang, T., Tsai, S., Zhai, Y., & Wang, J. (2020). An empirical study on greenhouse gas emission calculations under different municipal solid waste management strategies. *Applied Sciences (Basel, Switzerland)*, *10*(5), 1673. doi:10.3390/app10051673

Xue, T., Yuan, Y., Ahmed, Z., Moniz, K., Cao, G., & Wang, C. (2018). Proof of Contribution: A Modification of Proof of Work to Increase Mining Efficiency. *Proceedings - International Computer Software and Applications Conference*, (pp. 636–644). IEEE. 10.1109/COMPSAC.2018.00096

Yadav, A. K., Singh, K., Amin, A. H., Almutairi, L., Alsenani, T. R., & Ahmadian, A. (2023). A comparative study on consensus mechanism with security threats and future scopes: Blockchain. *Computer Communications*, *201*, 102–115. doi:10.1016/j.comcom.2023.01.018

Yang, H., Chen, G., & Wang, J. (2021). Microplastics in the marine environment: Sources, fates, impacts and microbial degradation. *Toxics*, *9*(2), 41. doi:10.3390/toxics9020041 PMID:33671786

Zhang, A. Y. (2018). Informal vs formal recycling; The dilemma of e-waste management. *Blavatnik School of Government*. https://www.bsg.ox.ac.uk/blog/informal-vs-formal-recycling-dilemma-e-waste-management

Zhang, P., Schmidt, D. C., White, J., & Dubey, A. (2019). Consensus mechanisms and information security technologies. *Advances in Computers*, *115*, 181–209. doi:10.1016/bs.adcom.2019.05.001

Zhang, Q., Jia, B., Li, T., & Li, W. (2024). Research on the sustainability of "greening" process in the Mu Us Sandy Land based on the spatiotemporal stability of ecological land. *PLoS One*, *19*(2), e0292469. doi:10.1371/journal.pone.0292469 PMID:38319930

Zhang, X., Sheng, Y., & Liu, Z. (2024). Using expertise as an intermediary: Unleashing the power of blockchain technology to drive future sustainable management using hidden champions. *Heliyon*, *10*(1), e23807. doi:10.1016/j.heliyon.2023.e23807 PMID:38226273

Zheng, Y., Yanful, E. K., & Bassi, A. S. (2005). A Review of Plastic Waste Biodegradation. *Critical Reviews in Biotechnology*, *25*(4), 243–250. doi:10.1080/07388550500346359 PMID:16419620

Zheng, Z., Fiddes, K., & Yang, L. (2021). A narrative review on environmental impacts of cannabis cultivation. *Journal of Cannabis Research*, *3*(1), 35. doi:10.1186/s42238-021-00090-0 PMID:34362475

Zhu, B., Wang, D., & Wei, N. (2022). Enzyme discovery and engineering for sustainable plastic recycling. *Trends in Biotechnology*, *40*(1), 22–37. doi:10.1016/j.tibtech.2021.02.008 PMID:33676748

Zuk, M., Podgórski, R., Ruszczyńska, A., Ciach, T., Majkowska-Pilip, A., Bilewicz, A., & Krysiński, P. (2022). Multifunctional Nanoparticles Based on Iron Oxide and Gold-198 Designed for Magnetic Hyperthermia and Radionuclide Therapy as a Potential Tool for Combined HER2-Positive Cancer Treatment. *Pharmaceutics*, *14*(8), 1680. doi:10.3390/pharmaceutics14081680 PMID:36015306

# Related References

To continue our tradition of advancing academic research, we have compiled a list of recommended IGI Global readings. These references will provide additional information and guidance to further enrich your knowledge and assist you with your own research and future publications.


Abbasnejad, B., Moeinzadeh, S., Ahankoob, A., & Wong, P. S. (2021). The Role of Collaboration in the Implementation of BIM-Enabled Projects. In J. Underwood & M. Shelbourn (Eds.), *Handbook of Research on Driving Transformational Change in the Digital Built Environment* (pp. 27–62). IGI Global. https://doi.org/10.4018/978-1-7998-6600-8.ch002

Abdulrahman, K. O., Mahamood, R. M., & Akinlabi, E. T. (2022). Additive Manufacturing (AM): Processing Technique for Lightweight Alloys and Composite Material. In K. Kumar, B. Babu, & J. Davim (Ed.), *Handbook of Research on Advancements in the Processing, Characterization, and Application of Lightweight Materials* (pp. 27-48). IGI Global. https://doi.org/10.4018/978-1-7998-7864-3.ch002

Agrawal, R., Sharma, P., & Saxena, A. (2021). A Diamond Cut Leather Substrate Antenna for BAN (Body Area Network) Application. In V. Singh, V. Dubey, A. Saxena, R. Tiwari, & H. Sharma (Eds.), *Emerging Materials and Advanced Designs for Wearable Antennas* (pp. 54–59). IGI Global. https://doi.org/10.4018/978-1-7998-7611-3.ch004

Ahmad, F., Al-Ammar, E. A., & Alsaidan, I. (2022). Battery Swapping Station: A Potential Solution to Address the Limitations of EV Charging Infrastructure. In M. Alam, R. Pillai, & N. Murugesan (Eds.), *Developing Charging Infrastructure and Technologies for Electric Vehicles* (pp. 195–207). IGI Global. doi:10.4018/978-1-7998-6858-3.ch010

Aikhuele, D. (2018). A Study of Product Development Engineering and Design Reliability Concerns. *International Journal of Applied Industrial Engineering*, *5*(1), 79–89. doi:10.4018/IJAIE.2018010105

Al-Khatri, H., & Al-Atrash, F. (2021). Occupants' Habits and Natural Ventilation in a Hot Arid Climate. In R. González-Lezcano (Ed.), *Advancements in Sustainable Architecture and Energy Efficiency* (pp. 146–168). IGI Global. https://doi.org/10.4018/978-1-7998-7023-4.ch007

Al-Shebeeb, O. A., Rangaswamy, S., Gopalakrishan, B., & Devaru, D. G. (2017). Evaluation and Indexing of Process Plans Based on Electrical Demand and Energy Consumption. *International Journal of Manufacturing, Materials, and Mechanical Engineering*, *7*(3), 1–19. doi:10.4018/IJMMME.2017070101

Amuda, M. O., Lawal, T. F., & Akinlabi, E. T. (2017). Research Progress on Rheological Behavior of AA7075 Aluminum Alloy During Hot Deformation. *International Journal of Materials Forming and Machining Processes*, *4*(1), 53–96. doi:10.4018/IJMFMP.2017010104

Amuda, M. O., Lawal, T. F., & Mridha, S. (2021). Microstructure and Mechanical Properties of Silicon Carbide-Treated Ferritic Stainless Steel Welds. In L. Burstein (Ed.), *Handbook of Research on Advancements in Manufacturing, Materials, and Mechanical Engineering* (pp. 395–411). IGI Global. https://doi.org/10.4018/978-1-7998-4939-1.ch019

Anikeev, V., Gasem, K. A., & Fan, M. (2021). Application of Supercritical Technologies in Clean Energy Production: A Review. In L. Chen (Ed.), *Handbook of Research on Advancements in Supercritical Fluids Applications for Sustainable Energy Systems* (pp. 792–821). IGI Global. https://doi.org/10.4018/978-1-7998-5796-9.ch022

Arafat, M. Y., Saleem, I., & Devi, T. P. (2022). Drivers of EV Charging Infrastructure Entrepreneurship in India. In M. Alam, R. Pillai, & N. Murugesan (Eds.), *Developing Charging Infrastructure and Technologies for Electric Vehicles* (pp. 208–219). IGI Global. https://doi.org/10.4018/978-1-7998-6858-3.ch011

Araujo, A., & Manninen, H. (2022). Contribution of Project-Based Learning on Social Skills Development: An Industrial Engineer Perspective. In A. Alves & N. van Hattum-Janssen (Eds.), *Training Engineering Students for Modern Technological Advancement* (pp. 119–145). IGI Global. https://doi.org/10.4018/978-1-7998-8816-1.ch006

Armutlu, H. (2018). Intelligent Biomedical Engineering Operations by Cloud Computing Technologies. In U. Kose, G. Guraksin, & O. Deperlioglu (Eds.), *Nature-Inspired Intelligent Techniques for Solving Biomedical Engineering Problems* (pp. 297–317). Hershey, PA: IGI Global. doi:10.4018/978-1-5225-4769-3.ch015

Atik, M., Sadek, M., & Shahrour, I. (2017). Single-Run Adaptive Pushover Procedure for Shear Wall Structures. In V. Plevris, G. Kremmyda, & Y. Fahjan (Eds.), *Performance-Based Seismic Design of Concrete Structures and Infrastructures* (pp. 59–83). Hershey, PA: IGI Global. doi:10.4018/978-1-5225-2089-4.ch003

Attia, H. (2021). Smart Power Microgrid Impact on Sustainable Building. In R. González-Lezcano (Ed.), *Advancements in Sustainable Architecture and Energy Efficiency* (pp. 169–194). IGI Global. https://doi.org/10.4018/978-1-7998-7023-4.ch008

Aydin, A., Akyol, E., Gungor, M., Kaya, A., & Tasdelen, S. (2018). Geophysical Surveys in Engineering Geology Investigations With Field Examples. In N. Ceryan (Ed.), *Handbook of Research on Trends and Digital Advances in Engineering Geology* (pp. 257–280). Hershey, PA: IGI Global. doi:10.4018/978-1-5225-2709-1.ch007

Ayoobkhan, M. U. D., Y., A., J., Easwaran, B., & R., T. (2021). Smart Connected Digital Products and IoT Platform With the Digital Twin. In P. Vasant, G. Weber, & W. Punurai (Ed.), Research Advancements in Smart Technology, Optimization, and Renewable Energy (pp. 330-350). IGI Global. https://doi.org/ doi:10.4018/978-1-7998-3970-5.ch016

Baeza Moyano, D., & González Lezcano, R. A. (2021). The Importance of Light in Our Lives: Towards New Lighting in Schools. In R. González-Lezcano (Ed.), *Advancements in Sustainable Architecture and Energy Efficiency* (pp. 239–256). IGI Global. https://doi.org/10.4018/978-1-7998-7023-4.ch011

Bagdadee, A. H. (2021). A Brief Assessment of the Energy Sector of Bangladesh. *International Journal of Energy Optimization and Engineering*, *10*(1), 36–55. doi:10.4018/IJEOE.2021010103

Baklezos, A. T., & Hadjigeorgiou, N. G. (2021). Magnetic Sensors for Space Applications and Magnetic Cleanliness Considerations. In C. Nikolopoulos (Ed.), *Recent Trends on Electromagnetic Environmental Effects for Aeronautics and Space Applications* (pp. 147–185). IGI Global. https://doi.org/10.4018/978-1-7998-4879-0.ch006

Bas, T. G. (2017). Nutraceutical Industry with the Collaboration of Biotechnology and Nutrigenomics Engineering: The Significance of Intellectual Property in the Entrepreneurship and Scientific Research Ecosystems. In T. Bas & J. Zhao (Eds.), *Comparative Approaches to Biotechnology Development and Use in Developed and Emerging Nations* (pp. 1–17). Hershey, PA: IGI Global. doi:10.4018/978-1-5225-1040-6.ch001

Bazeer Ahamed, B., & Periakaruppan, S. (2021). Taxonomy of Influence Maximization Techniques in Unknown Social Networks. In P. Vasant, G. Weber, & W. Punurai (Eds.), *Research Advancements in Smart Technology, Optimization, and Renewable Energy* (pp. 351-363). IGI Global. https://doi.org/10.4018/978-1-7998-3970-5.ch017

Beale, R., & André, J. (2017). *Design Solutions and Innovations in Temporary Structures*. Hershey, PA: IGI Global. doi:10.4018/978-1-5225-2199-0

Behnam, B. (2017). Simulating Post-Earthquake Fire Loading in Conventional RC Structures. In P. Samui, S. Chakraborty, & D. Kim (Eds.), *Modeling and Simulation Techniques in Structural Engineering* (pp. 425–444). Hershey, PA: IGI Global. doi:10.4018/978-1-5225-0588-4.ch015

Ben Hamida, I., Salah, S. B., Msahli, F., & Mimouni, M. F. (2018). Distribution Network Reconfiguration Using SPEA2 for Power Loss Minimization and Reliability Improvement. *International Journal of Energy Optimization and Engineering*, *7*(1), 50–65. doi:10.4018/IJEOE.2018010103

Bentarzi, H. (2021). Fault Tree-Based Root Cause Analysis Used to Study Mal-Operation of a Protective Relay in a Smart Grid. In A. Recioui & H. Bentarzi (Eds.), *Optimizing and Measuring Smart Grid Operation and Control* (pp. 289–308). IGI Global. https://doi.org/10.4018/978-1-7998-4027-5.ch012

Beysens, D. A., Garrabos, Y., & Zappoli, B. (2021). Thermal Effects in Near-Critical Fluids: Piston Effect and Related Phenomena. In L. Chen (Ed.), *Handbook of Research on Advancements in Supercritical Fluids Applications for Sustainable Energy Systems* (pp. 1–31). IGI Global. https://doi.org/10.4018/978-1-7998-5796-9.ch001

Bhaskar, S. V., & Kudal, H. N. (2017). Effect of TiCN and AlCrN Coating on Tribological Behaviour of Plasma-nitrided AISI 4140 Steel. *International Journal of Surface Engineering and Interdisciplinary Materials Science*, *5*(2), 1–17. doi:10.4018/IJSEIMS.2017070101

Bhuyan, D. (2018). Designing of a Twin Tube Shock Absorber: A Study in Reverse Engineering. In K. Kumar & J. Davim (Eds.), *Design and Optimization of Mechanical Engineering Products* (pp. 83–104). Hershey, PA: IGI Global. doi:10.4018/978-1-5225-3401-3.ch005

Blumberg, G. (2021). Blockchains for Use in Construction and Engineering Projects. In J. Underwood & M. Shelbourn (Eds.), *Handbook of Research on Driving Transformational Change in the Digital Built Environment* (pp. 179–208). IGI Global. https://doi.org/10.4018/978-1-7998-6600-8.ch008

Bolboaca, A. M. (2021). Considerations Regarding the Use of Fuel Cells in Combined Heat and Power for Stationary Applications. In G. Badea, R. Felseghi, & I. Aşchilean (Eds.), *Hydrogen Fuel Cell Technology for Stationary Applications* (pp. 239–275). IGI Global. https://doi.org/10.4018/978-1-7998-4945-2.ch010

Burstein, L. (2021). Simulation Tool for Cable Design. In L. Burstein (Ed.), *Handbook of Research on Advancements in Manufacturing, Materials, and Mechanical Engineering* (pp. 54–74). IGI Global. https://doi.org/10.4018/978-1-7998-4939-1.ch003

Calderon, F. A., Giolo, E. G., Frau, C. D., Rengel, M. G., Rodriguez, H., Tornello, M., ... Gallucci, R. (2018). Seismic Microzonation and Site Effects Detection Through Microtremors Measures: A Review. In N. Ceryan (Ed.), *Handbook of Research on Trends and Digital Advances in Engineering Geology* (pp. 326–349). Hershey, PA: IGI Global. doi:10.4018/978-1-5225-2709-1.ch009

Ceryan, N., & Can, N. K. (2018). Prediction of The Uniaxial Compressive Strength of Rocks Materials. In N. Ceryan (Ed.), *Handbook of Research on Trends and Digital Advances in Engineering Geology* (pp. 31–96). Hershey, PA: IGI Global. doi:10.4018/978-1-5225-2709-1.ch002

Ceryan, S. (2018). Weathering Indices Used in Evaluation of the Weathering State of Rock Material. In N. Ceryan (Ed.), *Handbook of Research on Trends and Digital Advances in Engineering Geology* (pp. 132–186). Hershey, PA: IGI Global. doi:10.4018/978-1-5225-2709-1.ch004

Chen, H., Padilla, R. V., & Besarati, S. (2017). Supercritical Fluids and Their Applications in Power Generation. In L. Chen & Y. Iwamoto (Eds.), *Advanced Applications of Supercritical Fluids in Energy Systems* (pp. 369–402). Hershey, PA: IGI Global. doi:10.4018/978-1-5225-2047-4.ch012

Chen, H., Padilla, R. V., & Besarati, S. (2021). Supercritical Fluids and Their Applications in Power Generation. In L. Chen (Ed.), *Handbook of Research on Advancements in Supercritical Fluids Applications for Sustainable Energy Systems* (pp. 566–599). IGI Global. https://doi.org/10.4018/978-1-7998-5796-9.ch016

Chen, L. (2017). Principles, Experiments, and Numerical Studies of Supercritical Fluid Natural Circulation System. In L. Chen & Y. Iwamoto (Eds.), *Advanced Applications of Supercritical Fluids in Energy Systems* (pp. 136–187). Hershey, PA: IGI Global. doi:10.4018/978-1-5225-2047-4.ch005

Chen, L. (2021). Principles, Experiments, and Numerical Studies of Supercritical Fluid Natural Circulation System. In L. Chen (Ed.), *Handbook of Research on Advancements in Supercritical Fluids Applications for Sustainable Energy Systems* (pp. 219–269). IGI Global. https://doi.org/10.4018/978-1-7998-5796-9.ch007

Chiba, Y., Marif, Y., Henini, N., & Tlemcani, A. (2021). Modeling of Magnetic Refrigeration Device by Using Artificial Neural Networks Approach. *International Journal of Energy Optimization and Engineering, 10*(4), 68–76. https://doi.org/10.4018/IJEOE.2021100105

Clementi, F., Di Sciascio, G., Di Sciascio, S., & Lenci, S. (2017). Influence of the Shear-Bending Interaction on the Global Capacity of Reinforced Concrete Frames: A Brief Overview of the New Perspectives. In V. Plevris, G. Kremmyda, & Y. Fahjan (Eds.), *Performance-Based Seismic Design of Concrete Structures and Infrastructures* (pp. 84–111). Hershey, PA: IGI Global. doi:10.4018/978-1-5225-2089-4.ch004

Codinhoto, R., Fialho, B. C., Pinti, L., & Fabricio, M. M. (2021). BIM and IoT for Facilities Management: Understanding Key Maintenance Issues. In J. Underwood & M. Shelbourn (Eds.), *Handbook of Research on Driving Transformational Change in the Digital Built Environment* (pp. 209–231). IGI Global. doi:10.4018/978-1-7998-6600-8.ch009

Cortés-Polo, D., Calle-Cancho, J., Carmona-Murillo, J., & González-Sánchez, J. (2017). Future Trends in Mobile-Fixed Integration for Next Generation Networks: Classification and Analysis. *International Journal of Vehicular Telematics and Infotainment Systems, 1*(1), 33–53. doi:10.4018/IJVTIS.2017010103

Costa, H. G., Sheremetieff, F. H., & Araújo, E. A. (2022). Influence of Game-Based Methods in Developing Engineering Competences. In A. Alves & N. van Hattum-Janssen (Eds.), *Training Engineering Students for Modern Technological Advancement* (pp. 69–88). IGI Global. https://doi.org/10.4018/978-1-7998-8816-1.ch004

Cui, X., Zeng, S., Li, Z., Zheng, Q., Yu, X., & Han, B. (2018). Advanced Composites for Civil Engineering Infrastructures. In K. Kumar & J. Davim (Eds.), *Composites and Advanced Materials for Industrial Applications* (pp. 212–248). Hershey, PA: IGI Global. doi:10.4018/978-1-5225-5216-1.ch010

Dalgıç, S., & Kuşku, İ. (2018). Geological and Geotechnical Investigations in Tunneling. In N. Ceryan (Ed.), *Handbook of Research on Trends and Digital Advances in Engineering Geology* (pp. 482–529). Hershey, PA: IGI Global. doi:10.4018/978-1-5225-2709-1.ch014

Dang, C., & Hihara, E. (2021). Study on Cooling Heat Transfer of Supercritical Carbon Dioxide Applied to Transcritical Carbon Dioxide Heat Pump. In L. Chen (Ed.), *Handbook of Research on Advancements in Supercritical Fluids Applications for Sustainable Energy Systems* (pp. 451–493). IGI Global. https://doi.org/10.4018/978-1-7998-5796-9.ch013

Daus, Y., Kharchenko, V., & Yudaev, I. (2021). Research of Solar Energy Potential of Photovoltaic Installations on Enclosing Structures of Buildings. *International Journal of Energy Optimization and Engineering, 10*(4), 18–34. https://doi.org/10.4018/IJEOE.2021100102

Daus, Y., Kharchenko, V., & Yudaev, I. (2021). Optimizing Layout of Distributed Generation Sources of Power Supply System of Agricultural Object. *International Journal of Energy Optimization and Engineering, 10*(3), 70–84. https://doi.org/10.4018/IJEOE.2021070104

de la Varga, D., Soto, M., Arias, C. A., van Oirschot, D., Kilian, R., Pascual, A., & Álvarez, J. A. (2017). Constructed Wetlands for Industrial Wastewater Treatment and Removal of Nutrients. In Á. Val del Río, J. Campos Gómez, & A. Mosquera Corral (Eds.), *Technologies for the Treatment and Recovery of Nutrients from Industrial Wastewater* (pp. 202–230). Hershey, PA: IGI Global. doi:10.4018/978-1-5225-1037-6.ch008

Deb, S., Ammar, E. A., AlRajhi, H., Alsaidan, I., & Shariff, S. M. (2022). V2G Pilot Projects: Review and Lessons Learnt. In M. Alam, R. Pillai, & N. Murugesan (Eds.), *Developing Charging Infrastructure and Technologies for Electric Vehicles* (pp. 252–267). IGI Global. https://doi.org/10.4018/978-1-7998-6858-3.ch014

Dekhandji, F. Z., & Rais, M. C. (2021). A Comparative Study of Power Quality Monitoring Using Various Techniques. In A. Recioui & H. Bentarzi (Eds.), *Optimizing and Measuring Smart Grid Operation and Control* (pp. 259–288). IGI Global. https://doi.org/10.4018/978-1-7998-4027-5.ch011

Deperlioglu, O. (2018). Intelligent Techniques Inspired by Nature and Used in Biomedical Engineering. In U. Kose, G. Guraksin, & O. Deperlioglu (Eds.), *Nature-Inspired Intelligent Techniques for Solving Biomedical Engineering Problems* (pp. 51–77). Hershey, PA: IGI Global. doi:10.4018/978-1-5225-4769-3.ch003

Dhurpate, P. R., & Tang, H. (2021). Quantitative Analysis of the Impact of Inter-Line Conveyor Capacity for Throughput of Manufacturing Systems. *International Journal of Manufacturing, Materials, and Mechanical Engineering, 11*(1), 1–17. https://doi.org/10.4018/IJMMME.2021010101

Dinkar, S., & Deep, K. (2021). A Survey of Recent Variants and Applications of Antlion Optimizer. *International Journal of Energy Optimization and Engineering, 10*(2), 48–73. doi:10.4018/IJEOE.2021040103

Dixit, A. (2018). Application of Silica-Gel-Reinforced Aluminium Composite on the Piston of Internal Combustion Engine: Comparative Study of Silica-Gel-Reinforced Aluminium Composite Piston With Aluminium Alloy Piston. In K. Kumar & J. Davim (Eds.), *Composites and Advanced Materials for Industrial Applications* (pp. 63–98). Hershey, PA: IGI Global. doi:10.4018/978-1-5225-5216-1.ch004

Drabecki, M. P., & Kułak, K. B. (2021). Global Pandemics on European Electrical Energy Markets: Lessons Learned From the COVID-19 Outbreak. *International Journal of Energy Optimization and Engineering, 10*(3), 24–46. https://doi.org/10.4018/IJEOE.2021070102

Dutta, M. M. (2021). Nanomaterials for Food and Agriculture. In M. Bhat, I. Wani, & S. Ashraf (Eds.), *Applications of Nanomaterials in Agriculture, Food Science, and Medicine* (pp. 75–97). IGI Global. doi:10.4018/978-1-7998-5563-7.ch004

Dutta, M. M., & Goswami, M. (2021). Coating Materials: Nano-Materials. In S. Roy & G. Bose (Eds.), *Advanced Surface Coating Techniques for Modern Industrial Applications* (pp. 1–30). IGI Global. doi:10.4018/978-1-7998-4870-7.ch001

Elsayed, A. M., Dakkama, H. J., Mahmoud, S., Al-Dadah, R., & Kaialy, W. (2017). Sustainable Cooling Research Using Activated Carbon Adsorbents and Their Environmental Impact. In T. Kobayashi (Ed.), *Applied Environmental Materials Science for Sustainability* (pp. 186–221). Hershey, PA: IGI Global. doi:10.4018/978-1-5225-1971-3.ch009

Ercanoglu, M., & Sonmez, H. (2018). General Trends and New Perspectives on Landslide Mapping and Assessment Methods. In N. Ceryan (Ed.), *Handbook of Research on Trends and Digital Advances in Engineering Geology* (pp. 350–379). Hershey, PA: IGI Global. doi:10.4018/978-1-5225-2709-1.ch010

Faroz, S. A., Pujari, N. N., Rastogi, R., & Ghosh, S. (2017). Risk Analysis of Structural Engineering Systems Using Bayesian Inference. In P. Samui, S. Chakraborty, & D. Kim (Eds.), *Modeling and Simulation Techniques in Structural Engineering* (pp. 390–424). Hershey, PA: IGI Global. doi:10.4018/978-1-5225-0588-4.ch014

Fekik, A., Hamida, M. L., Denoun, H., Azar, A. T., Kamal, N. A., Vaidyanathan, S., Bousbaine, A., & Benamrouche, N. (2022). Multilevel Inverter for Hybrid Fuel Cell/PV Energy Conversion System. In A. Fekik & N. Benamrouche (Eds.), *Modeling and Control of Static Converters for Hybrid Storage Systems* (pp. 233–270). IGI Global. https://doi.org/10.4018/978-1-7998-7447-8.ch009

Fekik, A., Hamida, M. L., Houassine, H., Azar, A. T., Kamal, N. A., Denoun, H., Vaidyanathan, S., & Sambas, A. (2022). Power Quality Improvement for Grid-Connected Photovoltaic Panels Using Direct Power Control. In A. Fekik & N. Benamrouche (Eds.), *Modeling and Control of Static Converters for Hybrid Storage Systems* (pp. 107–142). IGI Global. https://doi.org/10.4018/978-1-7998-7447-8.ch005

Fernando, P. R., Hamigah, T., Disne, S., Wickramasingha, G. G., & Sutharshan, A. (2018). The Evaluation of Engineering Properties of Low Cost Concrete Blocks by Partial Doping of Sand with Sawdust: Low Cost Sawdust Concrete Block. *International Journal of Strategic Engineering*, *1*(2), 26–42. doi:10.4018/IJoSE.2018070103

Ferro, G., Minciardi, R., Parodi, L., & Robba, M. (2022). Optimal Charging Management of Microgrid-Integrated Electric Vehicles. In M. Alam, R. Pillai, & N. Murugesan (Eds.), *Developing Charging Infrastructure and Technologies for Electric Vehicles* (pp. 133–155). IGI Global. https://doi.org/10.4018/978-1-7998-6858-3.ch007

Flumerfelt, S., & Green, C. (2022). Graduate Lean Leadership Education: A Case Study of a Program. In A. Alves & N. van Hattum-Janssen (Eds.), *Training Engineering Students for Modern Technological Advancement* (pp. 202–224). IGI Global. https://doi.org/10.4018/978-1-7998-8816-1.ch010

Galli, B. J. (2021). Implications of Economic Decision Making to the Project Manager. *International Journal of Strategic Engineering*, *4*(1), 19–32. https://doi.org/10.4018/IJoSE.2021010102

Gento, A. M., Pimentel, C., & Pascual, J. A. (2022). Teaching Circular Economy and Lean Management in a Learning Factory. In A. Alves & N. van Hattum-Janssen (Eds.), *Training Engineering Students for Modern Technological Advancement* (pp. 183–201). IGI Global. https://doi.org/10.4018/978-1-7998-8816-1.ch009

Ghosh, S., Mitra, S., Ghosh, S., & Chakraborty, S. (2017). Seismic Reliability Analysis in the Framework of Metamodelling Based Monte Carlo Simulation. In P. Samui, S. Chakraborty, & D. Kim (Eds.), *Modeling and Simulation Techniques in Structural Engineering* (pp. 192–208). Hershey, PA: IGI Global. doi:10.4018/978-1-5225-0588-4.ch006

Gil, M., & Otero, B. (2017). Learning Engineering Skills through Creativity and Collaboration: A Game-Based Proposal. In R. Alexandre Peixoto de Queirós & M. Pinto (Eds.), *Gamification-Based E-Learning Strategies for Computer Programming Education* (pp. 14–29). Hershey, PA: IGI Global. doi:10.4018/978-1-5225-1034-5.ch002

Gill, J., Ayre, M., & Mills, J. (2017). Revisioning the Engineering Profession: How to Make It Happen! In M. Gray & K. Thomas (Eds.), *Strategies for Increasing Diversity in Engineering Majors and Careers* (pp. 156–175). Hershey, PA: IGI Global. doi:10.4018/978-1-5225-2212-6.ch008

Godzhaev, Z., Senkevich, S., Kuzmin, V., & Melikov, I. (2021). Use of the Neural Network Controller of Sprung Mass to Reduce Vibrations From Road Irregularities. In P. Vasant, G. Weber, & W. Punurai (Ed.), *Research Advancements in Smart Technology, Optimization, and Renewable Energy* (pp. 69-87). IGI Global. https://doi.org/10.4018/978-1-7998-3970-5.ch005

Gomes de Gusmão, C. M. (2022). Digital Competencies and Transformation in Higher Education: Upskilling With Extension Actions. In A. Alves & N. van Hattum-Janssen (Eds.), *Training Engineering Students for Modern Technological Advancement* (pp. 313–328). IGI Global. https://doi.org/10.4018/978-1-7998-8816-1.ch015A

Goyal, N., Ram, M., & Kumar, P. (2017). Welding Process under Fault Coverage Approach for Reliability and MTTF. In M. Ram & J. Davim (Eds.), *Mathematical Concepts and Applications in Mechanical Engineering and Mechatronics* (pp. 222–245). Hershey, PA: IGI Global. doi:10.4018/978-1-5225-1639-2.ch011

Gray, M., & Lundy, C. (2017). Engineering Study Abroad: High Impact Strategy for Increasing Access. In M. Gray & K. Thomas (Eds.), *Strategies for Increasing Diversity in Engineering Majors and Careers* (pp. 42–59). Hershey, PA: IGI Global. doi:10.4018/978-1-5225-2212-6.ch003

Güler, O., & Varol, T. (2021). Fabrication of Functionally Graded Metal and Ceramic Powders Synthesized by Electroless Deposition. In S. Roy & G. Bose (Eds.), *Advanced Surface Coating Techniques for Modern Industrial Applications* (pp. 150–187). IGI Global. https://doi.org/10.4018/978-1-7998-4870-7.ch007

Guraksin, G. E. (2018). Internet of Things and Nature-Inspired Intelligent Techniques for the Future of Biomedical Engineering. In U. Kose, G. Guraksin, & O. Deperlioglu (Eds.), *Nature-Inspired Intelligent Techniques for Solving Biomedical Engineering Problems* (pp. 263–282). Hershey, PA: IGI Global. doi:10.4018/978-1-5225-4769-3.ch013

Hamida, M. L., Fekik, A., Denoun, H., Ardjal, A., & Bokhtache, A. A. (2022). Flying Capacitor Inverter Integration in a Renewable Energy System. In A. Fekik & N. Benamrouche (Eds.), *Modeling and Control of Static Converters for Hybrid Storage Systems* (pp. 287–306). IGI Global. https://doi.org/10.4018/978-1-7998-7447-8.ch011

Hasegawa, N., & Takahashi, Y. (2021). Control of Soap Bubble Ejection Robot Using Facial Expressions. *International Journal of Manufacturing, Materials, and Mechanical Engineering, 11*(2), 1–16. https://doi.org/10.4018/IJMMME.2021040101

Hejazi, T., & Akbari, L. (2017). A Multiresponse Optimization Model for Statistical Design of Processes with Discrete Variables. In M. Ram & J. Davim (Eds.), *Mathematical Concepts and Applications in Mechanical Engineering and Mechatronics* (pp. 17–37). Hershey, PA: IGI Global. doi:10.4018/978-1-5225-1639-2.ch002

Hejazi, T., & Hejazi, A. (2017). Monte Carlo Simulation for Reliability-Based Design of Automotive Complex Subsystems. In M. Ram & J. Davim (Eds.), *Mathematical Concepts and Applications in Mechanical Engineering and Mechatronics* (pp. 177–200). Hershey, PA: IGI Global. doi:10.4018/978-1-5225-1639-2.ch009

Hejazi, T., & Poursabbagh, H. (2017). Reliability Analysis of Engineering Systems: An Accelerated Life Testing for Boiler Tubes. In M. Ram & J. Davim (Eds.), *Mathematical Concepts and Applications in Mechanical Engineering and Mechatronics* (pp. 154–176). Hershey, PA: IGI Global. doi:10.4018/978-1-5225-1639-2.ch008

Henao, J., Poblano-Salas, C. A., Vargas, F., Giraldo-Betancur, A. L., Corona-Castuera, J., & Sotelo-Mazón, O. (2021). Principles and Applications of Thermal Spray Coatings. In S. Roy & G. Bose (Eds.), *Advanced Surface Coating Techniques for Modern Industrial Applications* (pp. 31–70). IGI Global. https://doi.org/10.4018/978-1-7998-4870-7.ch002

Henao, J., & Sotelo, O. (2018). Surface Engineering at High Temperature: Thermal Cycling and Corrosion Resistance. In A. Pakseresht (Ed.), *Production, Properties, and Applications of High Temperature Coatings* (pp. 131–159). Hershey, PA: IGI Global. doi:10.4018/978-1-5225-4194-3.ch006

Hrnčič, M. K., Cör, D., & Knez, Ž. (2021). Supercritical Fluids as a Tool for Green Energy and Chemicals. In L. Chen (Ed.), *Handbook of Research on Advancements in Supercritical Fluids Applications for Sustainable Energy Systems* (pp. 761–791). IGI Global. doi:10.4018/978-1-7998-5796-9.ch021

Ibrahim, O., Erdem, S., & Gurbuz, E. (2021). Studying Physical and Chemical Properties of Graphene Oxide and Reduced Graphene Oxide and Their Applications in Sustainable Building Materials. In R. González-Lezcano (Ed.), *Advancements in Sustainable Architecture and Energy Efficiency* (pp. 221–238). IGI Global. https://doi.org/10.4018/978-1-7998-7023-4.ch010

Ihianle, I. K., Islam, S., Naeem, U., & Ebenuwa, S. H. (2021). Exploiting Patterns of Object Use for Human Activity Recognition. In A. Nwajana & I. Ihianle (Eds.), *Handbook of Research on 5G Networks and Advancements in Computing, Electronics, and Electrical Engineering* (pp. 382–401). IGI Global. https://doi.org/10.4018/978-1-7998-6992-4.ch015

Ijemaru, G. K., Ngharamike, E. T., Oleka, E. U., & Nwajana, A. O. (2021). An Energy-Efficient Model for Opportunistic Data Collection in IoV-Enabled SC Waste Management. In A. Nwajana & I. Ihianle (Eds.), *Handbook of Research on 5G Networks and Advancements in Computing, Electronics, and Electrical Engineering* (pp. 1–19). IGI Global. https://doi.org/10.4018/978-1-7998-6992-4.ch001

Ilori, O. O., Adetan, D. A., & Umoru, L. E. (2017). Effect of Cutting Parameters on the Surface Residual Stress of Face-Milled Pearlitic Ductile Iron. *International Journal of Materials Forming and Machining Processes*, *4*(1), 38–52. doi:10.4018/IJMFMP.2017010103

Imam, M. H., Tasadduq, I. A., Ahmad, A., Aldosari, F., & Khan, H. (2017). Automated Generation of Course Improvement Plans Using Expert System. *International Journal of Quality Assurance in Engineering and Technology Education*, *6*(1), 1–12. doi:10.4018/IJQAETE.2017010101

Injeti, S. K., & Kumar, T. V. (2018). A WDO Framework for Optimal Deployment of DGs and DSCs in a Radial Distribution System Under Daily Load Pattern to Improve Techno-Economic Benefits. *International Journal of Energy Optimization and Engineering*, *7*(2), 1–38. doi:10.4018/IJEOE.2018040101

Ishii, N., Anami, K., & Knisely, C. W. (2018). *Dynamic Stability of Hydraulic Gates and Engineering for Flood Prevention*. Hershey, PA: IGI Global. doi:10.4018/978-1-5225-3079-4

Iwamoto, Y., & Yamaguchi, H. (2021). Application of Supercritical Carbon Dioxide for Solar Water Heater. In L. Chen (Ed.), *Handbook of Research on Advancements in Supercritical Fluids Applications for Sustainable Energy Systems* (pp. 370–387). IGI Global. https://doi.org/10.4018/978-1-7998-5796-9.ch010

Jayapalan, S. (2018). A Review of Chemical Treatments on Natural Fibers-Based Hybrid Composites for Engineering Applications. In K. Kumar & J. Davim (Eds.), *Composites and Advanced Materials for Industrial Applications* (pp. 16–37). Hershey, PA: IGI Global. doi:10.4018/978-1-5225-5216-1.ch002

Kapetanakis, T. N., Vardiambasis, I. O., Ioannidou, M. P., & Konstantaras, A. I. (2021). Modeling Antenna Radiation Using Artificial Intelligence Techniques: The Case of a Circular Loop Antenna. In C. Nikolopoulos (Ed.), *Recent Trends on Electromagnetic Environmental Effects for Aeronautics and Space Applications* (pp. 186–225). IGI Global. https://doi.org/10.4018/978-1-7998-4879-0.ch007

Karkalos, N. E., Markopoulos, A. P., & Dossis, M. F. (2017). Optimal Model Parameters of Inverse Kinematics Solution of a 3R Robotic Manipulator Using ANN Models. *International Journal of Manufacturing, Materials, and Mechanical Engineering*, *7*(3), 20–40. doi:10.4018/IJMMME.2017070102

Kelly, M., Costello, M., Nicholson, G., & O'Connor, J. (2021). The Evolving Integration of BIM Into Built Environment Programmes in a Higher Education Institute. In J. Underwood & M. Shelbourn (Eds.), *Handbook of Research on Driving Transformational Change in the Digital Built Environment* (pp. 294–326). IGI Global. https://doi.org/10.4018/978-1-7998-6600-8.ch012

Kesimal, A., Karaman, K., Cihangir, F., & Ercikdi, B. (2018). Excavatability Assessment of Rock Masses for Geotechnical Studies. In N. Ceryan (Ed.), *Handbook of Research on Trends and Digital Advances in Engineering Geology* (pp. 231–256). Hershey, PA: IGI Global. doi:10.4018/978-1-5225-2709-1.ch006

Knoflacher, H. (2017). The Role of Engineers and Their Tools in the Transport Sector after Paradigm Change: From Assumptions and Extrapolations to Science. In H. Knoflacher & E. Ocalir-Akunal (Eds.), *Engineering Tools and Solutions for Sustainable Transportation Planning* (pp. 1–29). Hershey, PA: IGI Global. doi:10.4018/978-1-5225-2116-7.ch001

Kose, U. (2018). Towards an Intelligent Biomedical Engineering With Nature-Inspired Artificial Intelligence Techniques. In U. Kose, G. Guraksin, & O. Deperlioglu (Eds.), *Nature-Inspired Intelligent Techniques for Solving Biomedical Engineering Problems* (pp. 1–26). Hershey, PA: IGI Global. doi:10.4018/978-1-5225-4769-3.ch001

Kostić, S. (2018). A Review on Enhanced Stability Analyses of Soil Slopes Using Statistical Design. In N. Ceryan (Ed.), *Handbook of Research on Trends and Digital Advances in Engineering Geology* (pp. 446–481). Hershey, PA: IGI Global. doi:10.4018/978-1-5225-2709-1.ch013

Kumar, A., Patil, P. P., & Prajapati, Y. K. (2018). *Advanced Numerical Simulations in Mechanical Engineering*. Hershey, PA: IGI Global. doi:10.4018/978-1-5225-3722-9

Kumar, G. R., Rajyalakshmi, G., & Manupati, V. K. (2017). Surface Micro Patterning of Aluminium Reinforced Composite through Laser Peening. *International Journal of Manufacturing, Materials, and Mechanical Engineering*, *7*(4), 15–27. doi:10.4018/IJMMME.2017100102

Kumar, N., Basu, D. N., & Chen, L. (2021). Effect of Flow Acceleration and Buoyancy on Thermalhydraulics of sCO2 in Mini/Micro-Channel. In L. Chen (Ed.), *Handbook of Research on Advancements in Supercritical Fluids Applications for Sustainable Energy Systems* (pp. 161–182). IGI Global. doi:10.4018/978-1-7998-5796-9.ch005

Kumari, N., & Kumar, K. (2018). Fabrication of Orthotic Calipers With Epoxy-Based Green Composite. In K. Kumar & J. Davim (Eds.), *Composites and Advanced Materials for Industrial Applications* (pp. 157–176). Hershey, PA: IGI Global. doi:10.4018/978-1-5225-5216-1.ch008

Kuppusamy, R. R. (2018). Development of Aerospace Composite Structures Through Vacuum-Enhanced Resin Transfer Moulding Technology (VERTMTy): Vacuum-Enhanced Resin Transfer Moulding. In K. Kumar & J. Davim (Eds.), *Composites and Advanced Materials for Industrial Applications* (pp. 99–111). Hershey, PA: IGI Global. doi:10.4018/978-1-5225-5216-1.ch005

Kurganov, V. A., Zeigarnik, Y. A., & Maslakova, I. V. (2021). Normal and Deteriorated Heat Transfer Under Heating Turbulent Supercritical Pressure Coolants Flows in Round Tubes. In L. Chen (Ed.), *Handbook of Research on Advancements in Supercritical Fluids Applications for Sustainable Energy Systems* (pp. 494–532). IGI Global. https://doi.org/10.4018/978-1-7998-5796-9.ch014

Li, H., & Zhang, Y. (2021). Heat Transfer and Fluid Flow Modeling for Supercritical Fluids in Advanced Energy Systems. In L. Chen (Ed.), *Handbook of Research on Advancements in Supercritical Fluids Applications for Sustainable Energy Systems* (pp. 388–422). IGI Global. https://doi.org/10.4018/978-1-7998-5796-9.ch011

Loy, J., Howell, S., & Cooper, R. (2017). Engineering Teams: Supporting Diversity in Engineering Education. In M. Gray & K. Thomas (Eds.), *Strategies for Increasing Diversity in Engineering Majors and Careers* (pp. 106–129). Hershey, PA: IGI Global. doi:10.4018/978-1-5225-2212-6.ch006

Macher, G., Armengaud, E., Kreiner, C., Brenner, E., Schmittner, C., Ma, Z., ... Krammer, M. (2018). Integration of Security in the Development Lifecycle of Dependable Automotive CPS. In N. Druml, A. Genser, A. Krieg, M. Menghin, & A. Hoeller (Eds.), *Solutions for Cyber-Physical Systems Ubiquity* (pp. 383–423). Hershey, PA: IGI Global. doi:10.4018/978-1-5225-2845-6.ch015

Madhu, M. N., Singh, J. G., Mohan, V., & Ongsakul, W. (2021). Transmission Risk Optimization in Interconnected Systems: Risk-Adjusted Available Transfer Capability. In P. Vasant, G. Weber, & W. Punurai (Ed.), *Research Advancements in Smart Technology, Optimization, and Renewable Energy* (pp. 183-199). IGI Global. https://doi.org/10.4018/978-1-7998-3970-5.ch010

Mahendramani, G., & Lakshmana Swamy, N. (2018). Effect of Weld Groove Area on Distortion of Butt Welded Joints in Submerged Arc Welding. *International Journal of Manufacturing, Materials, and Mechanical Engineering*, *8*(2), 33–44. doi:10.4018/IJMMME.2018040103

Makropoulos, G., Koumaras, H., Setaki, F., Filis, K., Lutz, T., Montowtt, P., Tomaszewski, L., Dybiec, P., & Järvet, T. (2021). 5G and Unmanned Aerial Vehicles (UAVs) Use Cases: Analysis of the Ecosystem, Architecture, and Applications. In A. Nwajana & I. Ihianle (Eds.), *Handbook of Research on 5G Networks and Advancements in Computing, Electronics, and Electrical Engineering* (pp. 36–69). IGI Global. https://doi.org/10.4018/978-1-7998-6992-4.ch003

Meric, E. M., Erdem, S., & Gurbuz, E. (2021). Application of Phase Change Materials in Construction Materials for Thermal Energy Storage Systems in Buildings. In R. González-Lezcano (Ed.), *Advancements in Sustainable Architecture and Energy Efficiency* (pp. 1–20). IGI Global. https://doi.org/10.4018/978-1-7998-7023-4.ch001

Mihret, E. T., & Yitayih, K. A. (2021). Operation of VANET Communications: The Convergence of UAV System With LTE/4G and WAVE Technologies. *International Journal of Smart Vehicles and Smart Transportation*, *4*(1), 29–51. https://doi.org/10.4018/IJSVST.2021010103

Mir, M. A., Bhat, B. A., Sheikh, B. A., Rather, G. A., Mehraj, S., & Mir, W. R. (2021). Nanomedicine in Human Health Therapeutics and Drug Delivery: Nanobiotechnology and Nanobiomedicine. In M. Bhat, I. Wani, & S. Ashraf (Eds.), *Applications of Nanomaterials in Agriculture, Food Science, and Medicine* (pp. 229–251). IGI Global. doi:10.4018/978-1-7998-5563-7.ch013

Mohammadzadeh, S., & Kim, Y. (2017). Nonlinear System Identification of Smart Buildings. In P. Samui, S. Chakraborty, & D. Kim (Eds.), *Modeling and Simulation Techniques in Structural Engineering* (pp. 328–347). Hershey, PA: IGI Global. doi:10.4018/978-1-5225-0588-4.ch011

Molina, G. J., Aktaruzzaman, F., Soloiu, V., & Rahman, M. (2017). Design and Testing of a Jet-Impingement Instrument to Study Surface-Modification Effects by Nanofluids. *International Journal of Surface Engineering and Interdisciplinary Materials Science*, *5*(2), 43–61. doi:10.4018/IJSEIMS.2017070104

Moreno-Rangel, A., & Carrillo, G. (2021). Energy-Efficient Homes: A Heaven for Respiratory Illnesses. In R. González-Lezcano (Ed.), *Advancements in Sustainable Architecture and Energy Efficiency* (pp. 49–71). IGI Global. https://doi.org/10.4018/978-1-7998-7023-4.ch003

Msomi, V., & Jantjies, B. T. (2021). Correlative Analysis Between Tensile Properties and Tool Rotational Speeds of Friction Stir Welded Similar Aluminium Alloy Joints. *International Journal of Surface Engineering and Interdisciplinary Materials Science*, *9*(2), 58–78. https://doi.org/10.4018/IJSEIMS.2021070104

Muigai, M. N., Mwema, F. M., Akinlabi, E. T., & Obiko, J. O. (2021). Surface Engineering of Materials Through Weld-Based Technologies: An Overview. In S. Roy & G. Bose (Eds.), *Advanced Surface Coating Techniques for Modern Industrial Applications* (pp. 247–260). IGI Global. doi:10.4018/978-1-7998-4870-7.ch011

Mukherjee, A., Saeed, R. A., Dutta, S., & Naskar, M. K. (2017). Fault Tracking Framework for Software-Defined Networking (SDN). In C. Singhal & S. De (Eds.), *Resource Allocation in Next-Generation Broadband Wireless Access Networks* (pp. 247–272). Hershey, PA: IGI Global. doi:10.4018/978-1-5225-2023-8.ch011

Mukhopadhyay, A., Barman, T. K., & Sahoo, P. (2018). Electroless Nickel Coatings for High Temperature Applications. In K. Kumar & J. Davim (Eds.), *Composites and Advanced Materials for Industrial Applications* (pp. 297–331). Hershey, PA: IGI Global. doi:10.4018/978-1-5225-5216-1.ch013

Mwema, F. M., & Wambua, J. M. (2022). Machining of Poly Methyl Methacrylate (PMMA) and Other Olymeric Materials: A Review. In K. Kumar, B. Babu, & J. Davim (Eds.), *Handbook of Research on Advancements in the Processing, Characterization, and Application of Lightweight Materials* (pp. 363–379). IGI Global. https://doi.org/10.4018/978-1-7998-7864-3.ch016

Mykhailyshyn, R., Savkiv, V., Boyko, I., Prada, E., & Virgala, I. (2021). Substantiation of Parameters of Friction Elements of Bernoulli Grippers With a Cylindrical Nozzle. *International Journal of Manufacturing, Materials, and Mechanical Engineering, 11*(2), 17–39. https://doi.org/10.4018/IJMMME.2021040102

Náprstek, J., & Fischer, C. (2017). Dynamic Stability and Post-Critical Processes of Slender Auto-Parametric Systems. In V. Plevris, G. Kremmyda, & Y. Fahjan (Eds.), *Performance-Based Seismic Design of Concrete Structures and Infrastructures* (pp. 128–171). Hershey, PA: IGI Global. doi:10.4018/978-1-5225-2089-4.ch006

Nautiyal, L., Shivach, P., & Ram, M. (2018). Optimal Designs by Means of Genetic Algorithms. In M. Ram & J. Davim (Eds.), *Soft Computing Techniques and Applications in Mechanical Engineering* (pp. 151–161). Hershey, PA: IGI Global. doi:10.4018/978-1-5225-3035-0.ch007

Nazir, R. (2017). Advanced Nanomaterials for Water Engineering and Treatment: Nano-Metal Oxides and Their Nanocomposites. In T. Saleh (Ed.), *Advanced Nanomaterials for Water Engineering, Treatment, and Hydraulics* (pp. 84–126). Hershey, PA: IGI Global. doi:10.4018/978-1-5225-2136-5.ch005

Nikolopoulos, C. D. (2021). Recent Advances on Measuring and Modeling ELF-Radiated Emissions for Space Applications. In C. Nikolopoulos (Ed.), *Recent Trends on Electromagnetic Environmental Effects for Aeronautics and Space Applications* (pp. 1–38). IGI Global. https://doi.org/10.4018/978-1-7998-4879-0.ch001

Nogueira, A. F., Ribeiro, J. C., Fernández de Vega, F., & Zenha-Rela, M. A. (2018). Evolutionary Approaches to Test Data Generation for Object-Oriented Software: Overview of Techniques and Tools. In M. Khosrow-Pour, D.B.A. (Ed.), Incorporating Nature-Inspired Paradigms in Computational Applications (pp. 162-194). Hershey, PA: IGI Global. https://doi.org/ doi:10.4018/978-1-5225-5020-4.ch006

Nwajana, A. O., Obi, E. R., Ijemaru, G. K., Oleka, E. U., & Anthony, D. C. (2021). Fundamentals of RF/Microwave Bandpass Filter Design. In A. Nwajana & I. Ihianle (Eds.), *Handbook of Research on 5G Networks and Advancements in Computing, Electronics, and Electrical Engineering* (pp. 149–164). IGI Global. https://doi.org/10.4018/978-1-7998-6992-4.ch005

Ogbodo, E. A. (2021). Comparative Study of Transmission Line Junction vs. Asynchronously Coupled Junction Diplexers. In A. Nwajana & I. Ihianle (Eds.), *Handbook of Research on 5G Networks and Advancements in Computing, Electronics, and Electrical Engineering* (pp. 326–336). IGI Global. https://doi.org/10.4018/978-1-7998-6992-4.ch013

Orosa, J. A., Vergara, D., Fraguela, F., & Masdías-Bonome, A. (2021). Statistical Understanding and Optimization of Building Energy Consumption and Climate Change Consequences. In R. González-Lezcano (Ed.), *Advancements in Sustainable Architecture and Energy Efficiency* (pp. 195–220). IGI Global. https://doi.org/10.4018/978-1-7998-7023-4.ch009

Osho, M. B. (2018). Industrial Enzyme Technology: Potential Applications. In S. Bharati & P. Chaurasia (Eds.), *Research Advancements in Pharmaceutical, Nutritional, and Industrial Enzymology* (pp. 375–394). Hershey, PA: IGI Global. doi:10.4018/978-1-5225-5237-6.ch017

Ouadi, A., & Zitouni, A. (2021). Phasor Measurement Improvement Using Digital Filter in a Smart Grid. In A. Recioui & H. Bentarzi (Eds.), *Optimizing and Measuring Smart Grid Operation and Control* (pp. 100–117). IGI Global. https://doi.org/10.4018/978-1-7998-4027-5.ch005

Padmaja, P., & Marutheswar, G. (2017). Certain Investigation on Secured Data Transmission in Wireless Sensor Networks. *International Journal of Mobile Computing and Multimedia Communications*, *8*(1), 48–61. doi:10.4018/IJMCMC.2017010104

Palmer, S., & Hall, W. (2017). An Evaluation of Group Work in First-Year Engineering Design Education. In R. Tucker (Ed.), *Collaboration and Student Engagement in Design Education* (pp. 145–168). Hershey, PA: IGI Global. doi:10.4018/978-1-5225-0726-0.ch007

Panchenko, V. (2021). Prospects for Energy Supply of the Arctic Zone Objects of Russia Using Frost-Resistant Solar Modules. In P. Vasant, G. Weber, & W. Punurai (Eds.), *Research Advancements in Smart Technology, Optimization, and Renewable Energy* (pp. 149-169). IGI Global. https://doi.org/10.4018/978-1-7998-3970-5.ch008

Panchenko, V. (2021). Photovoltaic Thermal Module With Paraboloid Type Solar Concentrators. *International Journal of Energy Optimization and Engineering*, *10*(2), 1–23. https://doi.org/10.4018/IJEOE.2021040101

Pandey, K., & Datta, S. (2021). Dry Machining of Inconel 825 Superalloys: Performance of Tool Inserts (Carbide, Cermet, and SiAlON). *International Journal of Manufacturing, Materials, and Mechanical Engineering*, *11*(4), 26–39. doi:10.4018/IJMMME.2021100102

Panneer, R. (2017). Effect of Composition of Fibers on Properties of Hybrid Composites. *International Journal of Manufacturing, Materials, and Mechanical Engineering*, *7*(4), 28–43. doi:10.4018/IJMMME.2017100103

Pany, C. (2021). Estimation of Correct Long-Seam Mismatch Using FEA to Compare the Measured Strain in a Non-Destructive Testing of a Pressurant Tank: A Reverse Problem. *International Journal of Smart Vehicles and Smart Transportation*, *4*(1), 16–28. doi:10.4018/IJSVST.2021010102

Paul, S., & Roy, P. (2018). Optimal Design of Power System Stabilizer Using a Novel Evolutionary Algorithm. *International Journal of Energy Optimization and Engineering*, *7*(3), 24–46. doi:10.4018/IJEOE.2018070102

Paul, S., & Roy, P. K. (2021). Oppositional Differential Search Algorithm for the Optimal Tuning of Both Single Input and Dual Input Power System Stabilizer. In P. Vasant, G. Weber, & W. Punurai (Eds.), *Research Advancements in Smart Technology, Optimization, and Renewable Energy* (pp. 256-282). IGI Global. https://doi.org/10.4018/978-1-7998-3970-5.ch013

Pavaloiu, A. (2018). Artificial Intelligence Ethics in Biomedical-Engineering-Oriented Problems. In U. Kose, G. Guraksin, & O. Deperlioglu (Eds.), *Nature-Inspired Intelligent Techniques for Solving Biomedical Engineering Problems* (pp. 219–231). Hershey, PA: IGI Global. doi:10.4018/978-1-5225-4769-3.ch010

Pioro, I., Mahdi, M., & Popov, R. (2017). Application of Supercritical Pressures in Power Engineering. In L. Chen & Y. Iwamoto (Eds.), *Advanced Applications of Supercritical Fluids in Energy Systems* (pp. 404–457). Hershey, PA: IGI Global. doi:10.4018/978-1-5225-2047-4.ch013

Plaksina, T., & Gildin, E. (2017). Rigorous Integrated Evolutionary Workflow for Optimal Exploitation of Unconventional Gas Assets. *International Journal of Energy Optimization and Engineering*, *6*(1), 101–122. doi:10.4018/IJEOE.2017010106

Popat, J., Kakadiya, H., Tak, L., Singh, N. K., Majeed, M. A., & Mahajan, V. (2021). Reliability of Smart Grid Including Cyber Impact: A Case Study. In R. Singh, A. Singh, A. Dwivedi, & P. Nagabhushan (Eds.), *Computational Methodologies for Electrical and Electronics Engineers* (pp. 163–174). IGI Global. https://doi.org/10.4018/978-1-7998-3327-7.ch013

Quiza, R., La Fé-Perdomo, I., Rivas, M., & Ramtahalsing, V. (2021). Triple Bottom Line-Focused Optimization of Oblique Turning Processes Based on Hybrid Modeling: A Study Case on AISI 1045 Steel Turning. In L. Burstein (Ed.), *Handbook of Research on Advancements in Manufacturing, Materials, and Mechanical Engineering* (pp. 215–241). IGI Global. https://doi.org/10.4018/978-1-7998-4939-1.ch010

Rahmani, M. K. (2022). Blockchain Technology: Principles and Algorithms. In S. Khan, M. Syed, R. Hammad, & A. Bushager (Eds.), *Blockchain Technology and Computational Excellence for Society 5.0* (pp. 16–27). IGI Global. https://doi.org/10.4018/978-1-7998-8382-1.ch002

Ramdani, N., & Azibi, M. (2018). Polymer Composite Materials for Microelectronics Packaging Applications: Composites for Microelectronics Packaging. In K. Kumar & J. Davim (Eds.), *Composites and Advanced Materials for Industrial Applications* (pp. 177–211). Hershey, PA: IGI Global. doi:10.4018/978-1-5225-5216-1.ch009

Ramesh, M., Garg, R., & Subrahmanyam, G. V. (2017). Investigation of Influence of Quenching and Annealing on the Plane Fracture Toughness and Brittle to Ductile Transition Temperature of the Zinc Coated Structural Steel Materials. *International Journal of Surface Engineering and Interdisciplinary Materials Science*, *5*(2), 33–42. doi:10.4018/IJSEIMS.2017070103

Robinson, J., & Beneroso, D. (2022). Project-Based Learning in Chemical Engineering: Curriculum and Assessment, Culture and Learning Spaces. In A. Alves & N. van Hattum-Janssen (Eds.), *Training Engineering Students for Modern Technological Advancement* (pp. 1–19). IGI Global. https://doi.org/10.4018/978-1-7998-8816-1.ch001

Rondon, B. (2021). Experimental Characterization of Admittance Meter With Crude Oil Emulsions. *International Journal of Electronics, Communications, and Measurement Engineering*, *10*(2), 51–59. https://doi.org/10.4018/IJECME.2021070104

Rudolf, S., Biryuk, V. V., & Volov, V. (2018). Vortex Effect, Vortex Power: Technology of Vortex Power Engineering. In V. Kharchenko & P. Vasant (Eds.), *Handbook of Research on Renewable Energy and Electric Resources for Sustainable Rural Development* (pp. 500–533). Hershey, PA: IGI Global. doi:10.4018/978-1-5225-3867-7.ch021

Sah, A., Bhadula, S. J., Dumka, A., & Rawat, S. (2018). A Software Engineering Perspective for Development of Enterprise Applications. In A. Elçi (Ed.), *Handbook of Research on Contemporary Perspectives on Web-Based Systems* (pp. 1–23). Hershey, PA: IGI Global. doi:10.4018/978-1-5225-5384-7.ch001

Sahli, Y., Zitouni, B., & Hocine, B. M. (2021). Three-Dimensional Numerical Study of Overheating of Two Intermediate Temperature P-AS-SOFC Geometrical Configurations. In G. Badea, R. Felseghi, & I. Aşchilean (Eds.), *Hydrogen Fuel Cell Technology for Stationary Applications* (pp. 186–222). IGI Global. https://doi.org/10.4018/978-1-7998-4945-2.ch008

Sahoo, P., & Roy, S. (2017). Tribological Behavior of Electroless Ni-P, Ni-P-W and Ni-P-Cu Coatings: A Comparison. *International Journal of Surface Engineering and Interdisciplinary Materials Science*, *5*(1), 1–15. doi:10.4018/IJSEIMS.2017010101

Sahoo, S. (2018). Laminated Composite Hypar Shells as Roofing Units: Static and Dynamic Behavior. In K. Kumar & J. Davim (Eds.), *Composites and Advanced Materials for Industrial Applications* (pp. 249–269). Hershey, PA: IGI Global. doi:10.4018/978-1-5225-5216-1.ch011

Sahu, H., & Hungyo, M. (2018). Introduction to SDN and NFV. In A. Dumka (Ed.), *Innovations in Software-Defined Networking and Network Functions Virtualization* (pp. 1–25). Hershey, PA: IGI Global. doi:10.4018/978-1-5225-3640-6.ch001

Salem, A. M., & Shmelova, T. (2018). Intelligent Expert Decision Support Systems: Methodologies, Applications, and Challenges. In T. Shmelova, Y. Sikirda, N. Rizun, A. Salem, & Y. Kovalyov (Eds.), *Socio-Technical Decision Support in Air Navigation Systems: Emerging Research and Opportunities* (pp. 215–242). Hershey, PA: IGI Global. doi:10.4018/978-1-5225-3108-1.ch007

Samal, M. (2017). FE Analysis and Experimental Investigation of Cracked and Un-Cracked Thin-Walled Tubular Components to Evaluate Mechanical and Fracture Properties. In P. Samui, S. Chakraborty, & D. Kim (Eds.), *Modeling and Simulation Techniques in Structural Engineering* (pp. 266–293). Hershey, PA: IGI Global. doi:10.4018/978-1-5225-0588-4.ch009

Samal, M., & Balakrishnan, K. (2017). Experiments on a Ring Tension Setup and FE Analysis to Evaluate Transverse Mechanical Properties of Tubular Components. In P. Samui, S. Chakraborty, & D. Kim (Eds.), *Modeling and Simulation Techniques in Structural Engineering* (pp. 91–115). Hershey, PA: IGI Global. doi:10.4018/978-1-5225-0588-4.ch004

Samarasinghe, D. A., & Wood, E. (2021). Innovative Digital Technologies. In J. Underwood & M. Shelbourn (Eds.), *Handbook of Research on Driving Transformational Change in the Digital Built Environment* (pp. 142–163). IGI Global. https://doi.org/10.4018/978-1-7998-6600-8.ch006

Sawant, S. (2018). Deep Learning and Biomedical Engineering. In U. Kose, G. Guraksin, & O. Deperlioglu (Eds.), *Nature-Inspired Intelligent Techniques for Solving Biomedical Engineering Problems* (pp. 283–296). Hershey, PA: IGI Global. doi:10.4018/978-1-5225-4769-3.ch014

Schulenberg, T. (2021). Energy Conversion Using the Supercritical Steam Cycle. In L. Chen (Ed.), *Handbook of Research on Advancements in Supercritical Fluids Applications for Sustainable Energy Systems* (pp. 659–681). IGI Global. doi:10.4018/978-1-7998-5796-9.ch018

Sezgin, H., & Berkalp, O. B. (2018). Textile-Reinforced Composites for the Automotive Industry. In K. Kumar & J. Davim (Eds.), *Composites and Advanced Materials for Industrial Applications* (pp. 129–156). Hershey, PA: IGI Global. doi:10.4018/978-1-5225-5216-1.ch007

Shaaban, A. A., & Shehata, O. M. (2021). Combining Response Surface Method and Metaheuristic Algorithms for Optimizing SPIF Process. *International Journal of Manufacturing, Materials, and Mechanical Engineering, 11*(4), 1–25. https://doi.org/10.4018/IJMMME.2021100101

Shafaati Shemami, M., & Sefid, M. (2022). Implementation and Demonstration of Electric Vehicle-to-Home (V2H) Application: A Case Study. In M. Alam, R. Pillai, & N. Murugesan (Eds.), *Developing Charging Infrastructure and Technologies for Electric Vehicles* (pp. 268–293). IGI Global. https://doi.org/10.4018/978-1-7998-6858-3.ch015

Shah, M. Z., Gazder, U., Bhatti, M. S., & Hussain, M. (2018). Comparative Performance Evaluation of Effects of Modifier in Asphaltic Concrete Mix. *International Journal of Strategic Engineering, 1*(2), 13–25. doi:10.4018/IJoSE.2018070102

Sharma, N., & Kumar, K. (2018). Fabrication of Porous NiTi Alloy Using Organic Binders. In K. Kumar & J. Davim (Eds.), *Composites and Advanced Materials for Industrial Applications* (pp. 38–62). Hershey, PA: IGI Global. doi:10.4018/978-1-5225-5216-1.ch003

Shivach, P., Nautiyal, L., & Ram, M. (2018). Applying Multi-Objective Optimization Algorithms to Mechanical Engineering. In M. Ram & J. Davim (Eds.), *Soft Computing Techniques and Applications in Mechanical Engineering* (pp. 287–301). Hershey, PA: IGI Global. doi:10.4018/978-1-5225-3035-0.ch014

Shmelova, T. (2018). Stochastic Methods for Estimation and Problem Solving in Engineering: Stochastic Methods of Decision Making in Aviation. In S. Kadry (Ed.), *Stochastic Methods for Estimation and Problem Solving in Engineering* (pp. 139–160). Hershey, PA: IGI Global. doi:10.4018/978-1-5225-5045-7.ch006

Siero González, L. R., & Romo Vázquez, A. (2017). Didactic Sequences Teaching Mathematics for Engineers With Focus on Differential Equations. In M. Ramírez-Montoya (Ed.), *Handbook of Research on Driving STEM Learning With Educational Technologies* (pp. 129–151). Hershey, PA: IGI Global. doi:10.4018/978-1-5225-2026-9.ch007

Sim, M. S., You, K. Y., Esa, F., & Chan, Y. L. (2021). Nanostructured Electromagnetic Metamaterials for Sensing Applications. In M. Bhat, I. Wani, & S. Ashraf (Eds.), *Applications of Nanomaterials in Agriculture, Food Science, and Medicine* (pp. 141–164). IGI Global. https://doi.org/10.4018/978-1-7998-5563-7.ch009

Singh, R., & Dutta, S. (2018). Visible Light Active Nanocomposites for Photocatalytic Applications. In K. Kumar & J. Davim (Eds.), *Composites and Advanced Materials for Industrial Applications* (pp. 270–296). Hershey, PA: IGI Global. doi:10.4018/978-1-5225-5216-1.ch012

Skripov, P. V., Yampol'skiy, A. D., & Rutin, S. B. (2021). High-Power Heat Transfer in Supercritical Fluids: Microscale Times and Sizes. In L. Chen (Ed.), *Handbook of Research on Advancements in Supercritical Fluids Applications for Sustainable Energy Systems* (pp. 424–450). IGI Global. https://doi.org/10.4018/978-1-7998-5796-9.ch012

Sözbilir, H., Özkaymak, Ç., Uzel, B., & Sümer, Ö. (2018). Criteria for Surface Rupture Microzonation of Active Faults for Earthquake Hazards in Urban Areas. In N. Ceryan (Ed.), *Handbook of Research on Trends and Digital Advances in Engineering Geology* (pp. 187–230). Hershey, PA: IGI Global. doi:10.4018/978-1-5225-2709-1.ch005

Stanciu, I. (2018). Stochastic Methods in Microsystems Engineering. In S. Kadry (Ed.), *Stochastic Methods for Estimation and Problem Solving in Engineering* (pp. 161–176). Hershey, PA: IGI Global. doi:10.4018/978-1-5225-5045-7.ch007

Strebkov, D., Nekrasov, A., Trubnikov, V., & Nekrasov, A. (2018). Single-Wire Resonant Electric Power Systems for Renewable-Based Electric Grid. In V. Kharchenko & P. Vasant (Eds.), *Handbook of Research on Renewable Energy and Electric Resources for Sustainable Rural Development* (pp. 449–474). Hershey, PA: IGI Global. doi:10.4018/978-1-5225-3867-7.ch019

Sukhyy, K., Belyanovskaya, E., & Sukhyy, M. (2021). *Basic Principles for Substantiation of Working Pair Choice*. IGI Global. doi:10.4018/978-1-7998-4432-7.ch002

Suri, M. S., & Kaliyaperumal, D. (2022). Extension of Aspiration Level Model for Optimal Planning of Fast Charging Stations. In A. Fekik & N. Benamrouche (Eds.), *Modeling and Control of Static Converters for Hybrid Storage Systems* (pp. 91–106). IGI Global. https://doi.org/10.4018/978-1-7998-7447-8.ch004

Tallet, E., Gledson, B., Rogage, K., Thompson, A., & Wiggett, D. (2021). Digitally-Enabled Design Management. In J. Underwood & M. Shelbourn (Eds.), *Handbook of Research on Driving Transformational Change in the Digital Built Environment* (pp. 63–89). IGI Global. https://doi.org/10.4018/978-1-7998-6600-8.ch003

Terki, A., & Boubertakh, H. (2021). A New Hybrid Binary-Real Coded Cuckoo Search and Tabu Search Algorithm for Solving the Unit-Commitment Problem. *International Journal of Energy Optimization and Engineering*, *10*(2), 104–119. https://doi.org/10.4018/IJEOE.2021040105

Tüdeş, Ş., Kumlu, K. B., & Ceryan, S. (2018). Integration Between Urban Planning and Natural Hazards For Resilient City. In N. Ceryan (Ed.), *Handbook of Research on Trends and Digital Advances in Engineering Geology* (pp. 591–630). Hershey, PA: IGI Global. doi:10.4018/978-1-5225-2709-1.ch017

Ulamis, K. (2018). Soil Liquefaction Assessment by Anisotropic Cyclic Triaxial Test. In N. Ceryan (Ed.), *Handbook of Research on Trends and Digital Advances in Engineering Geology* (pp. 631–664). Hershey, PA: IGI Global. doi:10.4018/978-1-5225-2709-1.ch018

Valente, M., & Milani, G. (2017). Seismic Assessment and Retrofitting of an Under-Designed RC Frame Through a Displacement-Based Approach. In V. Plevris, G. Kremmyda, & Y. Fahjan (Eds.), *Performance-Based Seismic Design of Concrete Structures and Infrastructures* (pp. 36–58). Hershey, PA: IGI Global. doi:10.4018/978-1-5225-2089-4.ch002

Vargas-Bernal, R. (2021). Advances in Electromagnetic Environmental Shielding for Aeronautics and Space Applications. In C. Nikolopoulos (Ed.), *Recent Trends on Electromagnetic Environmental Effects for Aeronautics and Space Applications* (pp. 80–96). IGI Global. https://doi.org/10.4018/978-1-7998-4879-0.ch003

Vasant, P. (2018). A General Medical Diagnosis System Formed by Artificial Neural Networks and Swarm Intelligence Techniques. In U. Kose, G. Guraksin, & O. Deperlioglu (Eds.), *Nature-Inspired Intelligent Techniques for Solving Biomedical Engineering Problems* (pp. 130–145). Hershey, PA: IGI Global. doi:10.4018/978-1-5225-4769-3.ch006

Verner, C. M., & Sarwar, D. (2021). Avoiding Project Failure and Achieving Project Success in NHS IT System Projects in the United Kingdom. *International Journal of Strategic Engineering*, *4*(1), 33–54. https://doi.org/10.4018/IJoSE.2021010103

Verrollot, J., Tolonen, A., Harkonen, J., & Haapasalo, H. J. (2018). Challenges and Enablers for Rapid Product Development. *International Journal of Applied Industrial Engineering*, *5*(1), 25–49. doi:10.4018/IJAIE.2018010102

Wan, A. C., Zulu, S. L., & Khosrow-Shahi, F. (2021). Industry Views on BIM for Site Safety in Hong Kong. In J. Underwood & M. Shelbourn (Eds.), *Handbook of Research on Driving Transformational Change in the Digital Built Environment* (pp. 120–140). IGI Global. https://doi.org/10.4018/978-1-7998-6600-8.ch005

Yardimci, A. G., & Karpuz, C. (2018). Fuzzy Rock Mass Rating: Soft-Computing-Aided Preliminary Stability Analysis of Weak Rock Slopes. In N. Ceryan (Ed.), *Handbook of Research on Trends and Digital Advances in Engineering Geology* (pp. 97–131). Hershey, PA: IGI Global. doi:10.4018/978-1-5225-2709-1.ch003

You, K. Y. (2021). Development Electronic Design Automation for RF/Microwave Antenna Using MATLAB GUI. In A. Nwajana & I. Ihianle (Eds.), *Handbook of Research on 5G Networks and Advancements in Computing, Electronics, and Electrical Engineering* (pp. 70–148). IGI Global. https://doi.org/10.4018/978-1-7998-6992-4.ch004

Yousefi, Y., Gratton, P., & Sarwar, D. (2021). Investigating the Opportunities to Improve the Thermal Performance of a Case Study Building in London. *International Journal of Strategic Engineering*, *4*(1), 1–18. https://doi.org/10.4018/IJoSE.2021010101

Zindani, D., & Kumar, K. (2018). Industrial Applications of Polymer Composite Materials. In K. Kumar & J. Davim (Eds.), *Composites and Advanced Materials for Industrial Applications* (pp. 1–15). Hershey, PA: IGI Global. doi:10.4018/978-1-5225-5216-1.ch001

Zindani, D., Maity, S. R., & Bhowmik, S. (2018). A Decision-Making Approach for Material Selection of Polymeric Composite Bumper Beam. In K. Kumar & J. Davim (Eds.), *Composites and Advanced Materials for Industrial Applications* (pp. 112–128). Hershey, PA: IGI Global. doi:10.4018/978-1-5225-5216-1.ch006

# About the Contributors

**Chandra Mohan**, Associate Professor of Chemistry, K R Mangalam University, Gurugram is a Senior Faculty with over 11 years of teaching and research experience, having expertise in Inorganic Chemistry. Dr. Chandra Mohan has obtained Doctorate (Chemistry) in the field of Chemical Sensors from Guru Gobind Singh Indraprastha University, Delhi, India and also received M.Phil in Chemistry from University of Delhi. His research work focused on heterocycle based transition metal complexes for biological and sensor applications, and clay based solid acid catalysis. He has published more than 60 research papers in Peer reviewed journals, 11 book chapters and 5 patent. He has also extended the privilege of being an organizing committee member in ICASE-2022, ASET-2021, NANO MALAYSIA-2019 International conferences. He serves as editorial board member of various international journals. He is an Associate Editor of INORGACHEM Journal, UAE. He serves as the Convener of International FDP and International Conference (AMGSE-2024) organized at K R Mangalam University.

**Shobhna Jeet** is an Associate Professor in Law at K.R. Mangalam University, Gurgaon. She has authored three books namely, "Cyber Crime and Cyber Terrorism: National and International Regime", and "Cyber Terrorism: A Threat to Nation". Her latest edited book is "Environmental Law Governance and Sustainable Development". She is in the editorial board of various national and international peer reviewed refereed journals. She was a Joint author and had shared her contribution in two chapters in the book "Right to Education: Imperative for Progress" edited by Prof. Mool Chand Sharma with a foreword from renowned Scientist Prof. Yash Pal.

**Saurav Dixit** currently holds the esteemed position of Dean of Research and Development at Lovely Professional University, located in Phagwara, India. His research interests include a wide range of topics such as construction productivity, synthesis and development, self-healing concrete, chemical and physical properties of fly-ash, additive manufacturing, advanced concrete production, and surface modification.

**Sónia Carabineiro** graduated in Applied Chemistry (branch of Biotechnology) at the NOVA School of Science and Technology (Universidade Nova de Lisboa), Portugal (FCT-UNL) and obtained her Ph.D. degree in Chemical Engineering (Catalysis) in 2001 at the same University. In 2000 she was an Invited Researcher at Leiden University, The Netherlands, and Post Doctoral Researcher from 2001 to 2003 at the same University. In 2004 she got a Post Doc grant at the Centre of Structural Chemistry of the University of Lisbon (CQE-IST). In 2007 she joined the Laboratory of Catalysis and Materials (LCM), part of Associate Laboratory LSRE-LCM, at the Department of Chemical Engineering (DEQ) of the Faculty of Engineering of the University of Porto (FEUP), as an Assistant Researcher, with a Ciencia 2007 grant from Foundation for Science and Technology (FCT). From 2013 to 2018 she was a

Principal Researcher (Investigador FCT grant) at the same lab. She was Integrated Researcher of CQE-IST and Invited (Visiting) Professor at Wuhan Textile University, China in 2019. She is now Assistant Professor at FCT-UNL (since 2020). She has also collaborations with other Universities of Portugal and abroad. She is member of the Editorial Board of ChemCatChem and Scientific Reports of the Editorial Advisory Board of Catalysis Today and of the Editorial Board of Nanomaterials and Catalysts, among others. She is one the females reviewer in the world with more revisions done (Publons data) and evaluator of several international projects (including ERC grants). She is one the females reviewer in the world with more revisions done (Publons data) and evaluator of several international projects (including ERC grants). She was included in the list of the worldwide most cited scientists (top 2% of Chemistry) published by Stanford University since 2017. She was also chosen as one of the Portuguese Women in Science featured in the 4th Edition of the Book "Mulheres na Ciência" from Ciência Viva, Portugal in 2023. Her research interests include: Catalysis by gold, nanostructured catalysts, mixed metal oxides, graphene, graphitic carbon nitride, carbon nanotubes, oxidation reactions (carbon monoxide, volatile organic compounds, hydrocarbons, alcohols), heterogenization of homogeneous catalysts, abatement of pollutants from water, soil and air.

**Payal Bansal** is working as Assistant Professor in the department of Fashion design and technology, Lovely Professional University, Phagwara, Punjab, India. She has received her master and Ph.D. degree from National Institute of Technology, Jalandhar, Punjab, India in 2021. Her research areas include garment technology, clothing comfort, fashion study, fabric design, apparel manufacturing, healthcare textiles, knitted fabric structures and technical textile. She is having a 8 year teaching experience (5 year teaching assistantship and 3 year teaching). She has published 22 number of publications (16 SCI/ Scopus and 6 UGC care listed), 2 patent published, 3 book chapters and organized one National, one International conferences. She has been recently involved in research projects also.

**Megersa Olumana Dinka** is a graduate with PhD from University of Natural Resources and Applied Life Science (Vienna) in 2010. He has MSc in Irrigation Engineering from ArbaMinch University (Ethiopia). He has about 18 years of experience as academician and 22 years of experience as a researcher. Currently, he is the head of Department of Civil Engineering Science at University of Johannesburg. He is specialized in the areas of water resource engineering specific to hydrology, hydraulics and water management aspects. He has taught various courses and modules at undergraduate and postgraduate levels successfully. Moreover, he also supervised a number of postgraduate students (31 M and 9 D) successfully. He has published more than 65 Journal articles, 1 book, 8 book chapters and 20 proceedings in international accredited publications.

**Mgha Garg**, BALLB, LLM, UGC- NET qualified working as an Assistant Professor in K.R. Mangalam University, Gurugram since April, 2022. She is having total 6.5 years of experience in academics. She has submitted her thesis titled" The Controversy of Surrogacy in Indian Perspective: An Analytical Study". She has contributed papers in UCG care listed, peer reviewed journals. She has also presented several papers in National and International Conferences/seminars. While working in various institutions Ms. Megha has handled various non academic assignments such as: Academic Incharge of Faculty of Law. Community Nodal Officer. Coordinator/Executive member of various intra faculty competitions like Moot court competitions, debate competitions, Webinars etc Coordinator of Legal Aid Committee, Student care center. Member of Examination Committee.

**Vishal Jain** is presently working as an Associate Professor at Department of Computer Science and Engineering, School of Engineering and Technology, Sharda University, Greater Noida, U. P. India. Before that, he has worked for several years as an Associate Professor at Bharati Vidyapeeth's Institute of Computer Applications and Management (BVICAM), New Delhi. He has more than 14 years of experience in the academics. He obtained Ph.D (CSE), M.Tech (CSE), MBA (HR), MCA, MCP and CCNA. He has authored more than 90 research papers in reputed conferences and journals, including Web of Science and Scopus. He has authored and edited more than 30 books with various reputed publishers, including Elsevier, Springer, Apple Academic Press, CRC, Taylor and Francis Group, Scrivener, Wiley, Emerald, NOVA Science and IGI-Global. His research areas include information retrieval, semantic web, ontology engineering, data mining, ad hoc networks, and sensor networks. He received a Young Active Member Award for the year 2012–13 from the Computer Society of India, Best Faculty Award for the year 2017 and Best Researcher Award for the year 2019 from BVICAM, New Delhi.

**Crstian Kaunert** is Professor of International Security at Dublin City University, Ireland. He is also Professor of Policing and Security, as well as Director of the International Centre for Policing and Security at the University of South Wales. In addition, he is Jean Monnet Chair, Director of the Jean Monnet Centre of Excellence and Director of the Jean Monnet Network on EU Counter-Terrorism ().

**Ai Kumar** is currently working as an Associate Professor in Department of Applied Sciences in Bharati Vidyapeeth's College of Engg. New Delhi. He has a teaching experience of more than 17 years to undergraduate and postgraduate students. He is supervising Ph.D. students at university level. He has obtained his PhD degree in Physics. He is Gold medalist in M.Sc. (Physics), Silver medalist in M.Phil. (Physics) in M.D.S.U. Ajmer. He has received academic excellence gold medal, awarded by D.A.V managing committee, Delhi in 2005. His Topic of PhD thesis was: Resonant Electron's Energy Diffusion in the Magnetosphere by the Whistler Mode VLF waves. He has 25 research publications to his credit and has presented papers in National, International conferences and workshops as well. His Research Areas are Wave – Particle Interaction & Whistler's waves, Magnetospheric Physics. He is a member of Rajasthan Physics Association. Three students have completed their Ph.D. in his supervision. Currently guiding three PhD. Students. 5 patent have been granted in his credit.

**Mhmmed Majeed** is a Senior Lecturer (PhD) at Tamale Technical University, Tamale-Ghana. He is the current Head of Department for Marketing. His current research interest includes branding, Hospitality and Tourism, social media in service organizations. Majeed holds Doctor of Business Administration (DBA), MPhil and MBA Marketing, Postgraduate Diploma, HND Marketing. He lectures part time in many Ghanaian public Universities and a reviewer to many Journals in Management, hospitality, and marketing. Majeed has also published in good journals like Journal of Hospitality and Tourism Insights (Emerald), Springer, Palgrave, Cogent Business & Management (Taylor & Francis), journal of African Business (Rutledge).

**Godwin Joseph** is currently pursuing a B.Tech in Nanotechnology in Amity University Dubai. Their passion for cutting-edge technology and its potential to address real-world challenges led them to delve into the field of nanotechnology. During their academic journey, Godwin Joseph authored a research paper titled "Fabrication of Silver Coated Iron Oxide Infused Hydrogel for Wastewater Treatment," which was presented at the International Conference on Innovation, Sustainability, and Applied

Sciences (ICISAS 2023). This work exemplifies their dedication to developing innovative solutions for environmental sustainability, particularly in the domain of wastewater treatment. In addition to their scholarly pursuits, Godwin Joseph demonstrated their analytical prowess and creative thinking by securing the 3rd prize at The Forensic Science Forum for Universities organized by the Dubai Police. Their paper, "Applications of Anticounterfeiting for Inks-Nano Based," showcased their ability to integrate nanotechnology with forensic science, addressing critical challenges in counterfeit detection and prevention. Furthermore, Godwin Joseph gained practical experience in water and wastewater analysis through an internship at Qualitas Material Testing Laboratories. This experience equipped them with valuable skills and insights into laboratory procedures and quality testing methodologies. Driven by a curiosity to explore the intersection of science and society, Godwin Joseph is dedicated to advancing knowledge and contributing to the development of practical solutions for global challenges. They are committed to pursuing excellence in research and innovation, with a vision to make a meaningful impact in their field and beyond.

**Cement Nangpiire** (PhD) is currently a Senior Lecturer at the Department of Procurement and Supply Chain Management at the SD-Dombo University of Business and Integrated Development Studies (SDD-UBIDS) Wa. He holds a PhD in Marketing and Strategy degree from the University of Minho-Portugal, and a Master of Business Administration (MBA) degree from the Maastricht School of Management in the Netherlands. Before then, he obtained a Bachelor's degree from the University of Ghana-Legon and holds a Diploma in Marketing from the University College of Management Studies-Ghana. His research interest is in Services Marketing and Supply Chain research, focusing on value co-creation, customer engagement, supply chain processes, procurement processes, market orientation, public sector marketing, hospitality and tourism marketing, and digital/social media marketing. He teaches undergraduate/graduate courses in Marketing Research, Services Marketing, Global Logistics Strategy, Social Marketing, Innovation and Entrepreneurship, Total Quality Management and Tourism/Hospitality Marketing

**Ja Nyika** is a lecturer at the Technical University of Kenya and a postdoctoral research fellow at the University of Johannesburg. She holds a doctorate degree in Science, Engineering and Technology from the University of South Africa and MSc. in Land and Water Management from the University of Nairobi, Kenya. In her research work, she specializes in hydrobiogeochemistry and has published more than 40 journal articles and book chapters in the discipline. She is also involved in teaching, supervising and mentoring undergraduate and postgraduate students.

**Prashant Panwar,** Assistant Professor of Law, currently works at K.R Mangalam University, where he teaches Law of Tort, Indian Constitution, and Indian Evidence Act. He is also the Co-Convener of Criminology and Victimology Cell and the Competitive Exam In-charge. Prashant has also worked as an Assistant Professor of Law at Shobhit University in Gangoh, Saharanpur, where he taught Contract Law, CPC, IPC, and Indian Evidence Act. Apart from teaching, Prashant has also conducted various webinars, seminars, workshops, and guest lectures on a range of legal topics. Prashant holds an LL.M. in Human Rights and Duties from the School of Law, Banaras Hindu University, and a BA.LL.B. (Hons.) from the School of Law, IMS Unison University, Dehradun. He has completed several certificate courses and has attended numerous faculty development programs, workshops, and organizational activities.

**SMNzmuz Sakib** was born in Bangladesh. He achieved his Primary School Certificate (PSC) from Cotton Research Station School, Sreepur, Gazipur and after several months of studying Class 6 in Al-haz Dhanai Bepari Memorial High School, Sreepur, Gazipur, he relocated to Jashore. He accomplished his Junior School Certificate (JSC) from Jagodishpur Mirzapur Ismail Secondary School, Jagodishour, Chaowgacha, Jashore and his Secondary School Certificate (SSC) from A. K. High School And College, Dhania, Dhaka-1236. He obtained his Higher Secondary Certificate (HSC) from Chowgacha Government College, Chowgacha, Jashore. He received a BSc in Business Studies, earning a CGPA of 4 out of 4 (1st class 1st) with 97.06% marks from School of Business and Trade, an online business school situated in Switzerland. He completed an MBA in Human Resources from International MBA Institute, an online business institute in Switzerland. He studied Bachelor of Laws (Hon's) or LLB(Hon's) at Dhaka International University, Dhaka, Bangladesh and studying LLM or Master of Laws in Bangladesh University of Professionals, Dhaka, Bangladesh. He has accomplished CPD verified advanced diploma and diploma programs i.e. Advanced Diploma in Tissue Engineering, Advanced Diploma in Genetic Engineering - Theory and Application, Diploma in Fashion Design, Diploma in Nutrition, Therapeutics and Health, Diploma in Biology, Diploma in Pharmacy Technician, Software Test Engineer Diploma (Advanced), Advanced Diploma in Soil Science and Technology, Diploma in Energy Economics, Energy Systems and Environmental Impact, Advanced Diploma in Political Ideologies, Diploma in Community Psychology, Diploma in Training of Trainers, Advance Diploma in Principles of Industrial Engineering, Advanced Diploma in Production & Operation Management, Advanced Diploma in Modelling and Analytics for Supply Chain Management, Business Administration Diploma (Foundation), Diploma in Effective Human Resource Administration, Diploma in Audio System Engineering, Diploma in ISO Standards - Integrated Management System (IMS), Diploma in Lean Manufacturing - Productive Management with Fundamental Tools, Advanced Diploma in Tourism and Hospitality Management, Diploma in Web Design, Diploma in Human Resources, and Diploma in Basic English Grammar, Diploma in River Engineering and has obtained certifications such as TESOL from Arizona State University, TEFL from Teacher Record, Scrum Master Professional Certificate (SMPC®), EF SET Certificate (C2 Level), Google IT Support Professional Certificate, Google Data Analytics Professional Certificate, Google Project Management Professional Certificate, Google Business Intelligence Professional Certificate, Google Digital Marketing & E-commerce Professional Certificate and IBM New Collar: Customer Engagement Specialist credential etc. As a multidisciplinary researcher, he has published several articles in the Asian Pacific Journal of Environment and Cancer, Simulacra, Journal of Innovation Information Technology and Application (JINITA), Waste Technology, and The IUP Journal of Electrical and Electronics Engineering. Several of his articles are published by Cambridge Open Engage, a publication of Cambridge University Press. Some of the titles of his published research works include "The Impact of Oil and Gas Development on the Landscape and Surface in Nigeria," "Assessing the Impact of Arctic Melting in the Predominantly Multilateral World System," "LiDAR Technology - An Overview," "Exploring the Intersection of Software Engineering and Mobile Technology from 2010 to 2021: A Review of Recent Research" "Comparing the sociology of culture in Bangladesh and India: Similarities and differences in Bangladeshi and Indian cultures" "Electrochemical Waste Water Treatment.", " Assessing enrichment and contamination of sediments in the effluent canal of the ore processing industry and Naviundu River in Lubumbashi, Democratic Republic of Congo" in "EQA (International Journal of Environmental Quality) – ISSN 2281-4485" and so on. Published chapters i.e. "Chapter 3: Artificial Intelligence Model for Analyzing the Buying Patterns of Customers" and "Chapter 11: Restaurant Sales Prediction Using Machine Learning" of the book titled "Handbook of Research on AI and Machine Learning Applications in Customer Support and Analytics", "Chapter 15: The Role of Innovation in Driving the Bioeconomy: The Challenges and Opportunities (pages 288-311)" of the book, "Handbook of Research on Bioeconomy and Economic Ecosystems", chapter, "Fixed Point Theory and Insurance Loss Modeling: An Unlikely Pairing" in the book "Advancement in Business Analytics Tools for Higher Financial Performance", " Mathematical Models and Formulas for Language Development and Disorders: A Collection of Hypothetical and Novel Approaches" in the book, "Childhood Developmental Language Disorders: Role of Inclusion, Families, and Professionals", " Salutogenic Marketing in the Elderly: Leveraging Digital Transformation in Geriatric Dentistry - Creating Positive and Meaningful Experiences for Older Adults" in "Geriatric Dentistry in the Age of Digital Technology", "Blockchain Technology for Smart Contracts: Enhancing Trust, Transparency, and Efficiency in Supply Chain Management" in "Achieving Secure and Transparent Supply Chains With Blockchain Technology", "Navigating the New Frontier of Finance, Art, and Marketing: A Look at Cryptocurrencies, NFTs, and Metaverse" in "Concepts, Technologies, Challenges, and the Future of Web 3" and so on.

**Buidr Singh** working as Professor at Sharda University, India. Also, Honorary Professor in University of South Wales UK and Santo Tomas University Tunja, Colombia. His areas of publications as Smart Healthcare, Medicines, fuzzy logics, artificial intelligence, robotics, machine learning, deep learning, federated learning, IoT, PV Glasses, metaverse and many more. He has 3 books, 139 paper publications, 163 paper presentations in international/national conferences and seminars, participated in more than 40 workshops/FDP's/QIP's, 25 courses from international universities of repute, organized more than 59 events with international and national academicians and industry people's, editor-in-chief and co-editor in journals, developed new courses. He has given talks at international universities, resource person in international conferences such as in Nanyang Technological University Singapore, Tashkent State University of Law Uzbekistan; KIMEP University Kazakhstan, All'ah meh Tabatabi University Iran, the Iranian Association of International Criminal law, Iran and Hague Center for International Law and Investment, The Netherlands, Northumbria University Newcastle UK,

**Ymboi Jonas** is a dedicated educator currently serving as a teacher in St. John's Integrated Senior High/Technical School, under Ghana Education Service, in Navrongo, Ghana. His academic journey reflects a commitment to both education and commerce. Yomboi holds a prestigious Master of Commerce in Accounting from the University for Development Studies, showcasing his deep understanding of financial principles and practices. His foundational academic qualifications include a Bachelor of Education in Accounting with Mathematics, indicating a strong background in the intersection of education and numerical disciplines. Additionally, he possesses a Diploma in Ghanaian Sign Language, emphasizing his versatility in communication methods. Beyond the confines of the classroom, Jonas extends his expertise by teaching part-time at the Community Health Training College. This dual commitment demonstrates his passion for imparting knowledge not only to high school students but also to those pursuing healthcare education. A keen researcher, Mr. Yomboi Jonas has focused his academic inquiries on areas of paramount importance. His research interests span across Social Media, where he explores the impact of digital platforms on society, as well as Public Sector Accounting and Financial Reporting, delving into the intricacies of financial management within the public domain. Through his multifaceted contributions to education and research, Mr. Yomboi Jonas emerges as a dynamic figure, bridging the realms of academia, commerce, and social inquiry. His dedication to both the theoretical and practical aspects of education positions him as a valuable asset in the academic landscape of Ghana.

# Index


## A

Agricultural Waste 5, 87, 148, 154-157, 159, 161, 165, 168, 170-172, 178, 180-181, 200-204, 206, 208-209, 212-217
Agricultural Waste Management 87, 155-156, 161, 165, 168, 170, 180, 200, 203-204, 208-209, 214, 216
Agro-Waste 155, 171, 180, 200, 213

## B

Bio-Economy 154
Blockchain 48, 94, 197-199, 219-233

## C

Circular Economy 3, 16, 18, 29-30, 33, 35, 37-38, 41-42, 45-46, 49-52, 72-73, 81, 84-85, 92, 94, 97, 103, 107, 123, 126-129, 132-134, 138, 158, 165, 168, 193-195, 200, 208, 227
Climate Change 3, 7, 11, 19, 22, 26, 31, 40, 51, 54, 63-64, 66, 73, 91, 98, 118, 121, 126, 136, 162, 172, 177, 181, 205-206, 220, 223
Closed-Loop Model 14-15
Collaborative Initiatives 34, 45, 47
Crop Residue 156, 170-176, 179-181, 202

## D

Deep Learning 182, 184-185, 187, 190, 192, 198
Development 2-5, 9-10, 16-22, 26, 29, 34, 36-38, 43-47, 49, 51-57, 59-70, 73, 81, 83, 85, 89-90, 93-94, 107, 109-110, 114-115, 117, 121-127, 129, 132-133, 136, 138, 141, 143, 145, 150-151, 155, 164-165, 167-168, 181, 183-184, 187-189, 192-198, 208, 210-215, 218-219, 225-229, 231-232
Disruptive 15, 94, 219-220

## E

Eco-Friendly Material 154
Economic Opportunity 72
Ecosystem 6-8, 15, 19, 26, 42, 55, 60, 62, 89, 96, 102, 134, 190, 199, 225, 227-228, 230
Education 7, 11, 18, 34, 42, 51, 53-65, 67-71, 73, 78, 81, 83, 85, 92, 95, 106, 116, 120, 141, 164, 167, 184, 186, 188-189, 199, 221, 225
Education in Sustainable Development 53
Environment 1-5, 9-11, 38, 43-44, 51-55, 57-58, 60-62, 65, 72, 86, 89, 91, 94, 96-98, 100, 103-107, 114, 117, 120-123, 126-127, 129, 131-133, 137-147, 150-151, 167-169, 175, 178, 180-181, 184, 189-192, 194, 203, 205-206, 208, 216, 218-221, 224-229, 231-233
Environmental Conservation 71-72, 74, 204
Environmental Health 90, 94, 106, 109-110, 119, 139
Environmental Safety 105, 140, 200
Environmental Sustainability 3-4, 10, 20-21, 37, 45, 58, 73, 82, 89, 107, 115, 121, 136, 139, 143, 156, 169, 183, 187, 195-196, 214, 216, 220, 225, 230
E-Products 96-99, 101-102, 104

## F

Food Wastage 124, 182, 185, 187, 191, 194-195
Formal E-Recycling 96, 104, 106

## G

Green 12, 15-16, 20, 34-35, 86, 94-95, 97, 99, 101-103, 105, 110, 117, 119-120, 125, 132, 134, 138-142, 144, 150, 155, 177, 179, 197-198, 200-201, 212-216, 218-221, 224-226, 228, 230-233
Greener Practices 96-98, 103
Greenhouse Crops 154

## H

Healthcare 18, 112, 117, 144, 192, 196, 198, 229
Human Health 3-5, 90-91, 106, 121, 133, 137, 143, 150-151, 168, 205-206, 215

## I

Influence 44, 64-68, 89, 97, 109, 111, 117, 135-136, 180, 184, 188, 192, 205, 219-220
IoT 48, 93-94, 157, 182-185, 187-192, 232-233

## N

Nano Biosensors 200, 210, 214
Nanomaterials 131-132, 134-137, 139, 143-145, 147, 149-152, 200, 209-210, 213-214, 217
Nanotechnology 125, 134-141, 143-144, 208-209, 211-217

## P

Partnerships for the Goals 37, 43
Plastic Waste Management 123, 125, 134-135, 137
Pollution 1-6, 11, 15, 17, 19-21, 29, 31, 38, 52-54, 61, 73, 87, 90-91, 98, 101, 104, 109-110, 113, 115-116, 118, 120, 122, 124, 126, 130, 132, 134-140, 169, 172-173, 176, 181, 194-195, 201, 205, 214, 217-218, 223, 231
Pollution and Waste Management 1
Public Health 38, 109-110, 119, 121, 142, 180, 187, 193, 201, 203, 205, 217

## R

Recyclable Leather 1
Recycle 18-19, 29, 42, 45-46, 78, 89-90, 96, 100, 102-103, 117-118, 125, 128, 130-131, 154
Resource Recovery 41, 71-73, 104
Rice-Wheat Cropping System 170-171, 180

## S

SDG 1 14-15, 17, 19, 37, 45-47, 115, 117-119, 182-183, 185, 188
SDG 2 115, 182, 184-185, 187
SDGs 14, 17-18, 62-65, 109-110, 115, 119, 183-184, 192, 197-198
Solid Waste 38-39, 94-96, 98, 100, 103, 109-111, 113-124, 203, 205, 207, 215
Stubble Burning 170, 172-174, 179-180
Sustainability 1-17, 19-24, 26-28, 31-35, 37-38, 40-47, 49-59, 61-70, 72-73, 81-86, 89-90, 93-100, 104-108, 115, 117, 119-121, 123-125, 132-133, 136-140, 143-144, 151, 154-158, 161, 164, 167-169, 182-185, 187, 190, 192-199, 202-203, 208, 213-214, 216, 218-223, 225-228, 230-233
Sustainable 1-24, 26-48, 51-53, 55-72, 74, 80-97, 99, 103-104, 107, 109-110, 114, 117-119, 121-126, 128-129, 131-133, 135-138, 140-148, 150-152, 154-158, 160, 162-168, 170, 176, 180-185, 187-190, 192-201, 203-208, 210-214, 216-228, 230-233
Sustainable Consumption 23, 46, 97, 182-185, 190, 198
Sustainable Fashion 7-10, 12-15, 19, 22, 24, 26-27, 29, 34, 36, 42, 52
Sustainable Practices 2, 5, 7, 11, 16, 18-20, 22-23, 31, 33, 37, 39, 42-44, 47, 50-51, 58, 71-72, 92, 94, 137, 154, 156, 164, 201, 224-227, 230
Sustainable Supply Chains 37

## T

Technology 4, 7, 11-12, 16, 35, 41, 61-62, 73, 86, 93-94, 99-106, 117, 122, 131, 139, 144, 157, 168-169, 176, 179, 181, 183, 188-189, 191-192, 194-199, 201, 205, 209-210, 215-216, 218-224, 226-233
Transparent 8-10, 119, 219-221

## U

Upcycling 14-21, 23, 29-31, 33-34, 36, 41, 51, 74, 91, 129, 134-135

## W

Waste Management 1, 15, 18, 30, 37-41, 43-50, 53-54, 58, 71-74, 77, 81-87, 89-95, 103, 106-110, 114-115, 117, 119, 121-128, 134-135, 137-139, 143, 154-156, 158, 160-161, 165, 167-168, 170, 177, 179-180, 182-185, 193-197, 199-201, 203-206, 208-209, 213-216
Waste Valorisation 109-110
Weed Control 71-72, 87
Weed Waste Management 71-74, 77, 81-85
WEEE 98, 101, 103-104, 107

**Publishing Tomorrow's Research Today**

IGI Global
Publishing Tomorrow's Research Today
www.igi-global.com

## Uncover Current Insights and Future Trends in
# Scientific, Technical, & Medical (STM)
## with IGI Global's Cutting-Edge Recommended Books

Print Only, E-Book Only, or Print + E-Book.
Order direct through IGI Global's Online Bookstore at **www.igi-global.com** or through your preferred provider.

ISBN: 9798369303689
© 2024; 299 pp.
List Price: US$ **300**

ISBN: 9798369314791
© 2024; 287 pp.
List Price: US$ **330**

ISBN: 9798369300442
© 2023; 542 pp.
List Price: US$ **270**

ISBN: 9798369308929
© 2024; 426 pp.
List Price: US$ **265**

ISBN: 9781668489383
© 2023; 299 pp.
List Price: US$ **325**

ISBN: 9781668465776
© 2023; 454 pp.
List Price: US$ **325**

**Do you want to stay current on the latest research trends, product announcements, news, and special offers?**
Join IGI Global's mailing list to receive customized recommendations, exclusive discounts, and more.
Sign up at: **www.igi-global.com/newsletters.**

Scan the QR Code here to view more related titles in STM.

**www.igi-global.com** Sign up at www.igi-global.com/newsletters facebook.com/igiglobal twitter.com/igiglobal linkedin.com/igiglobal

www.ingramcontent.com/pod-product-compliance
Lightning Source LLC
LaVergne TN
LVHW062146300926
849161LV00010B/2643
*9798369342640*